The Oxford Dictionary of English Grammar

The Oxford Dictionary of English Grammar

Sylvia Chalker
Edmund Weiner

OXFORD UNIVERSITY PRESS

Oxford University Press, Great Clarendon Street, Oxford OX2 6DP
Oxford New York
Athens Auckland Bangkok Bogota Bombay
Buenos Aires Calcutta Cape Town Dar es Salaam
Delhi Florence Hong Kong Istanbul Karachi
Kuala Lumpur Madras Madrid Melbourne
Mexico City Nairobi Paris Singapore
Taipei Tokyo Toronto Warsaw
and associated companies in
Berlin Ibadan

Oxford is a trade mark of Oxford University Press

Published in the United States by
Oxford University Press Inc., New York

First published 1994
First issued as an Oxford University Press paperback 1994
Reprinted with corrections 1996

British Library Cataloguing in Publication Data
Data available

Library of Congress Cataloging in Publication Data
Data available
ISBN 0–19–861314–8

5 7 9 10 8 6 4

Printed in Great Britain by
Biddles Ltd,
Guildford and King's Lynn

Contents

Introduction

Grammar, etymologically speaking, is related to *glamour*. Though few people might claim that grammar is glamorous in the modern sense, there is considerable interest in English grammar today and no shortage of grammar books, ranging from small basic books aimed at children or elementary-level foreign learners, through more advanced manuals to large scholarly works. The trouble is – they may be about the same language, but they do not always speak the same language. The very range of the grammar books on offer presents problems.

There are many ways of describing grammar, and a wealth of terminology. Some of it strikes the layman as jargon (*disjunct*, *matrix*, *pro-form*, *stative*); other words appear ordinary enough but conceal specialized meanings (*assimilation*, *comment*, *focus*, *specific*). Worse, the same terms, old or new – *comparison*, *formal*, *pronoun*, *reported speech*, *root*, *stress* – are used by different grammarians with different meanings.

Such difficulties are not entirely avoidable. Any subject of study needs specialist words. Different grammarians are entitled to analyse language in different ways, and fresh viewpoints may call for new terms. But while grammarians sometimes explain what they mean by a new or unusual term, it is rarer for them to point out that they are using an existing term in a different way. This is a cause of real confusion. Another problem is that new terms may in the end turn out simply to be alternatives for an old concept – a synonym in fact (e.g. *progressive*, *continuous*).

We have tried in this dictionary to indicate the range and variety of meanings that may lie behind a single term. The main emphasis is on the terminology of current mainstream grammar, but we have also included a considerable number of entries on the related areas of speech and meaning – more grandly known as phonetics and semantics. Users will also find some terms from generative grammar, which has greatly influenced mainstream grammar in recent years – but some of the more theoretical terminology of linguistics and semantics is excluded. We have also on the whole excluded outdated grammatical terminology, apart from a few traditional terms which may be familiar to the general reader.

The authors would like to thank Professor Flor Aarts, of the Katholieke Universiteit, Nijmegen, who read an early draft of the book: his comments, we believe, have led to many improvements, but the authors are alone responsible for any blemishes that remain. We would also like to thank our families for their support, encouragement, and, at times, forbearance.

SC
ESCW

London, Oxford 1993

Organization

1 *Entries* are strictly alphabetical. Thus:

agent
agentive
agentless passive
agent noun

2 Where two or more terms are *synonyms*, the definition appears under the preferred term, usually with a reference to the alternative term, e.g.

folk etymology
. . . Also called *popular etymology*.

and the other term is cross-referenced, e.g.

popular etymology
The same as FOLK ETYMOLOGY.

3 Where a word is not a grammatical term in itself but forms part of a *phrase* which is dealt with elsewhere, this is indicated, e.g.

act See SPEECH ACT.

4 Where a term is dealt with at the entry for some larger term, this is indicated, e.g.

formulaic subjunctive See SUBJUNCTIVE.

5 Where two or more terms are in a *contrastive* relationship, this is stated at the beginning of both entries, e.g.

abstract . . .
. . . Contrasted with CONCRETE.

concrete . . .
. . . Contrasted with ABSTRACT.

6 *See* at the end of (part of) an entry indicates that further information will be found at the entry indicated. Sometimes the user is referred to a closely related word, e.g.

coordinate . . .
. . . See COORDINATION.

At other times the reference is to a 'background' concept, e.g.

fall-rise . . .
. . . See INTONATION.

7 *Compare* at the end of (part of) an entry indicates that although the entry is complete, it may be useful to read entries for related or

overlapping terms, or terms with which this term could be confused, e.g.

> **abbreviated** . . .
> . . . Compare BLOCK LANGUAGE, REDUCED *clause.*

These particular entries show that certain types of language that might reasonably be described as *abbreviated* are in fact given special labels.

8 *Cross-references* to other entries are given in small capitals. A cross-reference to a phrase listed within an entry is given in a mixture of small capitals (for the entry headword) and italics (for the remainder of the phrase), e.g. DUMMY *element.*

9 Words are marked with *part-of-speech labels* (*n.* = noun, *adj.* = adjective, *v.* = verb) only when they are used as more than one part of speech.

10 Where part-of-speech labels are conjoined (e.g. (*n. & adj.*)), the definition is framed so as to cover both uses, with parentheses surrounding the part of the definition that applies to only one of the two uses, e.g.

> **ditransitive**
> (*n. & adj.*) (A verb) having two objects.

which is equivalent to:

> (*n.*) A verb having two objects. (*adj.*) Having two objects.

> **countable**
> (*n. & adj.*) (Designating) a noun with singular and plural forms.

which is equivalent to:

> (*n.*) A noun with singular and plural forms.
> (*adj.*) Designating a noun with singular and plural forms.

11 In certain entries for morphological terms, words and phrases quoted as examples are given *abbreviated dates* indicating their earliest known recorded appearance in English. In these, the number is that of the century and the preceding E, M, L mean 'early', 'mid', and 'late': 'E19' means '1800–1829', 'M19' means 1830–1869', and 'L19' means '1870–1899'. OE means 'Old English' (before 1150), ME 'Middle English' (1150–1349), and LME 'late Middle English' (1350–1469).

12 In many entries, *quotations* from works on language are given in order to illustrate the use of the word being defined. Only the author's name and the date of the work are cited: fuller details are given in the List of Works Cited.

13 *Derivatives* of a headword are listed undefined at the end of the entry if their meaning is plain once that of the parent word is known and if they are not found in special phrasal combinations.

A

A

ADVERBIAL as an ELEMENT of clause structure.

The symbol is used in some modern analyses of clause structure. See ADVERBIAL (1).

abbreviated

Shortened or contracted so that a part stands for the whole.

1 Designating language, or a clause or sentence, in which words inessential to the message are omitted and the grammar sometimes deviates from standard rules.

This is a very general term, since individuals will vary in how severely they abridge and exactly how they do it, when, for example, writing diaries or making lecture notes for private use.

Abbreviated sentences of a more predictable kind are a frequent feature of informal writing and conversation. Here the subject and part of the verb are often omitted.

Having a wonderful time here
See you soon
All news then
More tea? (= Would you like . . .?, Do you want . . .?)

Abbreviated language overlaps with ELLIPSIS, but has fewer 'rules'. Moreover, there is no need for the 'missing words' to be 'recoverable'.

Labels and printed instructions, too, often use abbreviated language; and here not only subjects but objects also are typically omitted, e.g.

Contains natural herb extracts
Avoid getting into the eyes

Other forms of abbreviated language appear in titles, notices, and newspaper headlines.

Compare BLOCK LANGUAGE, REDUCED *clause*.

2 **abbreviated clause**: the same as REDUCED *clause*.

3 **abbreviated form**: the same as CONTRACTION (2).

abbreviation

A shortened form of a word or phrase, standing for the whole.

This term is applied in three different ways.

abbreviation

(*a*) A string of letters—often spoken as such—formed from the initial letters of the (main) words of a phrase. Also called INITIALISM.

BBC (British Broadcasting Corporation)
CBI (Confederation of British Industry)
ERM (Exchange Rate Mechanism)
OTT (over the top)
PCW (personal computer word processor)
UK (United Kingdom)

Sometimes the letters represent syllables of a word.

ID (identity or identification card)
TB (tuberculosis)

(*b*) A word (sometimes called a CLIPPING) standing for the whole, retaining at least one syllable of the original word.

ad (advertisement) (M19)
demo (demonstration) (M20)
flu (influenza) (M19)
pub (public house) (M19)
phone (telephone) (L19)
sitcom (situation comedy) (M20)

Clippings vary in their level of formality; *mike* (microphone) and *wellies* (wellington boots) are at the informal end of the scale. Other abbreviations are acceptable in formal contexts, e.g. *bus* (omnibus), *maths* (US *math*) (mathematics); or their origin may even be virtually forgotten, e.g. *mob* (from Latin *mobile vulgus*).

(*c*) A written convention which is unpronounceable in its shortened form. This includes abbreviations of personal titles, e.g. *Col., Dr., Mrs., Sgt.*, etc. Also

St (street or saint)
Fr. (French)
Gk. (Greek)
etc. (etcetera)
kg (kilogram)
rpt. (repeat)
MS. (manuscript)

There are a few special written conventions for plurals:

pp. (pages)
ff. (following pages)
MSS. (manuscripts)

Written Latin abbreviations are sometimes read out in their English equivalents, but some are only pronounced as letter strings, e.g.

e.g. (*exempli gratia*) (for example, /ˌiː ˈdʒiː/)
i.e. (*id est*) (that is, /ˌaɪ ˈiː/)
cf. (*confer*) (compare, /ˌsiː ˈef/)
a.m. and *p.m.* (*ante* and *post meridiem*: /ˈeɪ ˌem/, /ˈpiː ˌem/).

Chemical formulae and other symbols can be regarded as a special type of abbreviation.

H_2O (water)
Fe (iron)
& (and)
+ (plus)
– (minus)

Compare ACRONYM, BLEND, CONTRACTION, INITIALISM.

ability

One of the semantic categories used in the classification of modal verbs.

The term is particularly applied to the dynamic meaning of *can* and *could.* It contrasts with other meanings of these verbs such as PERMISSION and POSSIBILITY.

ablative

(*n. & adj.*) (In older grammar.) (A case) that expresses meanings such as *by*, *with*, or *from.*

This case, occurring and originally named in Latin, is not relevant to English, where such meanings are expressed by prepositional phrases. The corresponding semantic categories include AGENT, INSTRUMENT, MEANS. The nearest equivalent in English to the *ablative absolute* of Latin is the ABSOLUTE CLAUSE.

Compare CASE.

ablaut

(In historical linguistics.) A particular type of alternation of different internal vowels between related words or forms. Also called *gradation*, *vowel alternation*.

This term was taken over from the German philologist and folklorist Jacob Grimm, who used it in his *Deutsche Grammatik*, first published in 1819. It is a phenomenon common to all the Indo-European languages, which retains a functional role in English in the formation of the past tense and participle of irregular verbs. Several of the Common Germanic ablaut series are still perceptible in the paradigms

ride	rode	ridden
drink	drank	drunk
speak	spoke	spoken
shake	shook	shaken
fall	fell	fallen

Compare MUTATION.

The term is occasionally extended to cover *umlaut* or MUTATION as well, but historically the two are different phenomena.

See VOWEL CHANGE.

absolute

1 Used to describe the uninflected form of a GRADABLE adjective or adverb (e.g. *kind, soon*) in contrast to the comparative and superlative forms (*kinder, kindest, sooner, soonest*). The same as POSITIVE (2).

2 Non-gradable. See GRADABLE.

3 Used of the *mine* series of pronouns, contrasted with the *my*-type of determiners.

4 (In older usage.) Designating an adjective or verb when standing outside certain usual constructions or syntactic relationships, as:

(*a*) designating an adjective used without a noun as a nominal (e.g. *the poor*);

(*b*) designating a normally transitive verb used intransitively (e.g. *Have you eaten?*); and

(*c*) designating a comparative or superlative form of an adjective used without specific mention of a relationship (e.g. *I only want the best*).

> 1931 G. O. CURME The absolute comparative is not as common as the absolute superlative . . *higher education*; a *better*-class cafe.

See also ABSOLUTE CLAUSE.

• **absolutely** (in older usage, as 4(*a*) above.)

> 1884 *New English Dictionary* In 'the public are informed', 'the young are invited', *public* and *young* are adjectives used absolutely.

absolute clause

A non-finite or verbless clause containing its own subject, separated from the rest of the sentence by a comma (or commas) and not introduced by a subordinator.

A verb, if used, can be an *-ing* or an *-en* form.

> The fight to board the train—*the women crushed against the doors, the children desperately clutching their mothers*—repeated itself at this provincial station
> *The platform empty once more*, I settled down for the night

Except for a few set phrases (*weather permitting*, *present company excepted*) absolute clauses tend to be formal and written. If the subject is a pronoun it must be in subject, not object, case (e.g. *I refusing to go, Nicholas went alone*) so absolute clauses are sometimes called *nominative absolutes*. (This contrasts with the 'ablative absolute' of Latin grammar, where the comparable noun is in the ablative case.)

abstract

Used mainly of nouns that denote an action, idea, quality or state; contrasted with CONCRETE.

The traditional division of common nouns into abstract and concrete nouns is semantic. It therefore cuts across the more strictly grammatical classification into UNCOUNT and COUNT nouns, and as a way of trying to deal with syntactic differences is unsatisfactory.

The abstract label does fit many uncount nouns (e.g. *Everybody needs advice/fun/luck*; not **an advice/*funs/*two lucks*). But abstract nouns also include count nouns (e.g. *We had an idea/another quarrel/better solutions*; not **We had idea/quarrel/better solution*). Other abstract nouns have both count and uncount uses (e.g. *several important discoveries/an important discovery*; *a voyage of discovery*).

accent

1 *Linguistics.* The mode of utterance peculiar to an individual, locality, or nation, as in 'he has a north country/Irish/Scottish/American/French/German accent'.

Accents in Britain may be regional or social, the latter related to educational and cultural background. Linguists insist that everyone speaks with an accent, and that the standard RP accent (see RECEIVED PRONUNCIATION) is just one among many. Accent refers only to pronunciation and is distinct from DIALECT.

See also IDIOLECT.

2 *Phonetics.*

(*a*) The same as STRESS.

(*b*) Stress (in its narrower sense) accompanied by pitch change.

Loosely, *accent* and *stress*, and their associated pairs of terms (*accented*, *stressed*, etc.) are used interchangeably. But some phoneticians distinguish between *accent*, defined as including PITCH change, and stress, which is due to the amount of force or energy used to produce a sound, but which does not include a pitch change. By this sort of definition, accent can only occur on a stressed syllable (whereas stress may not involve accent).

See PITCH, STRESS.

• **accentual**: relating to (phonetic) accent, particularly in the sense of word stress (rather than nuclear pitch).

> 1962 A. C. GIMSON The accentual patterns of words are liable to change. Considerable changes of this kind have taken place within the last three hundred years, in addition to the large-scale accentual shifts affecting French importations in ME. Thus, in the seventeenth century, and still in American English, a secondary accent with a strong vowel fell on the penultimate syllable of such words as *necessary*, *adversary*, *momentary*.

accentuation: the occurrence of accent (in the sense of pitch change).

> 1973 J. D. O'CONNOR Accentuation . . is a feature of the utterance, giving prominence to those parts which are semantically important; stress is a feature of the word and is just as much a part of its shape as the sequence of constituent phonemes is. The two are certainly related in those languages which have stress as a word feature, even though it is fixed, because the features of pitch which mainly constitute accentuation centre around the naturally stressed syllable of the word to be accented.

acceptability

Of a language form or an utterance: the quality of being judged by native speakers as normal or possible.

(*a*) Native speakers may disagree over whether a particular utterance is grammatically acceptable or not. An individual's judgement of acceptability may be affected by personal, regional, or social background, by perceptions of 'correctness', and so on. For example, judgements differ over the acceptability of:

? You ain't seen nothing yet.
? She was realizing there was a problem.
? The house was building for three years.
? We convinced them to go.
? Either Monday, Tuesday, or Wednesday would suit me.

(*b*) Linguists make a distinction between *acceptability* and GRAMMATICALITY, since sentences may be grammatically correct according to the rules, but unacceptable for some other reason. For example, a properly constructed, grammatically correct sentence could be so long that it becomes unacceptable because it is virtually impossible to understand. In this sense, acceptability is related to actual performance, while grammaticality is a feature of (more idealized) COMPETENCE.

Acceptability can extend to word formation. Thus, although the suffix *-ish*, meaning 'somewhat like', 'somewhat', combines with concrete nouns (e.g. *foolish*, *snobbish*, *kittenish*) and adjectives (e.g. *coldish*, *pinkish*), there could be degrees of acceptability as regards words newly formed with this suffix (e.g. ?*yuppyish*, **idiotish*, ?*trendyish*, ?**aquamarinish*).

• **acceptable.**

1962 A. C. GIMSON Any strongly rolled [r] sound, whether lingual or uvular, is not acceptable in RP.
1988 R. QUIRK Characters in Dickens can use *an't* or *ain't* for 'isn't' without any hint that such forms are other than fully acceptable.

accidence

(In older grammar.) The part of grammar that deals with the inflections of words; the way words change to indicate different grammatical meanings.

This category traditionally contrasts with SYNTAX. For example, the differences between

drive, *drives*, *driving*, *drove*, and *driven*

or between

driver, *driver's*, *drivers*, and *drivers'*

would come under accidence in a traditional grammar. In more modern grammar, the term has been superseded by INFLECTION, which, together with DERIVATION, is dealt with under MORPHOLOGY.

Apparently *accidence* was an alteration of *accidents* (plural), used around 1600 to mean 'the changes to which words are subject in accordance with the relations in which they are used', translating the Latin neuter plural *accidentia*; although it is possible that the latter was misunderstood as a feminine singular noun and rendered *accidence*.

accusative

(*n. & adj.*) (In older grammar.) The same as OBJECTIVE (1).

A traditional term, somewhat out of favour today as far as English is concerned.

acoustic phonetics

That branch of phonetics concerned with the way in which the air vibrates as sounds pass from speaker to listener.

Acoustic phonetics involves the measuring of sound features with instruments and electronic equipment that then present the information in visual form.

Compare ARTICULATORY *phonetics*, AUDITORY PHONETICS.

acquisition

Linguistics. The process of learning a language.

The term *child language acquisition* is used in descriptions of how children develop language ability.

See LANGUAGE ACQUISITION DEVICE.

acrolect

Sociolinguistics.

1 Originally, in a post-creole community, the social dialect most closely resembling the standard language from which the creole is derived.

2 The most prestigious or 'highest' social variety of a language.

The term is sometimes used in connection with mother-tongue English speakers. Thus standard British English with an RP accent may be considered an acrolect (unless the additional term HYPERLECT is introduced). It is also used of varieties of English in regions where English is a second (or third) language.

> 1977 J. T. PLATT I feel that in the case of Singapore English . . a very distinct non-British English acrolect is gradually emerging.

Compare BASILECT, HYPERLECT, MESOLECT.

• **acrolectal**.

acronym

Morphology.

1 Strictly, a word formed from (a) the initial letters of other words, or from (b) a mixture of initials and syllables.

E.g.

(*a*) *NATO* (=North Atlantic Treaty Organization)
NIMBY (=not in my back yard)
TINA (=there is no alternative)

(*b*) *radar* (=radio detection and ranging)
yuppie (=young urban professional + diminutive ending).

Sometimes included in the general term ABBREVIATION.

2 More loosely, an ABBREVIATION pronounced as a string of letters, especially letters that stand for the name of an organization or institution, e.g. *BBC, USA*.

This usage may be due to the fact that the specific term for this type of abbreviation (INITIALISM) is not widely known.

act See SPEECH ACT.

action

The process of acting or doing expressed by a verb.

Traditionally and loosely the term is used of any verb.

> 1884 *New English Dictionary* *Action of a verb, verbal action*: The action expressed by a verb; properly of verbs which assert *acting*, but conveniently extended to *the thing asserted by a verb*, whether action, state, or mere existence, as I *strike*, I *stand*, I *live*, I *am*.

Traditionally, transitive verbs used in an active tense are said to act upon their objects, which are given the role of 'patient'. But the concept is out of favour in modern grammatical theory, since obviously many verbs do not in any meaningful way imply actors, actions, or patients (e.g. *I heard screams*, *He's contracted hepatitis*).

actional

Of a passive verbal form: expressing a dynamic action rather than a state.

The usual meaning of most central passive verbs. Contrast STATAL.

action verb

A verb (also called an *event verb*) describing a happening that occurs in a limited time and has a beginning and an end. Contrasted with a STATE VERB.

e.g. *arrive, make, listen, walk*.

The terms *action* and *event*, used to describe verbs, are popular equivalents for DYNAMIC (sense 1) (similarly *state verb* is popularly substituted for *stative verb*). The alternative labels are not however strictly synonymous, since the verb in *I am growing old*, being in a progressive tense, must be described as *dynamic*, but less obviously denotes either an action or an event.

Compare DYNAMIC, STATIVE.

active

1 (*n. & adj.*) (Designating) the VOICE that attributes the ACTION of a verb to the person or thing from which it logically proceeds. As an adjective, often applied to clauses and sentences containing a verb in the active voice, and even to the verb itself.

The verbs in

> The sun *rises* in the east
> The early bird *caught* the worm

are in the active voice, in contrast to the PASSIVE voice. Many verbs, e.g. intransitive verbs, can occur only in the active.

1985 R. QUIRK et al. There are greater restrictions on verbs occurring in the passive than on verbs occurring in the active.

2 **active verb**: (in older usage) the same as ACTION VERB.

actor

The logical subject of a verb, particularly a dynamic verb.

The term is a semantic label, and roughly means the 'doer' of the verb, the person or other animate being that does the action. Broadly, the term can be used in relation to both transitive and intransitive verbs. It describes the subject of an active transitive verb and the by-AGENT of the corresponding passive sentence. Thus *the old lady* is the actor both in *The old lady swallowed a fly* (which can be described in terms of actor-action-goal) and in *The fly was swallowed by the old lady*); *actor* can also designate the grammatical subject of an intransitive verb (e.g. *Little Tommy Tucker* sings for his supper).

The term clearly makes more sense when restricted to a 'doer' that in a real sense initiates an action, than when applied to the subject of some 'mental process' verb (e.g. *She didn't like it*), or to a verb of 'being' (e.g. *She was old*). Some analysts therefore restrict the term and would exclude the old lady if her action was unintentional and involuntary.

Compare EXPERIENCER, SENSER.

adjectival

(*n. & adj.*) (A word, phrase, or clause) functioning as an adjective (including single word adjectives). e.g.

guide price
the *greenhouse* effect
the man *in the white suit*
an *I'm-all-right-Jack* attitude
Don't be so *holier-than-thou*

Some grammarians would loosely use the word *adjective* to describe all these, but it is sometimes useful to distinguish between true single-word adjectives and other words functioning adjectivally in a particular context.

Note that the terms *adjectival* and *adverbial* are not entirely comparable. *Adverbial* can denote one of the main elements in clause structure (the others being *subject*, *verb*, *object*, and *complement*); i.e., it is often a functional label. Adjectivals operate at a lower level, often as part of a noun phrase (which itself may function as subject or object). An adjectival may in some instances be the sole realization of complement (e.g. You look *hungry*), but the adjectival as such is not a functional element in clause structure.

•• **adjectival clause**: the same as ADJECTIVE CLAUSE (1).

adjectival noun: (in older usage) an adjective used as a noun.

e.g.

the *poor*
the *old*

Compare ABSOLUTE (4).

adjectival passive: the same as STATAL *passive*.

• **adjectivally**: in an adjectival manner, as an adjective.

adjectivalization

Morphology. The conversion of a member of another word class into an adjective; the use of such a word in an adjectival function.

The commonest way of forming an adjective from another part of speech is by adding an affix (e.g. *wealth*, *wealthy*; *fool*, *foolish*; *hope*, *hopeful*).

A well-known feature of English is the use of nouns in attributive position to modify other nouns (e.g. *greenhouse effect*, *holiday shop*, *wind instrument*. This usage too can be called *adjectivalization*, but such nouns do not take on the other characteristics of an adjective (**The effect is greenhouse*; **a more holiday shop*).

adjective

A major part of speech, traditionally defined as a describing word or 'a word that tells us something about a noun'.

In modern grammar *adjective* is usually defined in more grammatical terms. Formally, a CENTRAL adjective meets four grammatical conditions. It can

(1) be used attributively in a noun phrase (e.g. *an old man*)
(2) follow *be* or another copular verb and occur alone in a predicative position (e.g. *He looks old*)
(3) be premodified by intensifying words such as *very* (e.g. *He's very old*)
(4) have comparative and superlative forms (e.g. *an older person*, *most extraordinary*)

But not all adjectives pass all these tests. Adjectives with an absolute (i.e. ungradable) meaning fail (3) and (4) (e.g. **very unique*, **more unique than . . .*), while adjectives which are only attributive (e.g. *utter*) or only predicative (e.g. *afraid*) fail (2) and (1) respectively.

Adjectives other than attributive-only adjectives can sometimes take POST-POSITION:

People *impatient* with the slow progress of the talks

and for a few adjectives this position is obligatory:

the president *elect*, the body *politic*, the harbour *proper*

Adjectives used with indefinite compound pronouns must follow them:

nothing *special*, someone *silly*

Some adjectives are used with *the*, to function as a noun phrase. Used of people, the meaning is plural, 'people of that sort' in general:

the *great* and the *good*, the *poor*, the *disadvantaged*

but notice *the Almighty*. (A few participles also can, exceptionally, have singular meaning, e.g. *the accused*, *the deceased*.)

Other adjectives, prefaced by *the*, refer to abstract qualities, e.g.

the *bizarre*, the *grotesque*, the *occult*

Some are used nominally in set phrases:

in *public*, in *short*, for *better* or *worse*

Historically, adjectives were once called *noun adjectives* because they named attributes which could be added (Latin *adjectivus* from *adjicere* to add) to a *noun substantive* to describe it in more detail, the two being regarded as varieties of the class *noun* or 'name'.

1612 J. BRINSLEY *Q.* How many sorts of Nouns have you? *A.* Two: a Noun Substantive, and a Noun Adjective. A noun adjective is that cannot stand by itself, without the help of another word to be joyned with it to make it plain.

See also DEMONSTRATIVE *adjective*, QUALITATIVE *adjective*.

adjective clause

1 Commonly, another name for RELATIVE clause.

1932 C. T. ONIONS Adjective Clauses are introduced by Relative Pronouns.

2 A phrase, with an adjective head, which functions as a clause.

Keen to take part, he volunteered his services
The crowd, *angry now*, charged at the police

adjective complementation

A word or words added after an adjective to add to or complete the meaning in context.

Adjective complementation may be (*a*) a prepositional phrase

They were innocent *of the crime*
She is brilliant *at chess*

or (*b*) various kinds of clause

I am sorry *that you don't like it*
We were not very clear *why we had been asked*
You were mad *to tell them*
It's a very easy place *to find*
I've been busy *answering letters*

Such complementation may be obligatory, e.g.

She is fond of her mother (*She is fond)

or optional, e.g.

I am so glad (that you got the job)

adjective order

The order in which two or more adjectives come in attributive position.

When two or more adjectives premodify a noun, there is usually a 'natural' or a 'better' order for them. *Your wonderful new cream woollen jumper* is likely; **your woollen new cream wonderful jumper* is deviant. It has proved difficult to formulate comprehensive, satisfactory rules to describe the ordering, and there are often alternative possibilities, but in general the order is related to the semantic properties of the adjectives: inherent characteristics (e.g. *woollen*) are closest to the noun and subjective judgements (e.g. *wonderful*) are furthest from it.

A typical order is:

determiners (if any) (*your*)
'central' adjectives (*wonderful/expensive/useful*)
colour adjectives (*cream*)
inherent characteristics —origin (*Welsh*)
—material (*woollen*)
attributive noun describing purpose (*golfing* jacket)

adjective phrase

A phrase functioning adjectivally, and consisting of an adjective as HEAD plus, optionally, words before and/or after.

e.g.

It was *very difficult/too difficult to understand/simple enough.*

Compare ADJECTIVAL, ATTRIBUTIVE.

adjunct

1 (In older usage.) Any word or words expanding the essential parts of the sentence; an optional, inessential, element in a structure.

The term has been variously used for words or phrases considered to be of secondary importance, including vocatives, adjectives joined to nouns, and also some adverbs.

In Jespersen's terminology it describes the functional role of a SECONDARY joined to a PRIMARY by JUNCTION.

1933b O. JESPERSEN A secondary can be joined to a primary in two essentially different ways, for which we use the terms *Junction* and *Nexus*. As separate names for the secondary in these two functions we shall use the terms *Adjunct* and *Adnex*.

In this usage, the term *adjunct* typically applies to an adjective in attributive position (e.g. a *silly* person).

2 (In present-day use, more specifically.) An element functioning like an adverb, whether a single-word adverb, an adverb phrase, or an adverb clause; sometimes the same as ADVERBIAL.

Adjuncts in this sense are of course usually marginal to sentence structure and therefore omissible. Thus in *Sadly, my neighbour died two months ago*, both *sadly* and *two months ago* could be omitted, leaving a still grammatical, meaningful sentence.

3 (In some modern grammar.) A particular subclass of adverbial, contrasted with CONJUNCT, DISJUNCT, and (optionally) SUBJUNCT.

In this categorization, only adverbials functioning as an element of clause structure (represented as *A*) are adjuncts. By this definition, *clearly* is an adjunct in the first sentence below, but not in the second:

He spoke clearly and to the point
Clearly, I could be wrong

Adverbial adjuncts of this special type often refer to place, time, or process (including manner, means, instrument, and the agentive with the passive). In general, these adjuncts come in end position, after the verb (and object, if any). But other positions are possible, and *frequency adjuncts* usually come in mid position (e.g. they *usually* come).

Adjuncts are sometimes divided on more functional grounds into PREDICATION and SENTENCE adjuncts.

adnex

(In Jespersen's terminology.) The functional role of a SECONDARY when joined to a PRIMARY by NEXUS. See ADJUNCT.

This means a verb or some other part of the predication.

1933a O. JESPERSEN *The dog runs*, nexus: *runs* . . . is adnex to *dog*.
1933b O. JESPERSEN The adnex may be any word or combination of words which can stand as a predicative . . e.g. a prepositional group. Could she have believed herself in the way?

The term is not in general use today. Compare SECONDARY and see SENTENCE ADJUNCT.

adnominal

(*n. & adj.*) (A word or phrase) attached to and modifying a noun.

Adnominals can precede or follow their head noun, and with it they form a noun phrase, e.g.

our wonderful new home
a country cottage
my parents' flat
somewhere *to live*
a place *of one's own*

•• **adnominal relative clause**: an ordinary relative clause, whether defining or non-defining (so called because it functions in the same way as other adnominals):

houses *that we've looked at*
Sandringham House, *which we visited*

The label is not common, but is used to distinguish this type of clause from the so-called NOMINAL RELATIVE CLAUSE (when that is not included with the noun clause, as it is in many analyses).

advanced RP See RECEIVED PRONUNCIATION.

adverb

A major part of speech: a word that usually modifies or qualifies a verb (e.g. *spoke quietly*), an adjective (e.g. *really awful*), or another adverb (e.g. *very quietly*).

Adverbs form a notoriously mixed word class. Traditionally they are divided into various meaning-related categories, such as manner (e.g. *hurriedly*), place (e.g. *there*), time (e.g. *soon, often, tomorrow*), and degree (e.g. *very*). Some grammarians analyse them in much greater detail, and some have described categories that distinguish them according to different grammatical functions. See ADJUNCT, CONJUNCT, DISJUNCT, SUBJUNCT.

Loosely, in popular grammar, the term *adverb* is often used to cover ADVERB PHRASES and ADVERBIALS in general. More strictly, where the latter terms are used, *adverb* may be restricted to single words functioning adverbially.

See SENTENCE ADVERB.

adverb clause

Any clause (finite, non-finite, or verbless) functioning adverbially, that is, expressing notions such as time, reason, condition, concession, etc.:

I'll come *when I'm ready*
They succeeded *because they persevered*
Don't do it, *unless you're sure*
Although injured, he struggled on
While travelling, he contracted jaundice
Make it Thursday, *if possible*

In more traditional usage, only the finite clauses (i.e. the first three examples) would be included here.

Adverb (or *adverbial*) *clauses* are often classified on semantic grounds into such categories as clauses of TIME, PLACE, CONDITION, CONCESSION, PURPOSE, RESULT, COMPARISON, MANNER, and COMMENT.

adverbial

(*n.*) 1 (In some modern grammar.) One of the five elements of clause structure (symbolized by A), comparable to Subject (S), Verb (V), Object (O), and Complement (C).

In this sense, the adverbial may be a word, phrase, or clause:

You've done that *(very) well (indeed)*: SVOA
Hang your coat *on a hanger*: VOA
They arrived *the Sunday before last*: SVA
When you've quite finished, we can begin: ASV
Though disappointed, she said nothing: ASVO

The adverbial is the most optional element of clause structure. Subject and verb are usually essential, and many verbs make some sort of object or complement obligatory. Only a few verbs force the use of an adverbial (**Hang*/**put your coat*).

2 Any word, phrase, or clause used like an adverb (including the simple adverb alone) whether functioning as an element in clause structure (i.e. as an adjunct) or at some other level.

Compare:

She dresses simply
That was a simply dreadful thing to say

In the first sentence *simply* modifies the verb and is an adverbial element in the sentence (which has an overall pattern of SVA). In the second, *simply* is part of a noun phrase (*a simply dreadful thing to say*, which functions as a complement (making the sentence SVC).

(*adj.*) Of or pertaining to an adverb; functioning like an adverb.

1873 R. MORRIS In Elizabethan writers we find the adverbial *-ly* often omitted, as '*grievous* sick', '*miserable* poor'.

•• **adverbial clause**: the same as ADVERB CLAUSE.

adverbial group: the same as ADVERB PHRASE (1).

adverbial conjunction: see CONJUNCTION.

adverbial particle: see PARTICLE.

adverbial phrase: the same as ADVERB PHRASE.

• **adverbially**.

adverbialization

The process of forming an adverb from another part of speech.

Adding *-ly* to an adjective is the most obvious example of this process (e.g. *bad/badly*, *pretty/prettily*). Another affix with the same result is *-wise* (e.g. *jobwise* 'as regards jobs or a job').

adverb particle See PARTICLE.

adverb phrase

1 A phrase functioning as an adverbial in clause structure and containing an adverb as head:

He speaks *very quickly indeed*
We were able to use the pool *as often as we wished*

Notice the postmodification in the second example by a clause of comparison.

2 A phrase functioning as an adverbial, whether it contains an adverb or not:

We'll be with you *in a moment*
They arrived *last night*

These would be classified as adverbials in many grammars.

affected

Semantics. Influenced, acted upon: used to describe the typical semantic role of the direct object.

The direct objects in the following are said to have an *affected* role:

I love *Lucy*
Lucy Locket lost *her pocket*

GOAL, OBJECTIVE, and PATIENT are sometimes used in this way, but distinctions are sometimes made.

In some theories about semantic roles, the subjects of COPULAR verbs, and even of INTRANSITIVE verbs, are said to have an affected role, e.g.

Lucy was in the garden
Her hat lay on the seat beside her

Compare RECIPIENT.

affective

The same as ATTITUDINAL and EMOTIVE.

affirmative

Of a sentence or verb: stating that a fact is so; answering 'yes' to a question put or implied. Opposed to NEGATIVE.

In some grammars the term POSITIVE is used with the same meaning.

Compare POLARITY.

affix

Morphology. An addition to the root (or base form) of a word or to a stem in order to form a new word or a new form of the same word.

An affix added before the root is a PREFIX (e.g. *un*-natural, *over*-weight); one added at the end is a SUFFIX (natural-*ness*, weight-*less*-*ness*). Affixes may be derivational (garden*er*) or inflectional (garden*s*).

Some non-European languages also have *infixes*, additions inserted within the main base of the word. In English the way in which the plural *-s* comes in the middle of some compounds (e.g. *hangers-on*) or in a few cases a swearword can be inserted within a word (e.g. *abso-bloody-lutely*) are perhaps marginal examples of this phenomenon.

See also DERIVATION, FORMATIVE, INFLECTION, MORPHEME.

• **affixation**: (*Morphology*) the joining of an affix or affixes to (the root or stem of) a word.

affricate

Phonetics. (*n. & adj.*) (A consonant sound) that combines the articulatory characteristics of a PLOSIVE and a FRICATIVE; there is a complete closure in the vocal tract, so that the following release is a plosive, but the release is slow enough for there to be accompanying friction.

Two affricates are recognized in standard English: [tʃ], the voiceless sound heard at the beginning and end of *church*, in the middle of *feature* /ˈfiːtʃə/, and at the end of *catch*; and [dʒ], the voiced sound at the beginning of *gin* and *jam*, in the middle of *soldier* /ˈsəʊldʒə/, and the beginning and end of *judge*.

> 1988 J. C. CATFORD It thus appears that the [ʒ] in [dʒ] is indissolubly linked with the [d] and the sequence must thus be regarded as a single, unitary affricate phoneme.

Controversially, /tr/ and /dr/ are sometimes analysed as affricates rather than sequences of phonemes.

• **affricative**: another term for AFFRICATE, now rare.

agent

The doer of the action denoted by a verb.

The term is particularly applied to the 'doer', in semantic terms, of the action expressed by a passive verb because in such a case the grammatical subject is not the doer, as the subject of an active verb often is. The agent is often indicated with a *by* phrase:

The two were kidnapped *by masked terrorists*
The child was saved *by the dog*

The agent is typically animate. In popular grammar, *agent* contrasts principally with INSTRUMENT and MEANS; in Case Grammar, the *agent* (or AGENTIVE) case is one of a set.

Compare ACTOR.

agentive

Syntax & Semantics. (*n. & adj.*) (Designating) a noun, suffix, or semantic role that indicates an agent.

In the phrase *this great ceiling by Michelangelo*, *Michelangelo* is agentive (the agent) and *by Michelangelo* is an agentive phrase. In the words *doer*, *farmer*, and *lover* there is an agentive suffix, *-er* (which contrasts with the comparative suffix *-er* in *kinder*, *nicer*).

1964 J. VACHEK The comparative suffix *-er* does not imply the change of the word-category of the basic word, while the agentive *-er* necessarily does so.

Verbs in context can be described as agentive or *non-agentive*. An *agentive verb* posits an animate instigator of the action. Contrast

The postman banged on the door (agentive verb)
The door was banging in the wind (non-agentive verb)

In Case Grammar the *agentive case* is defined semantically (together with OBJECTIVE, DATIVE, etc.), and so the subject of an active verb is frequently (though not always) agentive (the agent). This is a specialized use.

1968 C. J. FILLMORE The cases that appear to be needed include: *Agentive* (A), the case of the typically animate perceived instigator of the action identified by the verb.

agentless passive

A passive clause or sentence in which no agent is mentioned.

The agent is not mentioned because either it is unimportant or unknown, e.g. *Rome was not built in a day*, or the identity of the agent is deliberately concealed, e.g. *Mistakes cannot afterwards be rectified.*

agent noun

A noun with the meaning 'one who or that which does (the action of the verb)', and formed by adding the suffix *-er* or *-or* to a verb.

e.g.

actor, instructor, manufacturer, teacher, worker

Some agent nouns are inanimate (e.g. *computer*, *shocker*) and some have no independent base (e.g. *author*, *butcher*).

agglutinative See ANALYTIC.

aggregate noun

A noun that is 'plural-only' in form (e.g. *outskirts, remains*) or functionally plural-only, though lacking an *-s* (e.g. *people, police*).

A few such nouns can take singular or plural verbs (e.g. *the works is/are in Birmingham*). This is the same as PLURALE TANTUM (b) and (c), i.e. excluding BINARY nouns (like *scissors*).

agree

Be in concord; take the same number, gender, person, etc. (as another element in the clause or sentence).

See AGREEMENT.

agreement

The agreeing of two (or more) elements in a clause or sentence; the taking of the same grammatical person, number, or gender; CONCORD.

In English, the most generally recognized agreement (or concord) is that between a subject and its verb. As verbs have few inflections, this mainly affects the third person of the present simple of lexical verbs, where the singular *-s* ending contrasts with the plural and with the other persons of the singular (e.g. *He* or *she works*, but *I, we, you, they work*). The verbs *be* and *have*, and progressive and perfect tenses formed with them (I *am* working, She *has* worked), also must agree with their subjects.

Prescriptivists favour strict grammatical agreement. But *notional concord*, where agreement follows the meaning, is a common feature of English, and acceptable to most grammarians:

Everybody knows this, don't they?
Neither of them approve (*more strictly* approves)
The committee have decided (. . . has . . .)
£10 is all I have

A minor type of verb agreement, called *proximity agreement* (or *proximity concord*), is the agreement of the verb with a closely preceding noun instead of the noun head that actually functions as subject of the sentence in question. Such agreement may be marginally acceptable when it supports notional concord, but is generally considered ungrammatical:

?No one except my parents care what happens to me
??A parcel of books have arrived for you

Number agreement also normally exists between subject and subject complement (e.g. *She is a pilot*; *They are pilots*) and between a direct object and an object complement (e.g. *I consider her a brilliant pilot*, *I consider them brilliant pilots*).

Number and gender agreement affects pronouns and determiners, e.g.

He has lost *his* umbrella
She blames *herself*
There were *many* problems and *much* heartsearching

air-stream mechanism

Phonetics. The system of air movement used in the production of speech sounds.

Most speech sounds, including all normal English phonemes, involve expelling air outwards in an egressive air stream. (Sounds made by drawing air into the body are INGRESSIVE.) Most sounds are also made with lung air (a pulmonic air stream), but other air mechanisms are possible: compare CLICK.

alethic

Necessarily and logically true.

The term, taken from modal logic, comes from the Greek word *alethes* 'true', and is concerned with the necessary truth of propositions. It is sometimes used in the analysis of modal verbs, though most grammarians include this meaning in EPISTEMIC. The distinction between alethic and epistemic modality, when it is made, is that alethic modality is concerned with logical deduction (e.g. *If she's a widow, her husband must have died*), whereas epistemic modality relates to confident inference (e.g. *They were married over fifty years—she must miss him*).

allograph

1 A particular printed or written form of a letter of the alphabet (or more technically of a GRAPHEME).

Thus a lower-case ⟨a⟩, a capital ⟨A⟩, an italic ⟨*a*⟩, and a badly scribbled letter *a* are all allographs of the same grapheme.

Compare GRAPH, PHONEME.

2 *Phonetics*. (Less usually.) One of a number of letters or letter combinations representing a phoneme.

Thus the ⟨f⟩ of *fish* and the ⟨gh⟩ of *cough* are both allographs of the phoneme /f/.

allomorph

Morphology. An alternant of a morpheme; any form in which a (meaningful) morpheme (in sense 1) is actually realized.

The English plural morpheme has three regular *allomorphs*: an /s/ sound as in cats /kæts/, a /z/ sound as in dogs /dɒgz/, and an /ɪz/ sound as in horses /ˈhɔːsɪz/. We can also talk of a *zero allomorph* in *sheep* (plural), and various irregular allomorphs in *mice*, *geese*, and so on.

Also called MORPHEMIC *variant*.

• **allomorphic**: of or pertaining to an allomorph.

Compare FORMATIVE, MORPH.

allophone

Phonetics. Any of the variants in which an (idealized) phoneme is actually realized.

Many allophones, that is actual articulations, are possible for any phoneme of a language, depending on individual people's pronunciation, but the main allophones of any particular language are conditioned by their relationship to the surrounding sounds. Thus in standard English the /l/ phoneme has a CLEAR sound when it precedes a vowel (as in *listen* or *fall in*); a somewhat DEVOICED sound when preceded by a voiceless plosive (as in *please*, *clue*), and a DARK sound when it occurs word-finally after a vowel (as in *fall down*) or when it is syllabic (as in *muddle*).

• **allophonic**: of or pertaining to an allophone. **allophonically**.

alternant

Any of the possible variants of a particular feature of the language.

1 *Phonology*. A particular vowel in one word viewed as a variant of the corresponding vowel in another word when the two words are connected by a process of derivation or word formation.

ei /iː/ in *deceive* and the second *e* /e/ in *deception* are alternants.

2 *Phonology*. Occasional term for ALLOPHONE. Also **allophonic alternant**.

3 *Morphology*. Another word for ALLOMORPH. Also **morphemic alternant**.

This label includes both phonologically conditioned alternants (e.g. plurals /s/, /z/, and /ɪz/, but also such 'grammatical' alternants as *-ed*, *-en* (as in *heated*, *frozen*) for the past participle morpheme.

• **alternance, alternation**: the existence of alternants and the relationship between them. **alternate**: vary between alternants.

> 1935b J. R. FIRTH Vowel alternance is also a very important morphological instrument in the strong conjugation of verbs. There are thirty vowel alternances for our babies to learn.
>
> 1974 P. H. MATTHEWS When the forms identifying a morpheme vary, a normal usage is to talk of an alternation between them; for example, the allomorphs of the English morpheme CATCH alternate between [kætʃ] and [kɔː].

alternative question

A type of question which names possible answers but does not leave the matter open.

One of the three main types of question, in contrast to *WH*-QUESTION and YES-NO QUESTION. An alternative question can begin with a *wh*-word, e.g.

Which do you `like—´coffee, ´tea, or `wine?

or may be more like a yes-no question, e.g.

Would you like ´coffee or `tea?

But it differs from both in intonation.

An alternative question implies that one, and only one, of the options is possible. Each option receives a separate pitch, and the final option is said with a fall, showing that the choice is complete.

alveolar

Phonetics. (*n. & adj.*) (Of or pertaining to) a consonant sound made by the front of the tongue against the hard ridge formed by the roots of the upper teeth.

The main alveolars in English are /t/ and /d/ (often dental consonants in other languages), which are alveolar plosives; /n/, an alveolar nasal; and /s/ and /z/, which are alveolar fricatives. The actual articulation of these alveolar phonemes is affected by adjacent sounds, so that not all their allophones are in fact alveolar.

• **alveolarity**: the quality of being alveolar; an instance of this.

Latin *alveolus* means 'a small cavity', and hence 'a tooth socket'.

ambient *it*

The same as DUMMY *it*.

ambiguity

Ability to be understood in two or more ways; a word or phrase that can be so understood.

(*a*) *Lexical ambiguity*. Some ambiguity is due to a word or phrase having more than one meaning; e.g. *I don't seem to have a chair* (was the lecturer complaining that she had nothing to sit on, or that there was nobody to introduce her?).

(*b*) *Grammatical ambiguity* (or *structural ambiguity*) has a variety of causes, including ellipsis (or uncertainty whether there is ellipsis) within a noun phrase; e.g.

He was wearing new red socks and boots (were the boots new and red too?)
They are advertising for teachers of French, German, and Russian (separate teachers of these languages, or people capable of teaching all three?)

As prepositional phrases can not only appear in noun phrases but also function adverbially in clause structure, there is often considerable doubt as to what they refer to:

These claims have been dismissed as mere bravado *by the police* (the context might make it clear that the meaning is 'dismissed by the police', but it sounds rather like bravado by the police).

Similarly *to*-infinitive clauses, which also have a multiplicity of functions, may be ambiguous:

Railmen defy union order *to stop coal shipments* (probably the railmen are carrying on working, but they could have decided to stop shipments in defiance of union orders).

Ellipted clauses of comparison are another frequent cause of ambiguity:

I had better taste in films than girls (was the writer's taste in films better than his taste in girls, or did he have better taste in films than girls had?)

In complex sentences, ambiguity can arise when not just a phrase but a whole finite clause is open to more than one interpretation. (Intonation may disambiguate if the sentences are spoken.)

I didn't go because it was my birthday (did the speaker not go at all, or go for some other reason?)
He said he wouldn't lend me the money and I couldn't go (did he say that I couldn't go, or could I not go as a consequence of not being lent the money?)
I'll tell you when they arrive (is it that I will inform you of their arrival, or that when they arrive I will tell you something?)

(*c*) A further type of ambiguity is caused by the fact that many English words can be interpreted as more than one part of speech; headline English is a rich source of such ambiguity:

Fine old houses are demanding mistresses (are they crying out for capable chatelaines, or are they proving themselves to be hard-to-please, bossy rulers?)
Peking bars escape from 'terror' purge (was the Chinese government taking measures to prevent dissidents from escaping, or had the city's drinking venues escaped some harsh new decree?)
He gave her dog biscuits

ambilingual

(*n. & adj.*) (Designating) a person who has complete mastery of two languages.

1959 J. C. CATFORD In everyday speech the word 'bilingual' generally refers to a person who has virtually *equal* command of two or more languages. If a special term is required for such persons of equal linguistic skill (which is very difficult to measure) I should prefer to call them 'ambilinguals'. Ambilinguals are relatively rare.

• **ambilingualism**.

Compare BILINGUAL.

American See GENERAL AMERICAN.

anacoluthon (Pronounced /ænəkə'lu:θɒn/. Plural **anacoluthons**, **anacolutha**.)

Syntactic discontinuity within a sentence; a sentence which either breaks off while incomplete or switches part-way through to a different syntactic structure; discontinuity of this kind as a general phenomenon.

Informal spoken language often contains anacoluthon, much of which may pass unnoticed by the listener, e.g.:

One of my sisters,—her husband's a doctor and he says if you take aspirin your cold will go in a week, but if you do nothing it will take seven days
It's a course which I don't know whether it will be any good
I thought that you were going—well, I hoped that you were going to help
Why don't you—it's only a suggestion—but you could walk

The term was formerly used in rhetoric but has been adopted into linguistics. In rhetoric, the general phenomenon was called *anacoluthia* and an individual instance *anacoluthon*.

• **anacoluthic. anacoluthically**.

analogy

Imitation of the inflections, derivatives, and constructions of existing words in forming inflections, derivatives, and constructions of other words.

Analogy normally governs the patterns of word-formation. Recent years for example have seen numerous new verbs with the prefixes *de-* (e.g. *deselect*) and *dis-* (e.g. *disinvest*) and nouns beginning with *Euro-* (e.g. *Eurocrat, Eurofare, Eurospeak*). Other new nouns have been formed with such well-established suffixes as *-ism* (e.g. *endism, handicapism*). New verbs almost always inflect regularly (e.g. *faxing, faxes, faxed*) by analogy with regular verbs.

In historical linguistics, the term *analogy* is used in connection with the tendency for irregular forms to become regular (e.g. *shape*, *shove*: past tense *shaped*, *shoved*, in the 14th century *shoop*, *shofe*; past participle *shaped*, *shoved*, in the 14th century *shopen*, *shoven*). Interestingly, irregular patterns are sometimes spread by analogy: for example, the historical past tense form of the verb *dive* is the regular *dived*, but the irregular *dove* arose in British dialects and American English during the nineteenth century; similarly *scarves*, *hooves*, and even *rooves* have tended to replace the historical and regular plurals *scarfs*, *hoofs*, and *roofs*. Analogy probably accounts for the recently developed pronunciation of *covert* (traditionally /ˈkʌvət/) to rhyme with *overt* and *dissect* (traditionally /dɪˈsekt/) to rhyme with *bisect* (each pair of words has a certain amount of shared meaning).

Over-regularization by analogy is seen in the early efforts at speaking English of both young children and foreign learners, who may well say *He goed*, *mouses*, *sheeps*, and so on.

Compare OVERGENERALIZATION.

analysable

Capable of being analysed; particularly used of words that can be broken down into constituent morphemes. Contrasted with **unanalysable**.

Compare *dis-interest-ed-ness* with *interest*, which cannot be analysed further into morphemes (although *in-* or *inter* are meaningful elements in other words, such as *inborn, incurable, interfaith*).

analyse

Distinguish the grammatical elements of (a sentence, phrase, or word).

There are many ways in which sentences, clauses, etc. can be analysed. Simple sentences or clauses may be analysed into subject and predicate, or into such elements as subject, verb, object, complement, and adverbial. Complex and compound sentences are often analysed into types of clauses, e.g. coordinate and subordinate; adverbial, nominal, and so on. A word may be analysed into its base and suffixes (e.g. *dis-interest-ed-ness*).

analysis

The process of breaking up sentences, phrases, and words into their constituent parts.

See also COMPONENTIAL ANALYSIS, CONTRASTIVE *analysis*, DISCOURSE *analysis*, IMMEDIATE CONSTITUENT, MULTIPLE *analysis*.

analytic

1 *Morphology*. Designating a language without (or with few) inflections. Also called ISOLATING.

In an analytic language, word order plays an important role in establishing meaning. In extremely analytic languages most words consist of single morphemes. Analytic languages contrast with *synthetic* languages, which rely heavily on changing the form of words, and *agglutinative* languages, in which words are built up from smaller words or units, each contributing a bit of grammatical meaning.

English, having few inflections, is more analytic than, say, Latin or German, but it has some synthetic characteristics (e.g. *happy/happier/happiest*; *time/times*) and some agglutinative features (e.g. *mis-understand-ing*, *bio-degrad-able*).

By analogy, the contrast between analytic and synthetic is sometimes applied to features of English. Thus multi-word verb tenses (e.g. perfect *have taken*) and periphrastic adjective comparisons (e.g. *more unusual*, *most unusual*) are analytic, while corresponding single word forms (e.g. *took*; *odder/oddest*) are synthetic.

2 *Semantics*. Designating a sentence that is necessarily true by virtue of the words themselves, without reference to any particular circumstances or situation, in contrast to SYNTHETIC (2).

E.g. *The sea is wet*, which is universally true; by contrast *The sea is cold* would need to be verified at a particular place and time, and is therefore not analytic.

anaphor

1 A word or phrase that refers back to an earlier word or phrase.

In *My cousin said he was coming*, *he* is used as an anaphor for *my cousin* (which is its ANTECEDENT). But more usually *he* would be described as a PRO-FORM substituting for *my cousin*.

2 **zero anaphor, null anaphor**: a 'space' where a word or words have been ELLIPTED (the latter being the more usual label).

anaphora

1 The use of a word or words as a substitute for a previous linguistic unit when referring back to the thing, person, happening, etc., denoted by the latter.

Pronouns and other PRO-FORMS are frequently used anaphorically to avoid repetition. But sometimes a noun is repeated, and then the identity of reference is usually shown by a marker of definiteness (*the*, *that*, etc.) in the later (*anaphoric*) reference:

Old Mother Hubbard
went to the cupboard
to get her poor dog a bone;
But when *she* got *there*,
the cupboard was bare,
and so *the* poor dog had none.

She and *there* refer back to *Old Mother Hubbard* and *the cupboard* (line 2). *The cupboard* (line 5) and *the poor dog* (line 6) refer back to lines 2 and 3.

The term anaphora is sometimes extended to include more indirect reference, e.g. *I've still got a book of nursery rhymes I had as a child, but the cover is torn.* Obviously *the cover* refers back to *the book*.

2 Loosely, a hypernym for both *anaphora* (sense 1) and CATAPHORA.

When the term is so used, the two types are then distinguished as *backwards*, *backward-looking*, or *unmarked order anaphora* (= (1) above) and *forwards*, *forward-looking*, *anticipatory*, or *marked order anaphora* (=cataphora).

• **anaphoric**: involving anaphora (*anaphoric ellipsis*: see ELLIPSIS). **anaphorically**.

1914 O. JESPERSEN *The little one* is used anaphorically if it means 'the little flower' or whatever it is that has just been mentioned.

Compare ANTECEDENT.

and-relations

A self-explanatory label for SYNTAGMATIC relations. Contrasted with *OR*-RELATIONS (i.e. PARADIGMATIC relationships).

Anglo-Saxon

The same as OLD ENGLISH.

animate

Denoting a living being.

The term is particularly used in the classification of nouns. Animate nouns (e.g. *girl*, *tiger*, etc.) refer to persons and animals in contrast to **inanimate** nouns (e.g. *girlhood, zoo*) referring to things, states, and ideas. In English this distinction is almost entirely a matter of meaning, not grammar, although there is rough correspondence in some personal and relative pronouns: *he, she, who* usually have animate reference; *it, which* are mainly used in connection with inanimate referents.

anomalous finite

A finite verb form capable of forming the negative by adding *-n't* and of expressing questions by inversion.

This category includes all the MODAL verbs; all uses of *be*; *do* as an auxiliary verb; and some uses of *have* (e.g. *I haven't enough money*, but not, for example, forms using *do*-support: *I don't have enough money*).

This is a somewhat dated term for dealing with the problems of *be*, *do*, and *have*, which sometimes are, and sometimes are not, auxiliary verbs.

Compare DEFECTIVE, IRREGULAR.

antecedent

A word or words to which a following word or phrase grammatically refers back.

Typically antecedents are noun phrases to which personal and relative pronouns refer, e.g.

> My brother telephoned to say he'd be late

My brother is the antecedent of *he*.

Such a grammatical relationship can exist even when the pronoun refers back not to the identical person or thing, but to a previous linguistic form, e.g.

> I've lost *my umbrella* and shall have to get a new *one*

Less obviously *do*, *so*, *do so*, *there*, *then*, and a few other pro-forms can refer back to antecedents which may be verbal, adverbial, or clausal, e.g.

> I *cried* more than I'd ever *done* before in my life
> You could *buy a yearly season ticket*, but I don't advise *doing so*
> '*Petrol prices are going up again.*' 'Who told you *that*?'

Loosely, despite the meaning of the word, the term *antecedent* is sometimes extended to refer to phrases that come later than their pro-forms, e.g.

> If you see *her*, will you give *Mary* a message for me?

Compare ANAPHORA, CATAPHORA, PRO-FORM.

anterior

Phonetics. Of a speech sound: made in the front part of the mouth. Contrasted with **non-anterior**.

Anterior consonants are made with some stricture in front of the palato-alveolar area. Thus LABIAL and DENTAL sounds can be so classified.

anterior time

Time preceding some other time referred to by tense or other means.

The term is sometimes used in describing the meaning of perfect tenses. For example, in

(i) You're too late. They'*ve already left*
(ii) This time next week, he'*ll have forgotten* all about it
(iii) I realized *I had lost* my key

the perfect tenses indicate time before points in (i) the present, (ii) the future, and (iii) the past. The third type of anterior time, as in (iii), is commonly called *before-past*.

anticipated dislocation See DISLOCATION.

anticipation

Psycholinguistics. A slip of the tongue by which a linguistic element is used earlier than it should be.

This is a term used by some psycholinguists for part of what many people would call a *Spoonerism*, e.g.

The cat popped on its drawers

anticipatory

1 Anticipating the 'real' subject or (more rarely) object, when the latter is postponed until later in the sentence by EXTRAPOSITION.

In the first sentence below, *it* is an anticipatory subject, and in the second *it* is an anticipatory object:

It is better *to have loved and lost* than never to have loved at all
I take *it that you agree with me*

There is considerable confusion in the usage of the several terms available to describe various functions of the word *it*. For some grammarians, *anticipatory it* (used with extraposition) and *preparatory it* are identical, but they distinguish this usage from DUMMY (or EMPTY or PROP) *it*. Others use all or some of these terms differently, or use one of them as an umbrella term.

See also INTRODUCTORY.

2 *Phonetics & Phonology*.

Of modification to speech articulation: occurring under the influence of following sounds. Also called REGRESSIVE; contrasted with PROGRESSIVE.

In everyday speech, the /d/ sound in *good boy* may be articulated more like the following /b/ sound; or the final /s/ of *this* in *this ship* may tend towards a 'sh' sound /ʃ/. These are examples of *anticipatory assimilation* (also sometimes called *regressive assimilation*).

3 **anticipatory anaphora**: see ANAPHORA (2).

antonym

Semantics. A word opposite in meaning to another.

For example, *good*, *thick*, *few*, and *life* are antonyms of *bad*, *thin*, *many*, and *death*. More accurately we should talk of a word that is opposite in some of its meanings to other words. For example the antonym of some meanings of *old* is *young*, and of others is *new*.

Some linguists distinguish various types of opposite meaning and reserve the term *antonym* for gradable opposites (e.g. *good/bad*), excluding both COMPLEMENTARY terms (e.g. *life/death*) and relational CONVERSES (e.g. *buy/sell*, *teach/learn*, *husband/wife*).

Compare HYPONYM, SYNONYM.

• **antonymous**: that is an antonym. **antonymy**: the relationship of opposite meaning that exists between pairs of words (itself the opposite of SYNONYMY).

Compare BINARY, COMPLEMENTARY, COMPLEMENTARITY.

apex

The TIP of the tongue. See also APICAL.

aphaeresis (Pronounced /æ'fɪərəsɪs/. Plural **aphaereses.**)

1 The omission of a sound at the beginning of a word, regarded as a morphological development.

The now unpronounced sounds at the beginning of *gnat, knight, psyche* are examples. This is now usually handled as a type of historical ELISION.

2 The omission of a syllable at the beginning of a word, as routinely occurs in (a) contractions or (b) clippings.

e.g. (a) I'll (=I (wi)ll), You've (=You (ha)ve), He'd better (=He (ha)d); (b) (omni)bus, (tele)phone.

This phenomenon would now be dealt with under *elision*, *contraction*, *clipping*, and so on.

This is a somewhat dated term, taken via historical linguistics from its original use in rhetoric.

aphesis (Pronounced /'æfəsɪs/.)

The gradual loss of an unstressed vowel at the beginning of a word (e.g. of *e-* from *esquire*, giving *squire*).

> 1885 *New English Dictionary* *Aphesis* . . . It is a special form of the phonetic process called *Aphaeresis*, for which, from its frequency in the history of the English language, a distinctive name is useful.

This term, which was introduced by J. A. H. Murray, editor of the *New English Dictionary* (the first edition of the *Oxford English Dictionary*) in 1880, was used in the diachronic study of English; in phonetics the phenomenon it covers would be treated as an aspect of ELISION.

• **aphetic**: (pronounced /ə'fetɪk/) pertaining to or resulting from aphesis.

apical

Phonetics. Made with the APEX (tongue tip).

The tip of the tongue is not normally involved in the formation of English speech sounds, though it is used in the articulation of a trilled [r].

Compare LAMINAL.

apocope (Pronounced /əˈpɒkəpɪ/.)

1 The omission of a sound at the end of a word.

This has happened historically in such words as *lamb*, *damn*, and happens currently in rapid or colloquial speech, e.g. *you an(d) me*, *fish an(d) chips*, *cup o(f) tea*. The more modern term covering this phenomenon is ELISION.

2 The omission of a syllable or syllables at the end of a word.

This has happened historically with the loss, since Old English times, of many verb inflections (e.g. OE *we lufodon*, ME *we loveden*, *we lovede*, ModE *we loved*; OE *sungen*, ModE *sung*). Today it happens as a type of CLIPPING (e.g. *auto(mobile)*, *des(irable) res(idence)*, *long vac(ation)*, *spag(hetti) bol(ognese)*, *trad(itional)*).

Like *aphaeresis* and *aphesis*, *apocope* is an old term from diachronic linguistics.

Compare APHAERESIS, SYNCOPE.

apodosis (Pronounced /əˈpɒdəsɪs/. Plural **apodoses**.)

The main clause in a complex sentence, particularly in a CONDITIONAL sentence.

e.g. *I would be upset* if they found out.

apo koinou (Pronounced /ˈæpəʊ ˈkɔɪnuː/.)

Applied to a construction consisting of two clauses which—unusually—have a word or phrase that is syntactically shared.

In *There's a man outside wants to see you*, *a man* is both the postponed subject of *-'s* (=*is*) and the subject of *wants*.

The term is not in general use today; such a construction would be called a BLEND or treated as deviant. But see CONTACT CLAUSE.

The term comes from Greek *apo koinou* 'in common'.

apostrophe

The sign ⟨'⟩ used in order to indicate (a) the omission of a letter or letters, as in *don't*, *thro'*, the *'90s*; and (b) the modern genitive or possessive case, as in *boy's, men's*.

The apostrophe is rarely used in abbreviations that are no longer felt to be contractions (e.g. *(in)flu(enza)*) or in colloquial abbreviations (e.g. *info(rmation)*). *O'clock* is an obvious exception.

The possessive apostrophe originally marked the omission of *e* in writing (e.g. *fox's*, *James's*), and was equally common in the nominative plural especially of proper names and foreign words (e.g. *folio's* = *folioes*); it was gradually disused in the latter, and extended to all possessives, even where *e* had not been previously written, as in *man's, children's, conscience' sake.*

In modern English the use of the apostrophe to mark ordinary plurals (e.g. *potato's*, *ice-cream's*) is generally regarded as illiterate and is disparagingly referred to as the 'greengrocer's apostrophe'. It is usually acceptable with the less usual plurals of letters and dates, e.g.

> Mind your *p's* and *q's*
> That is what people did in the *1960's/1960s*

The current rules for possessive apostrophes are:

> Add *'s* to a singular word (e.g. *the boy's statement*, *an hour's time*, *Doris's husband*, *her boss's address*)
> Also add *'s* to plural words that do not end in *s* (e.g. *the men's action*, *the people's will*)
> With plural words ending in *s* only add an apostrophe at the end (e.g. *ladies' shoes*, *the Lawsons' house*).

Names from ancient times ending with *s* do not necessarily follow these rules (e.g. *Socrates' death*).

It is an error to use the apostrophe with possessive pronouns (e.g. *hers, its, ours, theirs, yours*). *It's* means 'it is' or 'it has'. For 'belonging to it', the correct form is *its* (e.g. *The cat hasn't eaten its food*).

appellative

(In older grammar.) (*n.*) A common noun used as a name, particularly when addressing someone, e.g. *Mother*, *Sir*, *Doctor*.

In modern grammar, dealt with as a VOCATIVE.

(*adj.*) Of a word, especially a noun: designating a class, not an individual; common, in contrast to PROPER.

> 1755 S. JOHNSON As my design was a dictionary, common or appellative, I have omitted all words which have relation to proper names.

applied linguistics See LINGUISTICS.

apposition

A relationship of two (or more) words or phrases, especially noun phrases, such that the two units are grammatically parallel and have the same referent. e.g.

> *Our longest reigning monarch*, *Queen Victoria*, reigned from 1837 to 1901
> The second edition of *OUP's biggest dictionary*, *the Oxford English Dictionary*, was published in 1989

Grammarians vary in their use of the term *apposition*. In the narrowest definition (*full apposition*), both parts are noun phrases, they are identical in reference (i.e. in an *equivalence* relationship), and either part could be omitted, as in the examples above, without affecting the grammaticality or the essential meaning of the sentence.

More loosely, apposition may include pairs of units where not all these conditions apply (*partial apposition*). One part may be a clause (see APPOSITIVE CLAUSE). Or the relationship may be one of example, not identity, e.g.

A monarch, for example a twentieth-century monarch, may have limited powers

Or the omission of one part may result in an ungrammatical sentence, e.g.

A very important person is coming—the Queen (we could not say *is coming—the Queen*)

Apposition is predominantly non-restrictive (the second part adds information but is not essential), but restrictive apposition is possible (e.g. *the author Graham Greene*).

Grammarians also disagree about whether many such structures as the following show apposition or not:

the number thirteen
my sister Mary
the expression 'greenhouse effect'

and structures joined by *of* where the two parts share identity, as in *the city of Oxford, that fool of a man.*

appositional

Of or pertaining to apposition.

•• **appositional compound**: a compound which, semantically, is a hyponym of both of its constituent words.

e.g. manservant, drummer boy, oak tree.

appositive

(*n. & adj.*) (An element) standing in apposition to another element.

1990 S. GREENBAUM & R. QUIRK Appositives need not be noun phrases; compare:
She is *bigger* than her brother, *heavier*, that is.
Sixthly and *lastly*, I reject the claim on ethical grounds.
He *angered*, nay *infuriated*, his audience.

appositive clause

A finite clause often introduced by *that*, defining and postmodifying a noun phrase, and sharing identity of reference with it, e.g.

They had the idea *that everything would be all right in the end*

Or a similar, but non-defining clause:

They ignored Wendy's very sensible suggestion, (namely) *that the police should be told*

Such clauses are grammatically distinct from relative clauses. The *that* is a conjunction, not a pronoun, so these clauses have their own subject (and object), and *which* is never possible. The preceding noun is an abstract noun, such as *belief, fact, idea*.

The term *appositive clause* can be extended to non-finite clauses, e.g.

The request, *to send money*, shocked us
My work, *looking after these old people*, is rewarding

appropriacy

Sociolinguistics. The quality of being suitable to a given social situation.

The concept of appropriacy is linked to those of FORMAL versus INFORMAL, REGISTER, and so on.

• **appropriate**.

approximant

Phonetics. A sound made with an unimpeded airflow; contrasted with STOP and FRICATIVE.

Phoneticians group speech sounds in different ways. *Approximant* is used as a general term covering sounds made in various manners of articulation.

1971 P. LADEFOGED For the moment we may consider the approximant category to be simply a convenient general term to include what others have called semivowels, laterals, and frictionless continuants (as well as vowels).
1991 P. ROACH The term 'approximant' is usually used only for consonants.

Compare CONTINUANT.

arbitrary

Lacking any physical or principled connection.

Most language is arbitrary in this sense, however systematic its grammar. There is no inherent reason why a particular domestic animal should be *dog* in English, *chien* in French, *Hund* in German, *perro* in Spanish, and *cane* in Italian.

Compare ICONIC, ONOMATOPOEIC.

archaic

Of a word or grammatical structure: no longer in ordinary use, though retained for special purposes.

archaism

1 The use of words or grammar characteristic of an earlier period of the language.

2 An instance of such usage.

Some archaisms survive in special registers, such as legal or religous language, or they are familiar from literature, proverbs, and so on. Archaisms include

(*a*) individual words, e.g.

albeit
methinks
perchance

thou/thine/thee/thy
whence
whereat
ye

(*b*) verbal inflections, e.g.

goeth
knowest

and (*c*) some grammatical structures, e.g.

Our Father, *which art* in heaven (= *who are*)
He who hesitates is lost (= *Anyone who*)
All that glisters *is not* gold (= *Not everything that . . . is . . .*)
We *must away* (= *must go/leave/be off*)
Would that I could help (= *I wish*)
So be it
If it *please* your lordship
He has come but once in *these two years past*

argument

Semantics. A major element in a proposition.

The term is used in a technical sense from predicate calculus, as a part of logical semantics. Roughly speaking, an argument is the person, other animate being, or inanimate entity involved in the action of the verb.

1984 R. HUDDLESTON Semantic predicates may be classified according to the number of arguments they take. Thus "love" (the semantic predicate expressed by **love**) takes two; "sleep" (as in *Ed was sleeping*) takes one; "give" (as in *Ed gave me the key*) takes three.

The term hardly appears in mainstream grammar, where the phenomenon would be dealt with in terms of *intransitive*, *transitive*, *ditransitive verbs* etc.

article

A name for *the* (**definite article**) and *a, an* (**indefinite article**).

The articles are sometimes classified as a distinct part of speech. Earlier grammarians considered them to be a special kind of adjective.

1711 J. GREENWOOD There are two *articles*, *a* and *the*. These are really Nouns Adjective, and are used almost after the same Manner as other Adjectives. Therefore I have not made the Article (as some have done) a distinct Part of Speech.

In fact articles are much more like DETERMINERS than adjectives in their usage, and modern grammarians usually classify them as a subclass of determiner.

See also ZERO *article*.

articulate

Phonetics. Produce a speech sound.

See ARTICULATION.

articulation

Phonetics. The physical production of speech sounds.

Speech sounds are described in terms of both their PLACE and their MANNER of articulation. For example, English [p], [b], and [m] are bilabial sounds (made with both lips). In manner, [p] and [b] are plosives: that is, they are made by a complete closure followed by a release (an 'explosion') of air; [p] is voiceless and [b] is voiced. By contrast, [m] is articulated in a nasal manner, for though the lips are closed during its articulation, air escapes through the nose.

Compare COARTICULATION, DOUBLE ARTICULATION.

articulator

Phonetics.

1 A movable vocal organ.

> 1942 B. BLOCH & G. L. TRAGER The vocal organs are conveniently divided into two kinds: *articulators*, organs which can be moved more or less freely and can thus be made to assume a variety of positions; and *points of articulation*, fixed points or areas lying above the articulators, which these may touch or approach.

2 Any vocal organ, moving or not, involved in the production of speech sounds.

In this second usage, the moving organs may be termed *active articulators* and the others *passive articulators.*

articulatory

Phonetics. Of or pertaining to vocal articulation.

The International Phonetic Alphabet is largely based on articulatory features.

•• **articulatory phonetics**: the branch of phonetics concerned with the ways in which speech sounds are physically articulated.

> 1964 R. H. ROBINS Speech can therefore be studied in phonetics from three points of view. 1. It can be studied primarily as the activity of the speaker in terms of the articulatory organs and processes involved; this is called articulatory phonetics.

Compare ACOUSTIC PHONETICS, AUDITORY PHONETICS.

artificial language

A specially invented language, usually intended as a means of international communication.

Esperanto is perhaps the best known artificial language, but many others have been invented.

aspect

A category used in describing how the action of a verb is marked.

English is often considered to have two aspects: PROGRESSIVE (as in *I am/was writing to Robert*), which stresses action in progress (or incomplete action); and PERFECT or PERFECTIVE (as in *I have written to Robert*), which stresses completed action. This distinction of incomplete versus complete meaning is an oversimplification. More seriously, the analysis of 'perfect tenses' as showing aspect is itself disputed. See also PHASE (3).

Traditionally both aspects are treated as part of the tense system in English, and we commonly speak of tenses such as the present progressive (e.g. *We are waiting*) or even the past perfect progressive (e.g. *We had been waiting*), which combines two aspects. There is a distinction to be made, however, between tense and aspect. Tense is more concerned with past time versus present time and is based on morphological form (e.g. *write, writes, wrote*); aspect is concerned with duration, and in English is a matter of syntax, using parts of *be* to form the progressive, and of *have* to form the perfective.

• **aspectual**: of or pertaining to an aspect or aspects.

Aspects model, Aspects theory. See STANDARD *theory*.

aspirate

Phonetics.

(*n.*) (Pronounced /ˈæspərət/.) (In popular parlance) the sound of *h*.

This use is illustrated by the old joke:

A: I've got a 'orrible 'eadache.
B: What you need is a couple of aspirates.

In phonetic terms English /*h*/ is a voiceless glottal fricative.

The use of *aspirate* by grammarians of the past to mean 'fricative' (e.g. *th*, *ph*, etc.) is now obsolete.

(*v.*) (Pronounced /ˈæspəreɪt/.) Articulate (a sound) with an audible release of air.

In English the term is relevant to the description of plosives. The voiceless plosives /p/, /t/, and /k/ are aspirated in the final stage of their articulation when in initial position in a stressed syllable (e.g. *pun*, *two*, *careful*). In contrast, the same phonemes have little or no aspiration when initial in unstressed syllables (*p*erˈmission), when preceded by *s* (s*t*ory), or in final position, i.e. followed by silence (Bad lu*ck*!). See also ASPIRATION.

• **aspirated**: articulated with an audible release of air (contrasted with **unaspirated**).

Aspirated and *voiceless* articulations often occur together, but are distinct phenomena. *Voiced* and *voiceless* refer to the state of the vocal cords throughout the articulation of a phoneme; *aspirated* and *unaspirated* refer to the final release stage of plosion.

1971 P. LADEFOGED Aspirated and unaspirated refer to the state of the vocal cords during and immediately after the release of an articulatory stricture. In any aspirated sound the vocal cords are in the voiceless position during the release; in an unaspirated sound they are in a voiced position during this period.

aspiration

Phonetics. (Articulation accompanied by) an audible release of air.

1962 A. C. GIMSON The fortis series, /p,t,k/, when initial in an accented syllable, are usually accompanied by aspiration, i.e. there is a voiceless interval consisting of strongly expelled breath between the release of the plosive and the onset of a following vowel. When /l,r,w,j/ follow /p,t,k/ in such positions, the aspiration is manifested in the devoicing of /l,r,w,j/, e.g. in *please, pray, try, clean, twice, quick, pew, tune, queue.*

assertion

Pragmatics. A declaration that something is true.

In general, assertions tend to be made in grammatically positive sentences. Nevertheless, assertion is a pragmatic rather than a grammatical category, and other syntactical forms can make an assertion, e.g.

Isn't it hot today? (= it is hot today)
Wrong number! (= you have dialled/I got a wrong number)
If you believe that, you'll believe anything (= you cannot possibly believe that)

assertive

Of words and other sentence elements: typically used in positive statements; contrasted with NON-ASSERTIVE elements (typical of questions and negative statements).

Certain determiners, pronouns, and adverbs which, by reason of their meaning, are more usual in positive statements, are replaced by a corresponding set of *non-assertive* forms in negative and interrogative statements. Assertive forms include the *some* series of words (*some, someone, somewhere,* etc.) and adverbs such as *already*, which contrast with the non-assertive *any*-series and adverbs such as *yet*. Compare, for example:

I've already planted some spring bulbs

with:

I haven't planted any bulbs yet
Have you planted any yet?

The demarcation between assertive and non-assertive is not rigid. Although non-assertive forms are often impossible in positive statements (e.g. **I've already planted any*), assertive forms may be possible in questions and negative clauses, where they suggest a markedly positive meaning (e.g. *Have you planted some bulbs already?*). Non-assertive forms occasionally occur in positive statements (e.g. *Any* small bulbs would be suitable).

•• **assertive territory**: a term applied to the whole predication in positive statements, an area where assertive forms may be expected.

Similarly the predication of negative statements and questions is *non-assertive territory*. Non-assertive territory also includes clauses expressing other kinds of tentativeness, such as condition (e.g. *If anyone calls*, say I'm designing St Paul's).

Compare NEGATIVE.

- **assertiveness**: the fact of being assertive.

  1985 R. QUIRK et al. The contrast between assertiveness and non-assertiveness is basically a logical one.

assimilate

Phonetics. Make or become more similar in articulation (to an adjacent sound); (cause to) undergo assimilation.

1970 B. M. H. STRANG In the word *Tuesday* the opening sequence /tj/ can readily assimilate to /tʃ/ . . . Early NE [New English] had assimilated /dj/ to /dʒ/ and /tj/ to /tʃ/.

See ASSIMILATION.

assimilation

Phonetics. The effect on a speech sound of the articulation of other adjacent sounds; a kind of COARTICULATION.

This is a common feature of speech, though one that many native speakers are unaware of. In *anticipatory assimilation* (or *regressive assimilation*), the sound is influenced in its articulation by the following sound and not pronounced as it would be in isolation. For example, in some people's pronunciation of *width* the voiced /d/ has been assimilated to /t/ by the following voiceless /θ/, and in some people's pronunciation of *length*, the velar /ŋ/ has been assimilated to /n/ by the following dental /θ/.

In current speech, assimilation frequently occurs across word boundaries, as when *that case* becomes /ðæk keɪs/ or *this shop* becomes /ðɪʃ ʃɒp/ or *ten more* becomes /tem mɔː/.

A reverse type of assimilation (*progressive assimilation*) is found when a sound is changed by the influence of a previous one. This is an established and regular feature of the ending *-s* of verbs and nouns, which usually has a voiced /z/ sound (or /iz/ after all sibilants) but after voiceless sounds other than sibilants is /s/ (e.g. *taps*, *heats*, *dock's*, *griefs*, *Keith's*; compare *tabs*, *heeds*, *dog's*, *grieves*, *youths*, *eyes*, *seems*, *runs*, *dolls*, *pieces*, *daisies*.) Similarly the past tense *-ed* ending /d/ or /id/ is devoiced to a /t/ sound after a voiceless consonant other than *-t* itself (*roped, lacked, bussed, roofed, pushed* versus *robed, lagged, buzzed, grooved, rouged, hated, headed*).

In *coalescent assimilation* (or *reciprocal assimilation*), *there is a two-way influence. Historically this has occurred in words like soldier*, *picture*, or *fissure*, where the reconstructable earlier pronunciation ['souldjər], ['pɪktjuːr], ['fɪsjuːr] has become ['səʊldʒə], ['pɪktʃə], ['fɪʃə]. In current colloquial English, similar assimilation occurs in phrases such as *What d'you want?* /wɒtʃə wɒnt/ or *Could you?* /'kʊdʒuː/. This coalescent assimilation is also known as *yod coalescence*.

asyndetic

Not connected by conjunctions. Contrasted with **syndetic**.

The term is particularly applied to the coordination of words or clauses without an overt marker, as in

asyndeton

I came, I saw, I conquered.
Sad, white, frightened, her face was the picture of misery.

Such coordination is less usual than coordination with a conjunction and is therefore stylistically marked.

Compare PARATACTIC.

asyndeton

The omission of a conjunction. (Chiefly a rhetorical term.)

attachment rule

The 'rule' that the notional subject of a subjectless subordinate clause has the same referent as the subject of the superordinate clause.

This 'rule', sometimes called the *subject-attachment rule*, is a regular characteristic of English syntax, e.g.

Set loose by a freer economy and the easing of travel restrictions, they were selling fruit between one province and another

where it is the *they* of the main clause who were *set loose*.

There are exceptions where the rule need not apply, but failure to observe it often results in absurd MISRELATED constructions or HANGING PARTICIPLES.

attitudinal

Semantics. Relating to attitude or conveying an attitude.

A category of meaning (contrasted with COGNITIVE or REFERENTIAL meaning). Comparable terms sometimes used are AFFECTIVE or EMOTIVE.

A speaker or writer's choice of words may be affected by his or her attitude to the listeners or readers addressed. Thus *attitudinal varieties* of language (e.g. formal, informal, neutral, casual, slangy) contrast with regional or social varieties (determined largely by the speaker's origins), and with varieties affected by the 'field of discourse' (such as the journalistic, the literary, and so on). Attitudinal meaning is also conveyed by INTONATION in spoken English.

Compare CONNOTATIVE, INTERPERSONAL.

Many individual words are neutral as regards attitudinal meaning, but some are not. Compare *offspring, child, kiddy* or *terrorist, insurgent, guerilla, freedom fighter*.

•• **attitudinal disjunct**: the same as CONTENT DISJUNCT.

attitudinal past: the past tense used to express the speaker's attitude, usually a tentative one, rather than to refer to past time; e.g., *Did you want something?*, *I was hoping you could help*.

attitudinal prefix: a prefix that primarily describes an attitude, e.g. *anti*-social, *pro*-American.

attraction

The same as *proximity* AGREEMENT.

attribute

1 An adjective or noun preceding a noun and describing or expressing a characteristic of the noun; e.g. *new* in *the new library* or *power* in *power struggle*.

2 An adjective (phrase) or noun (phrase) acting as subject or object COMPLEMENT, and identifying or adding information about the subject or object, as *capable* in *She seems capable*, or *a capable person* in *We find her a capable person*.

The second meaning is less general than the first and also potentially confusing, although, etymologically, the sense is 'a word or phrase denoting an attribute of a person or thing' however that may be expressed within the sentence.

See ATTRIBUTIVE, PREDICATIVE.

attributive

1 Designating (the position in a sentence taken by) a noun or adjective that is an attribute (sense 1); contrasted with PREDICATIVE.

Some adjectives can only stand in attributive position (e.g. *former, inner, mere, lone, main, indoor*). Nouns used attributively cannot normally be transferred to predicative use (**the effect is greenhouse*, **this holiday is bank*), though a small number of predicative uses of nouns may be the historical result of such a transfer (e.g. *She is frightfully county*; *So Regency, my dear*).

Attributive adjectives are sometimes classified in more detail according to meaning. Subclasses include INTENSIFYING *adjective* (e.g. *pure invention, utter madness, total stranger*), RESTRICTIVE *adjective* (also called LIMITING or LIMITER ADJECTIVE) (e.g. *a certain person, the main trouble, that very day*), and adjectives related to nouns (*the criminal code, medical students*).

2 (In Systemic-Functional Grammar.) Designating any relationship in which an attribute is ascribed to an entity.

This may be a quality (e.g. *sensible* in *She is sensible*), a circumstance (e.g. *in the garden* in *She is in the garden*), or a possession (e.g. *a lovely garden* in *She has a lovely garden*).

• **attributively**.

auditory phonetics

The study of speech sounds from the point of view of the listener, concerned with the way the ears and brain process and perceive the speech sounds reaching them.

Compare ACOUSTIC PHONETICS, ARTICULATORY *phonetics*.

autosemantic

Semantics. (*n. & adj.*) (A word or phrase) that has meaning outside a context or in isolation.

1962 S. ULLMAN Full words are 'autosemantic', meaningful in themselves, whereas articles, prepositions, . . and the like are 'synsemantic', meaningful only when they occur in the company of other words.

Since most lexicographers manage to ascribe meanings to all words, and not merely to 'full' words, and even many full words change meaning according to context, the term is obviously somewhat relative.

Compare FULL *word.*

auxiliary

(*n. & adj.*) (A verb) used in forming tenses, moods, and voices of other verbs. In older grammar called *helping verb.*

(*a*) The verbs used for this purpose in English include *be*, *do*, *have*, and the MODAL verbs (sometimes called the *modal auxiliaries*).

An auxiliary cannot function as the only verb in a complete sentence. Apparent exceptions to this principle are in fact examples of ellipsis or substitution. Auxiliaries thus contrast functionally with MAIN verbs.

(*b*) On formal grounds an auxiliary can also be defined as a member of a class that is grammatically distinct from other verbs, notably in the way in which it forms questions by inversion of subject and verb (e.g. *Are you ready?*, *Can you help?*; but not **Want you to help?*) and negatives by simply adding *-n't* (e.g. *They aren't*, *We mustn't*, *She doesn't*; but not **She wantn't*). (See NICE PROPERTIES.) Auxiliary verbs thus contrast with LEXICAL verbs (*full verbs*), which form questions and negatives with (auxiliary) *be*, *do*, and *have*.

Both definitions fit the modal verbs. The verbs *be*, *do*, and *have*, however, present problems, since they can be used both in an auxiliary function and as the MAIN verb of a sentence, and *do* as a main verb uses *DO*-SUPPORT for questions and negatives, unlike *be* and *have*. In some grammatical models, therefore, the terms *auxiliary* and *main* are reserved for the functional role of verbs in sentence structure, and other terms are used for the formal classifications.

1985 R. QUIRK et al. We shall find good grounds for distinguishing 'auxiliary' and 'main', as functional terms, from the terms which define classes of word. Of these, there are three: MODAL verbs . . always function as auxiliaries; FULL verbs . . always function as main verbs; and PRIMARY verbs . . can function either as auxiliaries or as main verbs.

See also SEMI-AUXILIARY.

a-word

A word beginning with the syllable *a* and belonging to a class of words (some more like adjectives, the others more like adverbs) that mainly function predicatively.

Grammarians have variously classified *a-words* as adjectives and as adverbs, but few even among the adjective-like ones can be used in attributive position, e.g.

The children were afraid/alone/ashamed/awake

but not

*An afraid/alone/ashamed/awake child was crying

Some of the more adjective-like words can be used attributively if modified, for example:

You see before you a very ashamed person

and some can be modified by *much* (which typically goes with verbs), e.g.

I am very much afraid that . . .

The more adverb-like *a*-words can follow verbs of motion, e.g.

They've gone abroad/aground/away

Some *a*-adverbs have identical prepositions, e.g. *aboard, above, across, along, around,* etc., as in

We hurried along (adverb)
We hurried along the road (preposition)

There are also some words beginning with *a-* which are prepositions only (e.g. *amid*, *among*), but prepositions are usually excluded from the *a*-word category.

B

back

Phonetics. Of a speech sound: made in the back part of the mouth.

Vowel sounds are traditionally classified into *back*, CENTRAL, and FRONT vowels, the back vowels being made with the tongue humped towards the back of the mouth. The English back vowels are the vowels as pronounced by a standard speaker in:

hard [hɑːd]	hot [hɒt]
law [lɔː]	food [fuːd]

Among English consonants, the somewhat vowel-like h-sound /h/ and the non-standard GLOTTAL STOP /ʔ/ are both far back sounds, but there are no 'throaty' pharyngeal sounds like those that occur in Arabic and some other languages.

Compare UVULAR, VELAR.

back-formation

Morphology. The formation of a new word by the removal of (real or apparent) affixes etc. from an existing word; a word that is an instance of this.

A back-formation is revealed by the fact that the date of its first use is later than that of its apparent derivative. The majority of back-formations in English are verbs.

Examples in English are:

burgle (L19)	from *burglar* (M16)
caretake (L19)	from *caretaker* (M19)
housekeep (M19)	from *housekeeping* (M16)
liaise (E20)	from *liaison* (M17)
reminisce (E19)	from *reminiscence* (L16)
scavenge (M17)	from *scavenger* (M16, from earlier *scavager*)
shoplift (E19)	from *shoplifting* (L17)

backshift

(*n.*) The changing of tenses used in direct speech to past tenses when reporting indirectly what someone said or thought.

When using the past tense of a reporting verb (e.g. *He said*, *They thought*, *I remarked*) it is common to shift the tenses of the words spoken or thought into past tenses too. Thus

> 'I am sorry I haven't asked them yet. I will.'

may (with backshift) become

Mark said he was sorry he hadn't asked them yet, but he would.

Similarly a past tense (*'I asked them'*) may become a past perfect (*He said he had asked them*).

Backshift (sometimes known as the *sequence of tense rule*) is not, however, automatic. Importantly, if the time frame of the person writing or speaking now is the same as that of the original speaker whose words are being reported, tenses do not need to change:

He said he wouldn't } be around in the year 2050.
He said he won't }

This explains why the past tenses of direct speech are often not subject to backshift:

She told us that both her parents (had) died when she was ten (and not **when she had been ten*).

(*v.*) Change (a tense) in this way.

1985 R. QUIRK et al. The past subjunctive . . or hypothetical past . . is backshifted to hypothetical past perfective if there is a change in time reference.

Compare FREE *indirect speech*.

backwards anaphora See ANAPHORA.

bahuvrihi

Morphology. Of a compound noun: having the meaning 'a person or thing possessing a certain characteristic'.

The term is derived from a Sanskrit word, formed from *bahu* 'much' and *vrihi* 'rice', and literally meaning 'having much rice'. In English grammar it is usually applied to compounds with a non-literal meaning that cannot be deduced from the literal meaning of the separate parts. It is particularly used of somewhat pejorative words, in which some unflattering attribute stands for the person alleged to possess it. English compounds of this type include:

blockhead (M16)	hunchback (E18)
butterfingers (M19)	lazybones (L16)
egghead (E20)	loudmouth (M20)
fathead (M19)	paleface (E15)
highbrow (L19)	scatterbrain (L18)

The term is sometimes used in a wider sense to include other compounds in which the head element cannot be equated with the whole. For example, just as an *egghead* or a *scatterbrain* is a person, not a kind of head or brain, so a *hardback* is not a type of back, but a book with this characteristic.

Compare EXOCENTRIC.

bare infinitive

The infinitive of a verb without a preceding *to* particle.

Bare infinitives are thus identical with the BASE form of the verb. They are used after

the main modal verbs; e.g. *I must go, I shall return!* and can occur with semi-modals; e.g. *You needn't bother*
the verb *do* in questions and negatives; e.g. *Does he know? They didn't say*
verbs of perception; e.g. *We saw/heard them go, I felt it bite me* (these verbs can also be followed by *-ing* forms)
the verbs *make* and *let*; e.g. *Make/let them wait*
a few fixed expressions; e.g. *make do, make believe, (live and) let live, let go*
and in various other patterns; e.g. *I'd rather try than do nothing.*

A bare infinitive is only rarely interchangeable with a *TO*-INFINITIVE, and some of the verbs listed above need a *to*-infinitive when used in the passive, e.g. *We were made to stand.* The verb *help* is unusual, in that both the bare infinitive and the *to*-infinitive are often possible, e.g. *Please help (me) (to) do the washing-up.*

base

Morphology.

1 The basic or uninflected form of a verb. Also **base form.**

Go, like, and *sing* are bases or base forms, in contrast to *went, likes, sang,* which are not.

The base form of a verb functions (*a*) non-finitely

as the infinitive (e.g. *You must go*).

(*b*) finitely

as the imperative (e.g. *Listen!, Be quiet, Have a biscuit*)
as the present indicative tense for all persons other than the third person singular (e.g. *I always listen* as opposed to *He always listens*) (the verb *be* is an exception to this)
as the so-called present subjunctive (*They insisted that he listen*).

2 A basic element in word-formation. Also **base morpheme.**

(*a*) Usually the base is an irreducible 'core', to which one or more affixes are attached; e.g. *sing* + *-s* = *sings, great* + *-er* = *greater, great* + *-ly* = *greatly. Infectious* consists of a base *infect* + a suffix *-ious*; *indiscreet* consists of the negative prefix *in-* + the base *discreet.*

(*b*) Various problems arise however in the analysis of some less simple words, which sometimes do not consist merely of an affix or collection of affixes attached to a core word. For example, *unanswerable* consists of the negative prefix *un-*, the word *answer*, and the suffix *-able*. But we do not normally have a word **unanswer*; we can only attach *un-* to *answerable.*

A related problem is this: we can sometimes analyse a word as containing an affix, but what remains is not recognizable as a core word. *Gratuitous* apparently has the adjective suffix *-ous* (compare *pompous, monstrous, outrageous*), just as *gratuity* has a noun suffix *-y*, but we do not recognize *gratuit-* as a word.

Some linguists reserve the term *base* for units of these two types (*answerable* in the context of *unanswerable* and the incomplete form *gratuit-*). They may then refer to units such as *answer, sing, great* (as core words to which additions can be made as in 1 above) as ROOTS or STEMS. Thus *base* may be less 'basic' than *stem* or *root*.

This is, however, an area where terminology is confused.

1985 R. QUIRK et al. In some linguistic descriptions the minimal unit in morphology and word formation is called 'morpheme', with the further distinctions 'inflectional morpheme' (eg: plural *-s*), 'free morpheme' or minimal free form (eg: *pole*), 'bound morpheme' (eg: *un-*, *jeal-*), with the latter necessarily further subdivided between 'affixal morpheme' (eg: *un-*) and 'stem morpheme' (or 'root' or 'lexical morpheme', eg: *jeal-*). What we are calling BASE might in this framework be termed 'base morpheme'. It should be noted that linguists differ in their terminology for these distinctions, some reversing our use of STEM and BASE, others using 'root' for what in this book is called 'stem'.

3 **base component**: (in Generative Grammar) a major part of the grammatical apparatus, along with such other constituents as the TRANSFORMATIONAL *component*.

In the Aspects model, the *base component* contains the categorial and the lexical sub-components.

1980 E. K. BROWN & J. E. MILLER These two—the . . phrase structure rules and the lexicon—comprise what is referred to as the 'base component'.

Basic English

A variety of the English language, comprising a select vocabulary of 850 words, invented by C. K. Ogden, of Cambridge, and intended for use as a medium of international communication.

The word *basic* was a clever acronym for *British American Scientific International Commercial* (*English*). Ogden's book, *Basic English*, was published in 1931, and the idea enjoyed some vogue. But the language produced tended to be unnatural and un-English, an artificial language rather than simplified English.

Basic English is not to be confused with the computer language called BASIC, or Basic, an acronym for *Beginners' All-purpose Symbolic Instruction Code*.

basilect

Sociolinguistics.

1 Specifically, in a community in which a CREOLE has been current (a post-creole community), the social dialect that is furthest away from the standard language and nearest to the creole.

2 More generally, the least prestigious variety of a language. The term can be used to describe dialects of people speaking English as their mother-tongue, and may also be applied in communities where English is used as a second or third language.

A basilect is closer to the standard language of which it is a version than either a creole or a PIDGIN.

See ACROLECT, MESOLECT.

• **basilectal**: of, pertaining to, or characteristic of a basilect.

1977 J. T. PLATT Basilectal Singapore English is not a creole but what I have termed a 'creoloid'. It did not develop from a pidgin in the typical manner of a creole... There is a gap between the most basilectal form of Singapore English and the English-lexicon pidgins used in Singapore.

BBC English

Standard English, as supposedly spoken by professional BBC broadcasters.

In its early days the British Broadcasting Corporation encouraged a standard non-regional 'educated' accent among its broadcasters. The policy was established by the first managing director, John Reith, who sought 'a style or quality of English that would not be laughed at in any part of the country' and was implemented by the Advisory Committee on Spoken English, established by Reith in 1926 and succeeded, during the Second World War, by the BBC Pronunciation Unit. BBC policy has been considerably modified since the 1950s, and 'BBC English' is now only one of the accents heard from newsreaders, announcers, and other programme presenters.

Compare RP.

before-past See PAST PERFECT.

benefactive

(*n. & adj.*) (Indicating) the case or role taken by a noun (or noun phrase) referring to a person or animal that is intended to benefit from the action of the verb. Contrasted with RECIPIENT.

In an inflected language such as Latin, *benefactive* could describe the meaning of the dative case. In English this 'intended recipient' meaning is often indicated by a *for-* phrase. In

I bought her a present
I bought this for you
She got the poor dog a bone
He found himself a job

her, *you*, *the poor dog*, and *himself* have benefactive meaning. Some grammarians call objects that allow a prepositional construction with *for* '*benefactive objects*': so they distinguish between *her* in *I gave her a present* (recipient role) and *her* in *I bought her a present* (benefactive role).

Compare DATIVE.

beneficiary

Semantics. The person (or animal) who benefits from the action of the verb; see BENEFACTIVE.

bilabial

Phonetics (*n. & adj.*) (A speech sound) made with closed or nearly closed lips.

The English bilabials are /p/, /b/, and /m/, as in *pan*, *ban*, and *man*.
Compare LABIAL, LABIO-DENTAL.

bilateral

Phonetics. With the air released around both sides of the tongue.
A bilateral articulation is the normal articulation of LATERAL sounds in English. It contrasts with *unilateral* articulation, by which the air, unusually, is released around one side only.

bilingual

(*adj.*)
1 Able to speak two languages fluently.
2 Spoken or written in two languages (e.g. *a bilingual dictionary*).
(*n.*) A person who speaks two languages fluently.
• **bilingualism,** (rarely) **bilinguality**.
Compare AMBILINGUAL.

binary

1 Designating or relating to a pair of features in a language which are mutually exclusive, or the opposition between them.
(*a*) *Phonetics.* The contrasts between nasal and non-nasal or voiced and voiceless articulations are said to be *binary oppositions* or *binary features*. Such features are sometimes marked with a plus or minus sign. Thus /p/ is characterized as [-voice] and /b/ as [+voice].
Compare DISTINCTIVE FEATURE.
(*b*) Binary (dichotomous) contrasts are a notable feature of vocabulary, which contains many pairs of words of opposite meaning. The phenomenon is often dealt with under *antonymy*. Antonyms, however, include, or may be restricted to, gradable pairs (e.g. *good/bad*, *high/low*), whereas binary opposition, strictly speaking, characterizes pairs with an ungradable, all-or-nothing contrast (sometimes called *complementaries*), such as *alive/dead*, *married/single*, *human/non-human* (contrast the gradable *inhuman*). *Converse* relations (e.g. *buy/sell*, *husband/wife*) can also be considered binary.
(*c*) In Generative Grammar, both syntactic and semantic binary contrasts are made. Nouns, for example, can be characterized as + or – ANIMATE, + or – ABSTRACT. An alternative notation for [+ or –] is [±]. The system has considerable limitations, since by no means all syntax or vocabulary lends itself to this kind of either/or analysis.
Compare COMPONENTIAL ANALYSIS.
2 **binary noun**: the same as SUMMATION *plural*.

binding See GOVERNMENT-BINDING THEORY.

binomial

A phrase containing two parallel units joined by a conjunction, in which the order is relatively fixed. Sometimes called *irreversible binomial.*

e.g.

blood and thunder	ladies and gentlemen
heaven and hell	one and all
highways and byways	thick and thin
knife and fork	

Compare FIXED PHRASE.

biunique

Phonology. Consisting of or characterized by one-to-one correspondence between sound and phoneme.

The term is used in some theories of phonology that assert that there is a 'reversible' correspondence between phonemes and the sounds (ALLOPHONES) by which they are realized. But this idea is disputed.

•• **biuniquely. biuniqueness**.

blade

Phonetics. The tapering section of the front of the tongue, immediately behind the tip. Also called *lamina.*

In describing how speech sounds are articulated it is useful to label the speech organs in some detail. Tip, blade, and sides (rims) of the tongue articulate with the teeth in making the English *th*-sounds, /θ/ as in *theatre* and /ð/ as in *then.* Consonants primarily involving the blade are /t/, /d/, /s/, and /z/.

blend

A word, phrase, or construction formed by the merging of parts of two other linguistic elements.

(*a*) *Morphology.* Examples of *lexical blends* (also called *blend words, word blends*) are:

bit (= binary + digit) (M20)
brunch (= breakfast + lunch) (L19)
camcorder (= camera + recorder) (L20)
fantabulous (= fantastic + fabulous) (M20)
smog (= smoke + fog) (E20)
televangelist (= television + evangelist) (L20)
motel (= motor + hotel) (E20)

Note that while most blends are formed by joining a pair of words at the point where they have one or more letters or sounds in common (e.g. m*ot*or + h*ot*el), a few are not formed in this way (e.g. *brunch, camcorder*).

(*b*) Syntactic blends include such structures as

I would have liked to have done it (*I would have liked to do it* + *I would like to have done it*)

Neither claim impressed us, nor seemed genuine (*Neither claim impressed us or seemed genuine* + *The claims neither impressed us nor seemed genuine*)

I do not dare refuse ((modal) I *dare not* refuse + (ordinary verb) I *do not dare* to refuse)

This is a general term covering various types of structure, which might be regarded as merely stylistically awkward, or as grammatically dubious or anacoluthic. A dated term for a particular type of blend is APO KOINOU.

• **blending**: the process by which a blend is formed.

Compare CONTAMINATION.

block language

A type of structure different from normal clause or sentence structure, but often conveying a complete message.

Block language is found especially in notices and newspaper headlines. It sometimes consists of single noun phrases (e.g. *No exit*, *Essex's snappy reply to a negative image*). Other block language has a sort of abbreviated clause structure, with articles, auxiliary verbs, and other minor words omitted (e.g. *Tanks met by rain of stones*, *19 dockers dismissed unfairly*, *Jailed racing driver's bail request rejected*).

Bloomfieldian

(*n.*) An adherent of the linguistic theories of Leonard Bloomfield, the American linguist (1887–1949).

(*adj.*) Of, pertaining to, or characteristic of Bloomfield or his theories.

Bloomfield's book, *Language* (1933), became an influential textbook, particularly in the United States. The approach is associated with STRUCTURAL *linguistics* and with theories of behaviourist psychology, and the reaction against this position led to the development of GENERATIVE GRAMMAR in the 1950s and 1960s.

borrow

Morphology. Take (and often adapt) (a word) from another language.

borrowing

Morphology. The taking over of a word from a foreign language; a word so borrowed (also called a LOANWORD).

The term is somewhat misleading, since 'borrowed' words usually become a permanent, not a temporary, part of the borrowing language. Many borrowings are modified to bring them into line with the phonological rules of their new language.

As has often been remarked, the richness of the English vocabulary is in large part due to borrowing from many other languages of the world, sometimes in such a way as to allow fine denotative, connotative, or stylistic distinctions between semantically related or nearly synonymous words to grow up.

Loanwords attain different degrees of assimilation into the language. Some are totally assimilated to the native word-stock and are phonetically and orthographically integrated (e.g. *butter*, *fail*, *gas*, *umbrella*). Others are fully part of the English vocabulary, but retain traces of their foreign origin in their pronunciation, spelling, or inflection (e.g. *addendum*, *phenomenon*, *genre*, *faux pas*). A third group may be well assimilated in their form, but remain semantically tied to a foreign context (e.g. *matador*, *rajah*, *sampan*, *samurai*, *tundra*). Finally, there is a category (into which all loanwords must initially fall) of words which have not yet achieved general currency but occur in very limited contexts, such as during an English-speaker's stay in a foreign country, in news and current affairs, or in travel writing, books on foreign cuisine, anthropological works, etc. Examples would vary from one person's vocabulary to another's, but might include *desaparecido*, *intifada*, *peshmerga*, *tiramisu*, and *Waldsterben*.

Surprisingly few words have been borrowed into English from the neighbouring Celtic languages (Welsh, Gaelic, and Irish). *Bannock* and *crag* are among the few early borrowings from Old British; *coracle* and *flannel* came from Welsh later; *clan*, *slogan*, and *whisky* from Gaelic; and *banshee*, *galore*, and *shamrock* from Irish.

Borrowing from Latin has been constant from the very earliest times, and has always included quite central vocabulary items, such as *cheese*, *kiln*, *pillow*, and *tile*, borrowed before Old English was recorded. Later Latin loans tended to originate in a learned context but many have since become general (e.g. *focus*, *inflate*, *orbit*). Many Latin loanwords have come, virtually unchanged, through the intermediary of French (e.g. *condition*, *oracle*, *superior*) and in the same way, many borrowings from ancient Greek have come through Latin (e.g. *abyss*, *cemetery*, *history*), though some are direct loans (e.g. *acme*, *kudos*, *rhizome*).

The Scandinavian settlement in late Old English times had a marked effect on the English vocabulary: Danish- and English-speaking communities lived side by side for some time, so that penetration was deep and all-pervasive. Even form words, such as *they*, *them*, *their*, *though*, and *near*, were borrowed. Nearly all early Scandinavian loanwords are central items such as *cast*, *egg*, *law*, *take*.

French has contributed more than any other language to the English vocabulary, starting with the earliest post-Conquest loanwords (e.g. *castle*, *prison*, *war*). Borrowing at all levels of vocabulary was especially heavy during the later middle ages, greatly affecting the core vocabulary (e.g. *age*, *blue*, *chase*, *front*, *people*, *search*, and so on); more recent borrowing has been mostly at the learned and cultured level (e.g. *avant-garde*, *surrealism*).

Other important European sources of loanwords have been Dutch (e.g. *brandy*, *deck*, *hoist*), Low German (e.g. *hawker*, *smuggle*) and, of course, Italian (e.g. *motto*, *semolina*) and Spanish (*alligator*, *mosquito*).

Loanwords from outside Europe tended, in the earlier period of exploration, to come through other languages such as Dutch and Portuguese. As English-speaking settlements grew up, first in North America and then in other parts of the world, and as Britain imposed its political and commercial domination during the nineteenth century, direct borrowings came from a large number of other languages, e.g. *sheikh* (Arabic), *boomerang*, *kangaroo* (Australian aboriginal), *lychee* (Chinese), *taboo* (Tongan), *mocassin*, *skunk* (Algonquian), *judo*, *tycoon* (Japanese), *caddy*, *rattan* (Malay), *thug* (Hindi), *bungalow* (Gujarati), etc.

Owing to migration, ease of travel, mass communication, and similar factors, words of foreign origin abound in present-day English speech and writing, particularly in the fields of cookery, the arts, and politics. It is difficult to predict whether any given word will become part of the vocabulary in the long term.

bound

1 *Morphology*. Of a morpheme: normally occurring only in combination with another (*bound* or *free*) form; not FREE.

Bound morphemes (or *bound forms*) include inflections such as

> *-s*, *-ing*, and *-ed*

and affixes such as

> *de-*, *dis-*, *un-*, and *-ly*.

2 Of a clause: subordinate.

In some grammatical theory a *bound clause* is roughly the same as the subordinate clause of traditional grammar, but the terms *subordinate* or *dependent* are much more common.

3 See GOVERNMENT-BINDING THEORY.

bounded

Of the referent of a noun: capable of being conceived of as a separate unit.

• **boundedness.**

This terminology is not very general, but is intended to deal with the problem that many COUNT nouns (e.g. *cake*, *difficulty*) can also be interpreted as mass nouns.

> 1984 R. HUDDLESTON In *another cake*, *cake* has a bounded or individuated interpretation . . in *so much cake* it has an unbounded or mass interpretation.

When nouns are divided into either COUNT or UNCOUNT (usually the same as MASS), it is usual to describe nouns with membership of both categories in terms of *overlap* or *conversion*, the latter a recognized process of lexical morphology. The use of the term *boundedness* for noun meaning makes it possible to restrict the concept of COUNTABILITY to noun classes:

> 1984 R. HUDDLESTON An uncountable noun . . cannot sustain an individuated interpretation; the converse, however, does not hold. As the examples with *cake* . . show, particular instances of countable nouns can receive mass interpretations. It is precisely for this reason that I have treated countability and boundedness as distinct concepts.

bracket

Phonetics & Phonology. Each of the paired typographical marks used to enclose phonetic transcription.

The usual convention is to use square brackets [ ] for accurate phonetic values, and oblique strokes / / for phonemes.

bracketing

Linguistics. A method for showing the internal structure of a clause or sentence, using pairs of brackets.

At a simple level the technique may be useful. We might for example contrast

[He]	[looked up]	[the word]	[in his dictionary]
subject	verb	object	adverbial

with

[He]	[looked]	[up the chimney]
subject	verb	adverbial

More complicated sentences may involve brackets within brackets, which make the analysis difficult to read. For these TREE DIAGRAMS are often preferred.

Compare BRANCHING.

branching

Linguistics. (*n. & adj.*) (The connections between main and subordinate units) that can be symbolized by a branch (depending in a specified direction) within a TREE DIAGRAM.

The word sometimes occurs with prefixes, i.e. *left-branching*, *right-branching*, and *mid-branching* (also called *medial branching* and NESTING). Thus initial subordinate clauses are *left-branching* and final subordinate clauses are *right-branching*. Right-branching clauses tend to be easier to understand. Compare

I danced with a man
who danced with a girl
who danced with the Prince of Wales.

and (left-branching)

I danced with a man
a girl
the Prince of Wales danced with
danced with.

breaking

Phonetics. The process by which a pure vowel sound in certain contexts becomes a centring diphthong.

This process is particularly likely to happen before /l/ (closing a syllable). Thus words such as *feel* /fiːl/ or *meal* /miːl/ are frequently realized in RP as [fɪəl] and [mɪəl]. Historically, a similar process happened before /r/ (which, in word-final position, has been lost in RP); so *beer* was originally [biːr] which by breaking became [bɪər] (now /bɪə/).

The term originated in historical linguistics, where it denoted a similar change of vowels in Old English; it is a rendering of German *Brechung* 'breaking'.

breath

Phonetics. Expiration of air without any vibration of the vocal cords; contrasted with VOICE (2).

breathed

Phonetics. Of a speech sound: made without vibration of the vocal cords.

The term was used by the British phonetician Daniel Jones in his *An Outline of English Phonetics* 1918 (ninth edition, 1962). The more usual term today is VOICELESS.

Compare ASPIRATION.

breath-group

Phonetics. A word or succession of words, whether a sentence or part of a sentence, uttered without pause, in a single breath.

A somewhat dated term, roughly equivalent to TONE UNIT, *tone group*, or SENSE-GROUP in more modern analyses of intonation.

> 1962 D. JONES It is usual to employ the term breath-group to denote a complete sentence that can conveniently be said with a single breath, or, in the case of very long sentences, the longest portions that can conveniently be said with single breaths.

Briticism

A word or phrase that is of distinctively (modern) British origin, particularly in contrast to a different American equivalent.

British English

The variety of English used in Great Britain, as contrasted with those used in other English-speaking areas.

broad negative

A word which is almost negative in meaning and in its grammatical effect, or that is mainly used in a negative context.

The term is not in general use. It includes not only adverbs that are normally classified as SEMI-NEGATIVE (such as *barely, hardly, scarcely, seldom, rarely*, etc.) but also words such as *bother* and *necessarily* which normally appear in negative contexts and which some grammarians might call NON-ASSERTIVE.

broad transcription

A systematic method of representing in a rather general way (normally using the symbols of the INTERNATIONAL PHONETIC ALPHABET) how spoken language sounds; an example of this. Contrasted with NARROW transcription. Also called *broad notation*, *broad script*.

The term is used in two related ways:

(*a*) A purely PHONEMIC transcription. Here symbols are restricted to the phonemes of the language, i.e. the set of sounds that differentiate meaning. Thus in English the /t/ phoneme would be represented by this one symbol in all situations, regardless of the fact that the phoneme is realized by various allophones, e.g. being aspirated in a stressed initial position (*time*) and unaspirated after *-s* (*stay*), and ignoring also the fact that it may not always have alveolar articulation.

(*b*) A PHONETIC transcription, showing some articulatory details (e.g. aspiration), but not very many, in contrast to a narrow transcription showing a lot of detail.

A broad phonemic transcription (*a*) is generally felt to be simplest to use, but a knowledge of the allophonic systems of the language is needed if such a transcription is to be read aloud with even approximate accuracy.

by-form

Morphology. A collateral and sometimes less frequent form of a word.

This is an old-fashioned term from philology. It is generally used for a word form which has essentially the same origin as a related word but a distinct pronunciation and spelling, and which has had significant currency among speakers and writers of standard English. Examples in the *Oxford English Dictionary* include:

chaw (LME) besides *chew* (OE)
clift (LME) besides *cliff* (OE)
commonality (LME) besides *commonalty* (ME)
harrow (as in *the Harrowing of Hell*) (ME) besides *harry* (OE)

The *by-form* may be regarded as a subcategory of DOUBLET.

Compare HETERONYM (2).

C

C

1 COMPLEMENT as an ELEMENT in clause structure.

2 A symbol for a consonant in phonological structure.

calque

Morphology.

(*n.*) The same as LOAN TRANSLATION.

(*v.*) Form (a word or expression) as a loan translation (on a foreign word or expression).

> 1958 A. S. C. Ross M[oder]n E[nglish] *That goes without saying* is a translation-loan of (better, is calqued on) M[oder]n French *cela va sans dire.*

cardinal number

A number denoting quantity (*one, two, three*, etc.) in contrast to an ORDINAL number. Also called **cardinal numeral**.

cardinal vowel

Phonetics. One of a standard set of eighteen vowels, devised by the phonetician Daniel Jones (1881–1967) as a basis for describing the vowels of any language.

The system is mainly physiological. The vowels are described primarily in terms of tongue position and the amount of lip-rounding is specified.

There are eight primary vowels: four front vowels, defined according to the height of the front of the tongue, and four back vowels, where the height of the back of the tongue is relevant. The eight secondary cardinal vowels have the same tongue positions, but the lip-rounding or lip-spreading is different. Two further vowels are identified as depending on the centre of the tongue being raised.

The cardinal vowel system is conventionally presented in a stylized diagram, on which the actual vowels of a particular language can be superimposed. Thus the /iː/ sound in English *need* is a high (or close) front vowel, but not quite so high as the idealized cardinal vowel 1; while English /ɑː/ in *hard* is a low back or (open) vowel, near in sound to cardinal 5, the lowest or most open back vowel. [See diagrams p. 443]

Compare LIP POSITION.

case

1 The functional role of a noun or noun phrase in relation to other words in the clause or sentence.

2 The form of a word (shown by inflection) expressing this.

In English (unlike Latin, which has six cases) the only distinction of case in nouns is between the COMMON case (i.e. the ordinary base form for the singular: *boy*, with plural *boys*) and the GENITIVE (*boy's, boys'*). Even this analysis is disputed, since the genitive inflection can be added not only to a word, but also to a phrase (e.g. *the King of Spain's daughter*), which may not even end with a noun (e.g. *the man opposite's car*).

Among pronouns, six show distinctions between SUBJECT case and OBJECT case—*I/me, he/him, she/her, we/us, they/them, who/whom*. Apart from these few morphological distinctions, 'case' distinctions in English are shown by word order and the use of prepositions, and so the traditional case names (*nominative, accusative, dative* etc.) taken over from Latin are generally considered inappropriate.

case grammar

A theory about clause organization developed in the late 1960s by the American linguist Charles Fillmore (b. 1929), which analyses the semantic roles of nouns and noun phrases. The theory, along with other developments within generative grammar, grew to some extent out of dissatisfaction with early STANDARD THEORY.

In this analysis case is not a category of surface syntax (as in traditional grammar) but of meaning. Thus, in

> *The burglars* broke the door down

the grammatical subject is agent; but in

> *The jemmy* made little noise

it is instrument; in

> *The neighbours* heard nothing

it is dative; and in

> *The whole place* was a mess

it is locative.

The original six cases recognized were AGENT(IVE), the initiator of the action, INSTRUMENTAL, DATIVE (2), FACTITIVE (2), LOCATIVE, and OBJECTIVE. Subsequent adaptations and revisions introduced the EXPERIENCER case (formerly *dative*), RESULT (formerly *factitive*), SOURCE, GOAL (2), and PATIENT.

Case grammar is a useful contribution to the study of relationships between grammar and meaning, but is not now generally considered to offer universal solutions.

> 1976 F. R. PALMER The deeper the investigation, the more complex it seems to become . . . A particularly difficult problem is *My ear is twitching*. *My ear* could be agent since it is 'doing' the twitching, or experiencer, or even location (*I have a twitch in my ear*).

catachresis (Plural **catachreses.**)

'Wrong application of a term, use of words in senses that do not belong to them.' (H. W. Fowler, 1926.) An old-fashioned term, originally rhetorical.

Examples given by Fowler were the 'popular' use of *chronic* = severe, *asset* = advantage, *conservative* (as in 'conservative estimate') = low, *annex* = win, and *mutual* = common.

1589 G. PUTTENHAM Catachresis, or the Figure of abuse . . if for lacke of naturall and proper terme or worde we take another, neither naturall nor proper and do vntruly applie it to the thing which we would seeme to expresse.

• **catachrestic. catachrestically.**

cataphora

The use of a pronoun or other PRO-FORM to point forward to a later word, phrase, or clause; an example of this process. Sometimes called *forwards anaphora*, but usually contrasted with ANAPHORA.

Examples:

What I want to say is *this*. Please drive carefully.
If you see *him*, will you ask Bob to telephone me?
Here is *the news*. In the House of Commons, the Government had an overwhelming majority . . .

cataphoric

Of or involving cataphora.

1976 M. A. K. HALLIDAY & R. HASAN The presupposition may go in the opposite direction, with the presupposed element following. This we shall refer to as cataphora . . . The presupposed element may . . consist of more than one sentence. Where it does not, the cataphoric reference is often signalled in writing with a colon.

•• **cataphoric ellipsis**: see ELLIPSIS.

category

A CLASS of items with the same function; one of the characteristics of such a class.

This is a very general term, used in different ways by different grammarians. In some analyses, nouns, verbs, adverbs, etc. (i.e., parts of speech) are categories; in some, noun phrase, verb phrase, etc. are; in some, subject, predicate, etc. are. Sometimes the term is reserved for characteristics of such elements, e.g. ASPECT, TIME, CASE, GENDER, or PERSON.

In Scale-and-Category Grammar, four categories are contrasted with three (or four) scales. The categories are:

UNIT (five units are recognized—sentence, clause, group, word, and morpheme)
STRUCTURE (concerned with the free or bound relationships within a unit)
CLASS (concerned with such matters as classifying sentences into simple, compound, or complex)

SYSTEM (concerned with the range of choices in any particular area: for example, the person system offers three options.)

- **categorial**: relating to categories.

In early Generative Grammar the *categorial component* is the set of grammatical rules that forms an important part of the theoretical apparatus; and *categorial rules* (or *phrase structure rules*) are rewriting and expanding rules.

catenative

(*n. & adj.*) (A lexical verb) that is capable of linking with a following verb.

(*a*) Strictly, with a directly following dependent verb.

e.g.

want (to go); *begin* (walking); *go* (shopping); *get* (hurt).

Chance juxtapositions are not catenative. Contrast:

We stopped + to talk to the old man (= in order to talk)

with

We *stopped talking* to the old man (catenative)

or

You only helped me + to satisfy your own conscience (= in order to satisfy)

with

You only *helped me to answer* one question (catenative)

(*b*) (Popularly.) More widely, including the above types of verb and similar verbs with an object, e.g.

want (them to go); *watch* (them go/going); *have* (the house painted).

Some of these are classified in different ways in different grammars.

causal

Of a clause: expressing cause or reason. (An older term.)

e.g.

As you are not ready, we must go without you.

causative

Semantics. (*n. & adj.*) (A word, especially a verb) expressing causation.

The term is particularly used in connection with verbs. In classic semantic theory, the verb *kill* is a causative verb, meaning 'cause to die'. Other causatives include verbs of motion such as *place* or *put*, i.e. cause (something) to be (in a place), and more general verbs that have a result, e.g. *elect* (*They have elected my brother as chairman*).

In popular pedagogic grammar, *get* and *have* are the prime causative verbs, as in *Get your hair cut* or *We've had the house painted.*

The term is sometimes applied to other linguistic units. *Because of* can be described as a causative preposition; in *She died of a fever*, *of a fever* is a causative adverbial.

• **causatively**.

Compare CONATIVE.

cause

The fact of giving rise to an event or state, considered as one of the semantic categories used in the classification of adverb clauses and prepositions (e.g. *because of*). Often conflated with REASON.

central

1 Of, at, or forming, the centre.

•• **central determiner**: a determiner that must follow a PREDETERMINER (such as *all, both, such*), but precede a POSTDETERMINER (such as a number). Among the most frequent central determiners are *a/an* and *the*, possessives (*my, your, his, her, its, our, their*), and demonstratives (*this, these, that, those*). e.g.

such *a* nuisance
all *our* yesterdays
my two left feet
both *those* two criminals

central vowel: (*Phonetics*) a vowel made with the centre of the tongue raised towards the middle of the roof of the mouth, where the hard and soft palates meet.

In standard English (RP) the central vowels are:

/ʌ/ the sound in *hut, come, blood*
/ɜː/ the sound in *bird, nurse, worm*
/ə/ the sound at the beginning of *ago* and the end of *mother*
/ʊ/ the sound in *foot, put, wolf, could.*

2 Having the main features of a particular word class.

•• **central adjective**: an adjective that can be used in both ATTRIBUTIVE and PREDICATIVE positions.

central coordinator: see COORDINATOR.

central modal: a true MODAL VERB, contrasting with a SEMI-MODAL.

central passive: a true PASSIVE verb having a regular active counterpart, in contrast to a *semi-passive* or *pseudo-passive* verb.

central preposition: a preposition having all the characteristics of a preposition, in contrast to a MARGINAL *preposition*.

3 **central meaning**: see MEANING.

• **centrality**: the fact or quality of being central.

centralized

Phonetics. Of a vowel: articulated with the centre of the tongue raised more than the front or the back.

The term is generally reserved for the articulation of a normally front or back vowel nearer to the centre of the mouth than usual. In a broader sense most English vowels are centralized to some extent if measured against the CARDINAL VOWEL system.

centring diphthong

Phonetics. A diphthong that moves towards a central vowel position for its second element. Contrasted with CLOSING DIPHTHONG.

Standard RP has three centring diphthongs:

/ɪə/ as in *dear, here, idea*
/ʊə/ as in *tour, during*
/eə/ as in *fair, where, stare*

chain

(Designating) a relationship between two linguistic units such that they can occur together in a larger unit.

• • **chain and choice**: the syntagmatic and paradigmatic relationships that exist between linguistic units.

The contrast between *chain* and *choice* can be applied at various levels of language analysis. Thus, if we take the words *bat, cat, fat, hat,* etc., the letters *b*, *c*, *f*, and *h* are in a chain relationship with *-at*, but in a choice relationship with each other (and the same is true of the sounds to which they correspond).

At a higher level, if we wish to add one word to complete the following: *The cat . . . on the mat*, the chain relationship requires a verb, but the choice is wide (*is, jumped, lay, lies, sat, slept,* etc.).

Compare AND-RELATIONS, PARADIGM (2), SYNTAGM.

choice

(Designating) a relationship between two linguistic units such that only one can be used in a particular context.

See CHAIN. Compare OR-RELATIONS.

Chomskyan

(*adj.*) Of, pertaining to, or characteristic of (the theories of) the American linguist Noam Chomsky (b. 1928).

(*n.*) An adherent of Chomsky's theories.

Chomsky's *Syntactic Structures* (1957) and *Aspects of the Theory of Syntax* (1965) introduced Transformational-Generative Grammar and gave a radically new direction to linguistics. He has continued to develop his theories and has published extensively ever since: on language and the human mind; on phonology (his best-known contribution being *The Sound Pattern of English* (1968) with M. Halle); and also, critically, on American foreign policy and politics. His latest version of generative grammar is GOVERNMENT-BINDING THEORY (or *government and binding*).

circumstance

1 The state of affairs surrounding and affecting an action, event, etc.

The term is sometimes invoked in its general sense in a detailed analysis of subordinate clauses of CAUSE OR REASON. Thus the following sentences may be defined as showing a relationship of circumstance and consequence (rather than cause or reason plus effect):

Since you're so clever, why don't you do it yourself?
Seeing it's so late, we'd better take a taxi

2 (In Systemic-Functional Grammar.) A type of RELATIONAL *process* (in contrast to INTENSIVE and POSSESSIVE) used to distinguish various kinds of meaning relationships between different parts of a sentence.

• **circumstantial** (*n. & adj.*): (an element) expressing circumstance.

1985 M. A. K. HALLIDAY The relationship .. is one of time, place, manner, cause, accompaniment, matter or role. These are the circumstantial elements in the English clause.

The concept is a rather subtle one, with *circumstance* further divided into *circumstance as process* and *circumstance as participant*. When the circumstance is a process it may be realized by a verb as in:

The play *lasted* three hours
He *resembles* his grandfather

but typically, circumstantial elements are realized by adverbials and particularly by prepositional phrases, e.g.:

Dinner is *on the table*.

citation form

A word or other linguistic unit that is being cited in, and for the purpose of, discussion.

1985 R. QUIRK et al. *It* . . can only very rarely receive stress, for example when it is used as a citation form:
Is this word *IT*? [looking at a manuscript]

class

A group of linguistic items with shared characteristics. See OPEN *class*, CLOSED *class*, FORM *class*, WORD CLASS.

In Scale-and-Category Grammar, *class* is one of four major categories: see CATEGORY.

•• **class noun**: (less generally) the same as COUNT NOUN.

dual class membership: see DUAL (2).

class dialect See DIALECT.

classical plural See FOREIGN PLURAL.

classifier

1 An affix which shows the subclass to which a word belongs.

The term is not in general use in English grammar, but is sometimes applied to affixes, e.g. *un-* meaning 'not', 'the opposite of' (e.g. *unkind*, *unintentional*); *de-* and *dis-* (reversing an action, e.g. *decontaminate*, *disconnect*); *-let*, 'small' (e.g. *piglet*). The term is more useful with reference to a language such as Chinese, which has an obligatory system for marking semantic classes. Noun classifiers in such languages may indicate shape ('long', 'thin', 'sticklike'), size and colour, whether the referent is animate, etc.

2 A word (typically in the attributive position) which has the semantic role of identifying the class or kind of the following noun; in contrast to an EPITHET.

Classifiers include both adjectives (also called CLASSIFYING ADJECTIVES) and nouns, e.g.

a *medieval* castle, a *thatched* cottage, a *country* house.

classifying

Designating the use of *a/an* or ZERO article to indicate membership of a class.

In describing the meaning and usage of the ARTICLES many grammars contrast SPECIFIC and GENERIC (in addition to the better known distinction between definite and indefinite). However, although *the* + a singular count noun can clearly have generic meaning and refer to a class as a whole, as in *The black rhino is in danger of extinction*, the indefinite article cannot have this meaning (**A black rhino is in danger of extinction*). Some grammars therefore prefer the label *classifying* rather than *generic* for the use of *a/an* or zero article with various non-specific meanings:

A black rhino can be very dangerous
We cannot afford *a new car*
More people should train as *engineers*

classifying adjective

A term used in some analyses to describe a subgroup of adjectives:

1990 *Collins Cobuild English Grammar* **Classifying adjective**, an adjective used to identify something as being of a particular type; EG *Indian, wooden, mental*. They do not have comparatives or superlatives. Compare with qualitative adjective.

The term is related to CLASSIFIER (2) but is not the same, since the latter is defined partly in terms of position and includes nouns.

classifying genitive

A genitive which classifies the head noun, rather than showing possession. (Also called DESCRIPTIVE *genitive*.)

e.g. *a women's college* is a certain kind of college; similarly *child's play*, *a moment's thought*, *a stone's throw*.

clause

A grammatical unit operating at a level lower than a sentence but higher than a phrase.

(*a*) In traditional grammar, a clause has its own subject and a finite verb, and is part of a larger sentence. Thus *I was ten when I got my scholarship* consists of a MAIN clause (*I was ten*) and a SUBORDINATE clause (*when I got my scholarship*).

(*b*) Some modern grammar allows NON-FINITE and VERBLESS clauses (which would be categorized as phrases in more traditional analysis) so that the following, though containing only one finite verb, has four clauses:

My father travelled by two buses each day / to get there on time, / leaving home at 5.00 am / and usually returning after 10.00 pm.

Clauses are also defined functionally into three main types:

NOMINAL clause, functioning like a noun phrase
RELATIVE clause, functioning like an adjective, and
ADVERBIAL clause.

(*c*) Some modern grammar uses the clause, rather than the sentence, as the basis of structural analysis, so that in some instances clause and sentence are coterminous (e.g. *I was ten at the time*). More importantly, a clause-based analysis allows a more straightforward functional analysis into five possible elements of English clause structure: SUBJECT, VERB, OBJECT, COMPLEMENT, and ADVERBIAL, with the verb element as the most essential and the adverbial the most mobile.

See also *THAT*-CLAUSE, *WH*-CLAUSE.

• • **clause complex**: a compound or complex sentence, viewed as a 'complex' of clauses.

• **clausal**: of or relating to a clause.

clear *l*: see LATERAL.

cleft sentence

A sentence derived from another by dividing the latter into two clauses, each with its own finite verb, so as to place emphasis on a particular component of the original sentence.

The form is *It* + part of the verb *be* + . . . + *who*/*that*. For example:

Bob always plays golf on Sundays

could be reworded as any of the following cleft sentences:

It is Bob who always plays golf on Sundays (i.e., not his brother)
It is golf (that) Bob always plays on Sundays (not tennis)
It is on Sundays that Bob always plays golf (not Mondays)

In these sentences the chief focus comes at the end of the main clause (the first clause) and the subordinate clause contains information which is assumed to be 'known' and is therefore less important.

In another type of sentence, sometimes included as *cleft*, but more carefully distinguished as *pseudo-cleft*, the focus of information can unequivocally come at the very end. A pseudo-cleft sentence starts with a nominal relative clause (beginning with *what*) as subject:

What Bob plays on Sundays is golf

Unlike the cleft sentence, the pseudo-cleft can give focus to the verb element:

What Bob does on Sundays is (to) play golf.

click

Phonetics. A type of stop consonant made with ingressive (i.e. sucked-in) air.

In English, clicks are found only as extralinguistic sounds, such as those made to express disapproval (the sound conventionally written *tut-tut*) or to encourage horses to move on and hurry up.

Clicks are not made with the normal pulmonic air stream mechanism, but with a velaric mechanism, by which the back of the tongue is pressed tightly against the velum (soft palate) and then moved backwards and forwards. Clicks are found as regular speech sounds in some African languages, such as Xhosa.

cline

A continuum, a series of gradations.

A term applied in various areas of grammar and phonetics where there are no clear-cut contrasts. Many closely similar grammatical structures lie along a cline of acceptability, from fully acceptable instances, through those about which native speakers disagree, to unacceptable instances:

My parents persuaded me to try again.
?They convinced me to try again.
*They insisted me to try again.

Compare GRADIENCE.

clipping

Morphology. The formation of a new word by shortening an existing one; an example of this. A type of ABBREVIATION.

e.g.

(omni)bus, exam(ination), (in)flu(enza), (tele)phone

Compare REDUCTION.

clitic

(*n. & adj.*) (A word) that cannot stand on its own in a normal utterance.

The term is abstracted from the words denoting the two kinds of clitic: PROCLITIC (attached to a following word) and ENCLITIC (dependent on a preceding word). Although the articles (*a/an*, *the*) are proclitics and, arguably, so is *do* when reduced to a consonant only in *d'you know?*, the term *proclitic* is little used with reference to English Grammar.

close

1 *Phonetics*. Of a vowel: made with the tongue high in the mouth; contrasted with OPEN (1).

In English /iː/ as in *feet* or *sea* is a fairly close front vowel, and /uː/ as in *food, group, rude, move* is a close back vowel. *Close* vowels are sometimes called *high* vowels.

Compare OPEN, HALF-CLOSE, HALF-OPEN.

2 **close juncture**: see JUNCTURE.

3 **close-rounded**: see LIP POSITION.

closed

1 Of a word class: to which new words are rarely or never added; in contrast to OPEN (2).

The closed classes—also called *closed systems*—in English are the articles, pronouns and determiners, modal verbs, prepositions, and conjunctions. Very occasionally a new word may be added to a closed class: *plus* (once only a preposition and a noun) began in the mid-twentieth century to be used colloquially as a conjunction, e.g. *He's handsome plus he's rich*. But on the whole closed classes do not allow newcomers; for example, though in many ways it would be desirable for English to possess a singular unisex pronoun, none of the many words suggested has become part of ordinary usage.

Contrast MINOR and MAJOR word class.

2 Of a syllable: ending with a consonant sound; in contrast to OPEN (3).

3 Of a conditional clause or sentence: the same as HYPOTHETICAL; in contrast to OPEN (4).

closing diphthong

Phonetics. A DIPHTHONG which glides towards a closer sound.

This includes all the diphthongs ending in /ɪ/ and /ʊ/, and contrasts with CENTRING DIPHTHONG.

closure

Phonetics. A closing of the air passage by some part of the vocal organs in the production of certain speech sounds. Also called *constriction*.

A complete closure is a feature of plosives, affricates, and nasals. Most other consonants are produced with incomplete or partial closure.

See STRICTURE and compare CONTINUANT.

cloze

Of or pertaining to a method of testing the difficulties presented by a passage to a reader's comprehension by requiring the reader to supply single words that have been omitted. Usually in **cloze test, cloze passage**.

An abbreviation of *closure*. The method was first adopted in psychology. Strictly speaking, the words should be removed at regular intervals (e.g. every sixth or seventh word). When the words are removed irregularly, the result is 'modified' cloze.

It is a favourite method for testing foreign learners, since grammar as well as vocabulary can be tested. In some tests the candidate has ..... supply the ..... words deleted ..... the passage; in ...., any ..... word ..... acceptable.

cluster See CONSONANT CLUSTER.

coalescence

Phonetics. A process whereby two separate speech sounds merge to form a single new phoneme. (Also called *coalescent assimilation* or *reciprocal assimilation*.)

• **coalescent**: participating in or resulting from coalescence.

These terms are particularly applied to the process (**yod coalescence**) in which /t/, /d/, /s/, /z/ merge with /j/ and become /tʃ/, /dʒ/, /ʃ/, /ʒ/ respectively. Historically this has happened in such words as

question /ˈkwestʃ(ə)n/	ambition /æmˈbɪʃ(ə)n/
soldier /ˈsəʊldʒə/	occasion /əˈkeɪʒ(ə)n/
pressure /ˈpreʃə/	measure /ˈmeʒə/

In present-day speech similar coalescent variants are heard (*a*) in certain words, e.g.

intuition /ɪntjuːˈɪʃ(ə)n/ or /ɪntʃuːˈɪʃ(ə)n/
grandeur /ˈgrændjə/ or /ˈgrændʒə/
duel /ˈdjuːəl/ or /ˈdʒuːəl/

and (*b*) across word boundaries, e.g.

/ˈkʊdʒuː/ as an alternative to /ˈkʊdjuː/ for *Could you?*

Except where historically established, coalescence tends to be regarded as colloquial or non-standard.

coarticulation

Phonetics. (Also called *double articulation*.)

1 The simultaneous use of two strictures of equal rank in the articulation of a sound.

'Equal rank' implies for example two stops or two fricatives. This kind of articulation, phonemically significant in some languages, is not specially important in English, although a coarticulation of a voiceless plosive (/p/, /t/, or /k/) with a glottal stop /ʔ/ is sometimes heard.

Nasalized vowels (made with two open 'strictures') are sometimes also included in this category; for example the vowel of *man*, coming between two nasal consonants, may be nasalized. But in other analyses NASALIZATION is treated as a secondary articulation.

2 More broadly, a term including unequal double articulations involving a SECONDARY articulation, which in English usually occur under the influence of neighbouring sounds.

The term thus overlaps with ASSIMILATION, but whereas assimilation is concerned with modifications and changes to phonemes as a phonological phenomenon, coarticulation, as the word implies, is more concerned with the mechanics of speech sounds:

> 1991 P. ROACH It is important to realise that the traditional view of assimilation as a change from one phoneme to another is naive; modern instrumental studies in the broader field of *coarticulation* show that when assimilation happens one can often show how there is some sort of combination of articulatory gestures.

code

1 *Sociolinguistics.* (A general term for) a language, dialect, or speech variety.

Code can mean any of two or more distinct languages (in a situation where more than one is available to a speaker): a bilingual Welsh/English speaker could be said to have two codes. But the term is particularly favoured by those wishing to avoid the possibly pejorative overtones of the word *dialect*. Thus a person who frequently changes from, say, a regional variety of English to standard English is said to be operating with two codes. See CODE-*switching*.

A sociological theory put forward in the early 1970s contrasted an *elaborated* code and a *restricted* code. Speakers of the former, it was claimed, used more complicated sentence structure and a larger vocabulary and conveyed their meaning more explicitly than people operating with a restricted code. The two codes were said to characterize middle-class and working-class speech. The theory aroused both interest and argument.

•• **code-switching**: changing from one language code to another, according to where one is, who one is talking to, which SPEECH COMMUNITY one is identifying with, and so on.

2 See NICE PROPERTIES.

cognate

1 (*n. & adj.*) (A word or language) related in form to another word or language because both are derived by direct descent from the same source.

French, Italian, and Spanish are cognate languages, being all derived from Latin. Latin *mater*, German *mutter*, and English *mother* are cognate words or cognates.

2 **cognate object**: an object related in form and meaning to the verb it is used with.

e.g.

fight the good *fight*
sing a *song* of sixpence
smile a shy *smile*

cognitive

Semantics. Of meaning: relating objectively to facts and the denotations of words, in contrast to ATTITUDINAL meaning.

The analysis of different types of meaning is far from simple, and different semanticists make different distinctions. Cognitive meaning is similar to DESCRIPTIVE or REFERENTIAL meaning, in contrast to subjective, personal meaning.

Compare DENOTATIVE, IDEATIONAL.

coherence

The set of relationships within a TEXT that link sentences by meaning. The term contrasts with COHESION.

Coherence often depends on shared knowledge, implication, or inference. A dialogue such as

A: You weren't at the meeting yesterday.
B: My daughter's ill.

shows coherence. A's statement can be understood as a question, and B's statement can be understood as an explanation. But if B had replied with a rather different statement, e.g. 'Marmalade is a kind of jam', the conversation would normally lack coherence.

cohesion

1 Grammatical or lexical relationships that bind different parts of a TEXT together, in contrast to COHERENCE.

e.g.:

A: You weren't at the meeting yesterday.
B: No, I'm sorry I wasn't there. How did it go?

Here *there* substitutes for *at the meeting*, and *it* refers to *the meeting*.

2 The 'uninterruptability' of a word.

It is a defining characteristic of a WORD in English that normally no extra element can be inserted within it. About the only exception to this that is ever cited is the possibility of inserting a swearword into *abso- . . - lutely*.

Compare INFIX.

cohesive

Grammatically linking together different parts of a TEXT or UTTERANCE.

When pronouns and other words are used to refer back or forward to other words in a text etc., they are sometimes described as *cohesive devices*.

Compare ANAPHORA, CATAPHORA.

• **cohesively. cohesiveness**: another name for COHESION.

co-hyponym See HYPONYM.

coinage See NEOLOGISM, NONCE, WORD FORMATION.

collective

Short for COLLECTIVE NOUN.

collective noun

1 A noun that refers to a group of individual people or animals, and which in the singular can take either a singular or plural verb: *army, audience, committee, family, herd, majority, parliament, team*, etc.

The choice of singular or plural verb—and corresponding pronouns and determiners—depends on whether the group is considered as a single unit or a collection of individuals;

e.g.:

The audience, *which was* a large one, *was* in *its* place by 7 pm

The audience, *who were* all waving *their* arms above *their* heads, *were* clearly enjoying *themselves*

The use of a plural verb with a grammatically singular noun of this type is commoner in British English than in American. But even when followed by a plural verb, such a noun still takes a singular determiner (e.g. *This family are all accomplished musicians*).

2 Loosely (notionally defined), any noun referring to a group—including nouns that can only (in the sense used) take a plural verb: *cattle, clergy, people, police*.

See AGGREGATE, GROUP NOUN, PLURALE TANTUM.

colligation

A grouping of words, based on their functioning in the same type of syntactic structures, in contrast to a semantic relationship (or COLLOCATION).

The class of verbs of perception such as *hear, notice, see, watch* enters into *colligation* with the sequence of object + either the bare infinitive or the *-ing* form; e.g.

We	heard	the visitors	leave/leaving
	noticed	him	walk away/walking away
	heard	Pavarotti	sing/singing
	saw	it	fall/falling

The term is far less general than the contrasting term COLLOCATION.

• **colligate** (cause to) be in colligation (with another word).

collocable

Of a word or phrase: typically able to COLLOCATE with another word or phrase.

1961 Y. OLSSON Although '*have* a look' and '*take* a look' are both collocable with *at*, '*have* a look' is alone in collocating with *for*.

• **collocability**.

collocate

v. (Pronounced /'kɒləkeɪt/.) (Cause to) co-occur with (another word) so as to form a collocation.

1951 J. R. FIRTH In the language of Lear's limericks . . *person* is collocated with *old* and *young*.

n. (Pronounced /'kɒləkət/.) A word that collocates with another.

Any word is to some extent restricted in its usage by virtue of both its meaning and its word-class. But many are much more restricted and can occur only with a limited set of other words or be used only in a particular type of structure. The prepositions used with certain nouns, adjectives, and verbs are often fixed, e.g.:

adherence to	by chance	on foot	under the auspices of
similar to	inconsistent with		
account for	consist of	long for	rely on

adherence and *similar* are said to collocate with *to*; *adherence* and *to* or *similar* and *to* are said to be collocates.

collocation

The habitual juxtaposition of a particular word with other particular words; an instance of such a juxtaposition.

The technical sense in linguistics was introduced by J. R. Firth, although the word had been loosely applied in linguistic contexts previously.

1951 J. R. FIRTH I propose to bring forward as a technical term, meaning by 'collocation', and to apply the test of 'collocability'.

Collocation is a type of SYNTAGMATIC relationship between words. Two kinds should be distinguished:

•• **grammatical collocation**: a type of construction where a verb, adjective, etc. must be followed by a particular preposition, or a noun must be followed by a particular form of the verb (e.g. *account for*, *afraid of*; *the foresight to do it* (not **of doing it*).

lexical collocation: a type of construction where particular nouns, adjectives, verbs, or adverbs form predictable connections with each other (e.g. *cancel a* (luncheon) *engagement* or *break off an engagement* (to be married), not normally **withdraw*, **revoke*, or **discontinue an engagement*; compare also such collocations as *take advantage of*).

Special cases of collocation (e.g. *come a cropper*, *kith and kin*), in which one of the elements is predictable from another, merge into IDIOM or CLICHE.

collocational

Relating to collocations.

E.g. *collocational differences* (between American and British English); *collocational possibilities, collocational range, collocational restrictions.*

Compare SEMANTIC RESTRICTION.

colloquial

Belonging or proper to ordinary conversation; not formal or literary. Compare INFORMAL.

In ordinary everyday language, especially between speakers who know each other well, a casual style of speech is both frequent and appropriate. *Are you doing anything tomorrow evening?* as a preliminary to an invitation is probably more suitable than *Have you an engagement for tomorrow evening?* Colloquial speech is not substandard, nor is it the same as SLANG.

Compare REGISTER.

• **colloquialism**: a colloquial word or phrase; the use of such words or phrases. **colloquially**: in the language of ordinary conversation.

colon See PUNCTUATION.

combination

A SYNTAGMATIC relationship between words or other language units, as opposed to a PARADIGMATIC or contrastive relationship; an instance of this.

At phoneme level /s/ can enter into such combinations as *sap, pass, wasp*. At word level, *combination* denotes a frequently occurring sequence of two (or occasionally more) words. At word level some usage restricts the term to a sequence that functions virtually as a single word (e.g. *oak tree, prime minister*); usually in contrast to a COLLOCATION, which may be discontinuous and phrasal (e.g. *advantage* was *taken of* him). In other usage they are virtually interchangeable.

combinatorial

Able to combine with other linguistic units.

> 1968 J. LYONS In general, any formal unit can be defined (i) as being distinct from all other elements which contrast with it, and (ii) as having certain combinatorial properties.

combinatory

Relating to (a) combination or combinations; collocational.

> 1986 M. BENSON et al. The BBI Combinatory Dictionary of English gives essential grammatical and recurrent word combinations, often called collocations.

•• **combinatory coordination**: coordination in which the two elements have a grammatically joint meaning, in contrast to SEGREGATORY coordination.

Henry and Margaret met is an example of combinatory coordination, since the only possible interpretation is that the two subjects are in combination: they met each other. The sentence cannot be segregated into **Henry met* and **Margaret met*.

combining form

A bound form (or bound morpheme) used in conjunction with another linguistic element in the formation of a word.

The term is usually used in a narrower sense than BOUND *morpheme* to refer to forms that contribute to the particular sense of words, e.g.

arch- chief, pre-eminent:	*archduke, arch-enemy*
geo- earth:	*geography, geology*
-(o)cracy rule:	*meritocracy, theocracy*
-(o)logy study:	*archaeology, zoology*

Combining forms contrast with prefixes and suffixes that adjust the sense of a base (e.g. *un-*, *ex-*) or change the word-class of the base (e.g. *-ation*, *-ize*).

comitative

Having the meaning 'in company with' or 'together with'.

The term is primarily useful in the description of languages that have a particular case for this meaning, but it is sometimes applied to English phrases on the pattern *with* + animate noun. In *I went there with my cousin*, *with my cousin* has a comitative function.

comma See PUNCTUATION.

command

A sentence with the form or function of an order.

There is considerable looseness and conflicting usage in the description of sentence types. Since a command in the sense understood by the layman is often expressed syntactically by an IMPERATIVE (e.g. *Speak up! Leave me alone! Stop teasing the cat*) the terms are sometimes used interchangeably. However, what is in meaning effectively a command or order can be expressed grammatically in other ways (e.g. *You will do as I say*; *Could you make less noise?*). Conversely, grammatical imperatives can have other pragmatic functions, such as requests (e.g. *Please give generously*) or invitations (*Have some more coffee*). Some grammarians therefore use separate sets of words for sentence forms and sentence functions. *Command* may therefore be confined to either form or function only; or two completely different terms may be preferred.

See DIRECTIVE, IMPERATIVE.

In Government-Binding Theory, *constituent command* (*c-command* for short) is used as a technical term.

comment

Semantics. That part of a sentence which says something about the TOPIC. Also called FOCUS or RHEME by some linguists. The comment often coincides with the predicate.

See TOPIC.

comment adjunct

(In some models.) A subcategory of adjuncts (adverbials) that other grammarians mainly label *disjuncts.*

e.g. frankly, no doubt

The category also overlaps with COMMENT CLAUSES (e.g. *to be frank*).

See MODAL ADJUNCT.

comment clause

A parenthetical clause, only loosely connected with the rest of the sentence. Comment clauses may be finite or non-finite, and include many cliches and conversation fillers, e.g.

you know	generally speaking
you see	as I said
to be frank	

common

General, not specially marked; in which a (usual) distinction is not made.

•• **common case**: the unmarked case of a noun (e.g. *boy*, *week*), which is used for subject and object function and after prepositions; in contrast to the marked GENITIVE case (e.g. *boy's*, *week's*). Compare BASE.

common core: the basic grammar and vocabulary shared by all varieties of the English language (a rather indefinable concept).

common gender (occasionally **common sex**): the characteristic shared by many animate nouns of making no distinction of gender (examples are the nouns *baby, person, horse, sheep*).

Occasionally it refers more narrowly to nouns that denote animate beings when sex is irrelevant and that *it* or *which* can be used to refer to (e.g. *A baby cannot feed itself*). Compare DUAL GENDER.

common noun: a noun which is not the name of any particular person, place, thing, quality, etc.; opposed to PROPER NOUN.

Common nouns are further classified grammatically into COUNT and UNCOUNT nouns and semantically into ABSTRACT and CONCRETE.

communicative

1 Of or pertaining to communication.

• **communicative competence**: a speaker's ability to understand the implications of utterances and to appreciate what language is appropriate in different situations; in contrast to GRAMMATICAL competence. It is of course a very practical kind of social competence, and therefore contrasts also with the idealized COMPETENCE of generative grammar.

communicative dynamism: variation in the importance of different parts of an utterance in conveying information.

The concept has been a basic one for the Prague School. A similar concept underlies the theories of THEME and rheme, and TOPIC and comment.

communicative function: the purpose and meaning of an utterance, whatever its FORM.

Thus an interrogative sentence (e.g. *Isn't it a lovely day?*) may have the communicative function of an exclamation.

2 Of teaching method etc.: that emphasizes functions and meaning rather than grammatical forms alone.

Communicative teaching methods have led in some quarters to learners developing fluency in talking ('communication') at the expense of grammatical accuracy, but this was not the intention of the early proponents of the approach, and there has been a shift of emphasis back towards grammar.

comparative

(*adj.*)

1 Of a gradable adjective or adverb form, whether inflected (essentially, by the addition of *-er* to the positive form) or periphrastic (by the use of *more*): expressing a higher degree of the quality or attribute denoted by the base form.

e.g.

better, happier, sooner, more beneficial, more energetically

•• **comparative degree**: the middle degree of comparison, between POSITIVE and SUPERLATIVE.

comparative element: a comparative phrase in a main clause that is followed by a comparative clause.

e.g. It was *colder/more exhilarating* than we expected.

2 *Linguistics.* Concerned with the similarities and differences between different languages or varieties.

• **comparative linguistics**: the study of language change over time or of the historical connections between different languages.

This discipline made great advances in the nineteenth century. Related terms are *comparative grammar*, (*comparative*) *philology*, and *historical linguistics*. *Comparative grammar* and *comparative linguistics* can also refer to the study and comparison of two modern languages, but this is often distinguished as CONTRASTIVE *linguistics*.

(*n.*)

(An adjective or adverb that is in) the comparative degree.

comparative clause

A clause expressing some kind of comparison.

There is considerable variation in the labelling of different kinds of clauses and sentences expressing ideas of COMPARISON. Narrowly, the term *comparative clause* can be restricted to a clause following a main clause containing a comparative form, e.g.:

It was colder *than we expected*
It was more/less expensive *than last year*

But usually clauses expressing EQUIVALENCE, beginning with *as* (and following *as* or *so* + a positive adjective or adverb) are included, e.g.:

It was (not) as cold *as it was last year*

In popular grammar, a comparative clause is a type of subordinate clause; in stricter models, it is an EMBEDDED clause (postmodifying the preceding adjective or adverb).

Comparative clauses are frequently ellipted, sometimes until nothing but the conjunction and one word remain, e.g.

It's not as cold *as* (it was) *yesterday*
You know them better *than I* (do)

Interestingly, not only do some comparative clauses lack a subject, but in some no subject can be inserted, e.g.

It was much more expensive *than was anticipated*

See COMPARISON.

compare

Form the COMPARATIVE and SUPERLATIVE degrees of (an adjective or adverb).

Compare GRADABLE.

comparison

1 The act or an instance of comparing one thing with another.

This very general term can be used to cover any grammatical means of comparing things. In comparative clauses containing *than* or *as*, whatever is represented by the adjective or adverb in the comparative element is sometimes called the *standard* of comparison, and the *basis* of the comparison is whatever or whoever the subject is being measured against. Thus in

It's colder than/not as cold as yesterday

the standard is coldness, and the basis (for comparing today's temperature) is yesterday.

Comparison may include expressions of sufficiency and excess, e.g.

They did not arrive early *enough* to help
They arrived *too* late to help

2 The action of forming the comparative (or comparative and superlative) of an adjective or adverb.

The three *degrees of comparison* are POSITIVE, COMPARATIVE, and SUPERLATIVE.

comparison clause

A clause containing some kind of comparison.

This label may include the COMPARATIVE CLAUSE and also the type of clause introduced by *as if/as though*:

He looked as if/as though he'd seen a ghost.

Or it may exclude the comparative clause and be used in contrast with them. (To confuse the issue further, clauses introduced by *as if* and *as though* are often analysed as MANNER clauses).

competence

Linguistics. The internalized knowledge of the rules of a language that native speakers have; contrasted with their actual PERFORMANCE.

The distinction between competence and performance is an important element in Generative Grammar theory. Native speakers' competence makes them aware of all possible ambiguities and enables them to generate an infinite number of 'correct' sentences. By contrast, performance, what a speaker actually says, may at times be ungrammatical or confused—and is also subject to constraints, such as length, that do not affect idealized competence.

See COMMUNICATIVE *competence*. Compare LANGUE.

complement

1 One of the five elements of clause structure, along with Subject, Verb, Object, and Adverbial.

Typically complements of this type 'complete' the verb *be* or another linking verb, and are either adjective phrases or noun phrases, e.g.:

My brother-in-law is *very clever*
He's *a brain surgeon*

This type of complement is called a *subject* or *subjective complement*, because the complement refers back to the subject. An *object* or *objective complement* refers back to the object, e.g.

I consider tranquillizers *dangerous*
They make some people *addicts*

2 More widely, any element needed to 'complete' an adjective, preposition, verb, or noun; an example of COMPLEMENTATION.

Complements of adjectives include prepositional phrases, e.g.

fond *of chocolate*

and clauses, e.g.

sorry *that you are ill*
sorry *to hear your news*

Complements of prepositions (also called OBJECTS of prepositions) are usually noun phrases, e.g.

out of *order*
in *the bag*
over *the moon*

The complement of a verb, in this wider sense, is a very unspecific term, and can include not only complements in sense (1), but also adverbials, objects, non-finite verbs, and entire sentence predicates apart from the verb itself.

1961 R. B. LONG *His sister is buying antiques* will always be understood to have *is buying* as predicator and *antiques* as complement; *his hobby is buying antiques* . . to have *is* as predicator and *buying antiques* as complement.

The words following a head noun, particularly an abstract noun in some types of noun phrase, are sometimes analysed as the complement of a noun (e.g. They were annoyed by her refusal *to answer*). The label may be useful where a noun is incomplete by itself (e.g. *They deplored her lack of remorse*: **They deplored her lack* is impossible). But usually such complements are dealt with under a wider term such as *postmodification*.

complementarity

Semantics. A relationship of oppositeness between pairs of words, such that to deny one is normally to assert the other, and vice versa.

See COMPLEMENTARY.

complementary

(*n. & adj.*) (Designating) a word that with another word forms a pair of mutually exclusive opposites.

Such complementaries (sometimes considered a type of ANTONYM) are usually ungradable either-or terms, e.g.

alive/dead, married/single.

complementary distribution

(Especially in *Phonology*.) A distribution of two or more similar or related linguistic units in a mutually exclusive way.

The term is particularly used of the context-conditioned allophones of the same phoneme. In standard British English the most obvious example of this is the complementary distribution of the CLEAR and DARK allophones of /l/. See ALLOPHONE.

The term is inapplicable to dissimilar or unrelated sounds. For example, English /h/, which occurs only in syllable-initial position, and /ŋ/ (the sound of *ng* in sing), which is only syllable-final, never occur in the same phonetic context, but are evidently not allophones and are not usefully described as in complementary distribution.

Compare DISTRIBUTION, DISTINCTIVE FEATURE.

complementation

1 The addition of a COMPLEMENT to a linguistic unit.

1985 R. QUIRK et al. We reserve the term COMPLEMENTATION (as distinct from *complement*) for the function of a part of a phrase or clause which follows a word, and completes the specification of a meaning relationship which that word implies.

•• **ditransitive complementation**: see DITRANSITIVE.

2 *Linguistics*. Complementary distribution (a rare use).

1948 E. A. NIDA The forms *I* and *me* generally occur in complementation: *I* occurs in preverbal subject position, *me* in postverbal object position and after prepositions.

complementizer

A word or morpheme marking some types of complement.

1976 R. HUDDLESTON Finite complements are typically introduced by *that* (though it is often omissible); infinitival ones are sometimes introduced by *for*, as in *For John to be late was rare*; and in gerundives the subject NP is often in the possessive/genitive case, as in *She resented John's getting the job* (the case morpheme has a variety of phonological realizations and is generally known as *Poss*). The elements *that*, *for* and *to*, *Poss* and *ing* are known as complementizers.

The term is particularly used in Transformational-Generative Grammar. It is narrower in meaning than *conjunction* or *subordinator*.

complex

(*adj.*) Consisting of at least two unequal parts; often in contrast to either SIMPLE or COMPOUND.

•• **complex conjunction**: Two- or three-word conjunction (e.g. *in that*, *providing that*, *as soon as*).

complex preposition: a two- or three-word preposition (e.g. *out of*, *because of*, *prior to*, *on behalf of*), in contrast to an ordinary one-word preposition (e.g. *from*, *before*).

complex sentence: a sentence containing at least one subordinate clause, in addition to its main (matrix) clause, e.g. *When you've quite finished, we can begin*. Compare COMPOUND SENTENCE.

complex stem: see STEM.

complex verb phrase: any verb phrase (except a one-word one), including perfect and progressive tenses (e.g. *have forgotten*, *is hoping*) and phrases containing modals (e.g. *should apologize*, *must remember*).

complex word: generally, a word consisting of at least two parts, usually a base and one or more bound morphemes.

Thus *im-polite*, *rude-ness* contrast both with unanalysable words, long or short, (e.g. *dog*, *hippopotamus*) and with COMPOUND words. However, there is very considerable variation in the way different linguists deal with morphology, and usage of terminology is correspondingly fluid. The word *blackboard*, for example, a *compound* in many definitions (*black* + *board*), is classed as a *complex word* in another model, and even as a *compound stem*.

(*n.*) **clause complex**: see CLAUSE.

complex transitive verb

1 A verb that takes a direct object plus an object complement.

I.e. a verb in an SVOC structure:

Let's *paint* the town red
They *made* him leader

2 More widely, a verb in any structure in which the object noun phrase alone is not 'acted upon' by the verb, but the object and what follows it are in a sort of 'subject-predicate' relationship as regards meaning; e.g.

We watched him leave/leaving (i.e. He left)
I knew him to be a crook (i.e. He was a crook)
They made him pay (i.e. He paid)
I saw him arrested (i.e. He was arrested)

There are however considerable differences of analysis here: some of these verbs would be considered ordinary transitive (monotransitive) verbs in some grammars, or dealt with as catenatives.

3 (In other models.) A verb that takes an obligatory adverbial in an SVOA pattern; e.g.

She put the car in the garage
He threw himself into the role

Compare DITRANSITIVE.

component

1 Any one of the major parts of a generative grammar.

Different sets of components have been proposed at different times, for example a *base component* (see BASE (3)), a set of transformational rules (the *transformational component*), a *phonological component*, and a *semantic component*. In more recent theory the term *module* is preferred.

See TRANSFORMATIONAL GRAMMAR.

2 *Semantics.* A small unit of meaning that forms part of the total meaning of a word and that may be shared by other words. See COMPONENTIAL ANALYSIS.

componential analysis

Semantics. The analysis of linguistic elements into fundamental properties or attributes; especially a method of breaking down words into bundles of meanings.

Classic examples cited usually concern terms for people and animals that can be shown to have or lack certain attributes, and these are often indicated by a binary notation of plus or minus signs. Thus *stallion* or *boar* could be shown as [+male] [+adult] [–human] etc.

This sort of technique is theoretically applicable to the entire vocabulary, and a study of prepositions proposes 'locative path locative' as the componential definition of *beyond*, whereas *past* is defined as 'path locative proximity'. But the validity of the technique has been criticized.

Compare DISTINCTIVE FEATURE.

composition

1 An older term for COMPOUNDING.

1926 H. W. FOWLER Composition—How words are fused into compounds.

2 The way in which language is composed of units which incorporate other units.

1968 J. LYONS The relationship between the five units of grammatical description . . is one of *composition*. If we call the sentence the 'highest' unit and the morpheme the 'lowest', we can arrange all five units on a scale of *rank* (sentence, clause, phrase, word, morpheme), saying that units of higher rank are composed of units of lower rank.

compound

(*n.*) A word formed by combining two or more bases (or free morphemes); a compound word.

(*adj.*) Formed by combining two or more units; especially, consisting of two or more parts of equal value, in contrast to COMPLEX and SIMPLE.

Among *compound words*, *compound nouns* and *compound adjectives* are particularly common, e.g.

bookcase, handlebar, laptop, mind-set, windshield, fact-finding, home-made, south-facing, tax-free, etc.

Some adverbs (e.g. *somehow*, *hereby*) are also described as compound, in contrast to simple adverbs (e.g. *just*, *only*).

In general, compound words contrast with simple words, and with words formed by DERIVATION or INFLECTION. At a more technical level, terminology is by no means agreed, and terms such as *compound root* and *compound stem* are differentiated in ways that seem idiosyncratic, while even the distinction between *compound* and COMPLEX may be blurred. See COMPOUNDING.

•• **compound lexeme**: see LEXEME.

compound preposition: now more usually called COMPLEX preposition.

compound sentence: a sentence containing two or more coordinate clauses.

Compare COMPLEX sentence, and see COMPOUND-COMPLEX sentence.

compound stem: see STEM.

compound subject, compound object: now more usually called *coordinated subject*, *coordinated object*.

compound verb: an older term for MULTI-WORD VERB.

(*v.*) Combine (bases, or one base with another) so as to form a compound word.

• **compounding**: the process of forming compound words by joining at least two independent bases together; contrasted with the other main type of word-formation, DERIVATION. Also called COMPOSITION.

compound-complex sentence

A sentence containing at least two coordinated clauses (making it compound) plus at least one subordinate clause (making it complex).

compound tense

(In older grammar.) A 'tense' combining parts of the verbs *be* or *have*, or both, plus a lexical verb (*been waiting*, *have wondered*).

In some older grammar, compound tenses also included questions and negatives with *do/does/did* and 'future' tenses with *will/would* and *shall/should*.

All such tenses would now be dealt with as complex VERB PHRASES.

compression

Phonology. The reduction of two syllables into one within a word.

Compression is related to ELISION. Thus the RP pronunciation of *medicine* as a two-syllable word /'medsən/ shows established compression due to the elision of the middle syllable. Similarly, a word like *rational* may have three or two syllables, /'ræʃənəl/ or /'ræʃnəl/.

Compression can additionally involve phonetic change. Thus the word *influence* may be pronounced /'ɪnflʊəns/ or reduced by compression to /'ɪnflwəns/, with the change of /ʊ/ to /w/.

computational linguistics See LINGUISTICS.

conation

'Trying' (as a meaning expressed in language).

This term is sometimes applied to a verb construction such as *manage to do*, *fail in doing*.

The term is derived from Latin *conari* to endeavour.

conative

Semantics. (*n. & adj.*) (Designating) language the function of which is to persuade.

concession

A term used occasionally in theories of meaning to refer to the use of language to cause or persuade others to do what the speaker wants (as with COMMANDS). *Conative meaning* thus overlaps with INSTRUMENTAL function.

The term is derived from Latin *conari* to endeavour.

concession

The act or an instance of conceding, admitting.

Concession is one of the meaning categories used in the analysis of adverbial clauses. See CONCESSIVE.

> 1985 R. QUIRK et al. Clauses of concession are introduced chiefly by *although* or its more informal variant *though*.

concessive

Expressing concession.

A *concessive clause* (or *clause of concession*) is usually introduced by a *concessive conjunction*, e.g. *although, though, whereas, while*. These clauses are sometimes classified along with clauses of CONTRAST and sometimes distinguished. They include finite, non-finite, and verbless clauses, e.g.

Although he was angry }
Although feeling angry } he did not raise his voice
Although angry }

> 1985 R. QUIRK et al. Concessive clauses indicate that the situation in the matrix clause is contrary to expectation in the light of what is said in the concessive clause.

Concessive prepositions include *despite, in spite of, for* (obligatorily followed by *all*), *notwithstanding*, e.g.

For all his protestations, nobody believed him.

There are also *concessive conjuncts*, e.g. *anyhow, anyway, however, nevertheless, still, though, yet, in any case, all the same*, and many more, e.g.

He was angry; he did not raise his voice, though

conclusive

Of a verb or its meaning: implying a resulting change of state, progress towards some goal, or possibly some single action or transitional action or event (leading to a definite end). Contrasted with *non-conclusive*.

The term, which can only apply to dynamic verbs, is not in popular use, and cuts across such better known distinctions as DURATIVE and PUNCTUAL.

Contrast *conclusive* meaning in:

The situation has improved (durative; it is better now)
I am redecorating the kitchen (durative; it will end up redecorated)
The concert's beginning (punctual; it will soon have begun)

with *non-conclusive* meaning in

It was raining (durative; no conclusion implied)

The children are watching TV (durative; no conclusion implied)
Why are you smiling? (punctual; no conclusion implied)

concord

The same as AGREEMENT.

concrete

Of a noun: denoting a physical object: a person, an animal, or an observable, touchable thing; contrasted with ABSTRACT.

Originally the term was applied by logicians and grammarians to a quality viewed as *concreted* or adherent to a substance, and so to the word expressing this, namely the adjective, in contrast to the quality as mentally *abstracted* from substance and expressed by an *abstract* noun; hence *white* (paper, horse, etc.) was the concrete quality while *whiteness* was the abstract quality. Later *concrete* was extended also to nouns involving attributes, such as *fool, sage, hero,* etc. Finally it was applied by grammarians to all nouns that are not abstract, thus parting company with, and even contradicting, the older use in logic.

Compare COUNT.

condition

Something upon the fulfilment of which something else depends.

Condition is one of the meaning categories used in the analysis of adverbial clauses. A distinction is often made between *open condition* and *hypothetical condition.*

An open condition (also called *real condition*) is neutral. The condition may or may not be true, and therefore the proposition of the main clause may or may not be true, e.g.

If it rains tomorrow, we won't go.
If Bob's there already, he'll have heard the news.

A hypothetical condition (also called *unreal, counterfactual*, or *rejected condition*) implies that the speaker does not think that the condition will be, is, or has been fulfilled, and therefore the proposition is either in doubt or untrue, e.g.

If he made a bit more effort, he might get somewhere.
If you hadn't told me, I'd never have guessed.

The commonest subordinators introducing clauses of condition are *if* and *unless*. Others are: *on condition (that), providing that, provided (that)*. *If* and *unless* can introduce non-finite and verbless clauses, e.g.

If in doubt, say nothing, unless advised otherwise.

Most conditional clauses posit a *direct condition*. An *indirect condition* occurs when there is a *logical gap* in the overt meaning between the two parts of a conditional sentence. For example, the stated outcome in the following does not depend on the fulfilment of the *if*-clause:

You look tired, if you don't mind my saying so
If you're going in July, it will be raining

See RHETORICAL.

conditional

(*adj.*) (Used for) expressing a condition, e.g. *conditional clause, conditional conjunction, conditional phrase*, etc.

Traditional grammarians often label *should/would* + infinitive and *should/would* + perfect infinitive as *conditional tenses*. But this analysis is somewhat out of favour today.

(*n.*) A sentence containing a conditional clause (a *clause of condition*).

Simplified grammar books for foreign learners often classify conditionals into three structure or meaning types according to the tense forms used:

First conditional	If X DOES [present simple], Y *will* DO . . .
Second conditional	If X DID [past tense], Y *would* DO . . .
Third conditional	If X *had* DONE, Y *would have* DONE . . .

e.g.

If I see them, I'll tell them
If I saw them, I'd tell them
If I had seen them, I would have told them

This analysis is however a misleading oversimplification, as many other tense combinations are normal, e.g.

If you had paid attention, you would know
If you listen, you learn things

conditioned

Phonology. Automatically altered by reason of the phonological context.

The most obviously *conditioned variants* in English are ALLOPHONES and ALLOMORPHS. Thus clear and dark *l* are not in FREE VARIATION, but are conditioned by phonological context (e.g. clear *l* in *let* but dark *l* in *welfare*). Similarly the allomorphs of regular past tenses are conditioned (e.g. *wished* /-t/, *wined* /-d/, and *wanted* /-ɪd/), as is the *a/an* distinction (e.g. *a mother*, *an aunt*).

• **conditioning**.

congruence

Linguistics. Correspondence between grammatical and semantic classification.

Although NOTIONAL grammar has proved unsatisfactory by itself, there are many consistent relationships between forms and meanings. Many abstract nouns, for example, are non-count; most nouns denoting people require *who*, not *which*, and so on.

congruent

1 Exhibiting CONGRUENCE.

2 Of meaning: literal; not metaphorical.

> 1985 M. A. K. HALLIDAY If something is said to be metaphorical, there must also be something that is not; and the assumption is that to any metaphorical expression corresponds another, or perhaps more than one, that is 'literal'—or, as we shall prefer to call it, CONGRUENT.

conjoin

(*v.*) Join (two or more usually equal units); coordinate.

The term is favoured by generative grammarians, where *conjoining* roughly corresponds to COORDINATION in traditional grammar. Thus *conjoined clauses* result in what is often a compound sentence in traditional grammar.

(*n.*) (In some specialized analysis.) An item coordinated with another.

This could be a coordinated clause or phrase or a single word, e.g.

by hook or *by crook*
boys and *girls*
red, *white*, and *blue*

conjoint

(In some specialized analysis.) A coordinated structure, containing two or more conjoins (e.g. *by hook or by crook*, *boys and girls*).

The parts (*conjoins*) are thus distinguished from the whole (*conjoint*): a specialist distinction.

conjugate

Give the different inflected forms of (a verb).

This traditional term is not now considered applicable to English. It is more appropriately used in connection with inflected languages like Latin or French that have forms varying in accordance with number and person.

conjugation

1 A connected scheme of all the inflectional forms of a verb; a division of the verbs of a language according to the general differences of inflection.

> 1841 R. G. LATHAM The Praeterite Tense of the Weak Verbs is formed by the addition of *d* or *t* . . . The Verbs of the Weak Conjugation fall into Three Classes.

2 A presentation of the various inflected forms of the verb.

Neither sense is in current use for English grammar.

conjunct

1 An adverbial with a joining (or connective) function, often that of joining a clause or sentence to an earlier clause or sentence. Popularly called CONNECTOR.

Conjuncts have a variety of meanings, including (a) listing, (b) reinforcement, (c) result, and (d) concession, e.g.

(*a*) *First of all*, I'd like to thank all those people . . .
(*b*) *Moreover* (or *Above all*), I owe a debt of gratitude to . . .
(*c*) I would like, *therefore*, to . . .
(*d*) I must *nevertheless* point out . . .

In some analyses of adverbials conjuncts contrast with ADJUNCTS, DISJUNCTS, and (sometimes) SUBJUNCTS. In analyses that do not use these labels or make different distinctions, conjuncts (roughly as here defined) may be called *conjunctive/connective/discourse/linking adjuncts* (or *adverbs*) or *structural conjunctives*.

2 Another word for a CONJOIN, i.e. a unit joined to another by a coordinating conjunction.

conjunction

1 A word used to join clauses, words in the same clause, and sometimes sentences.

The conjunction is one of the generally recognized word classes (parts of speech). Two main types are generally distinguished:

(*a*) *coordinating conjunction* (also called COORDINATOR) joining units of 'equal' status:

free *and* easy, poor *but* honest, speak now *or* for ever hold your peace.

(*b*) *subordinating conjunction* (also called SUBORDINATOR) introducing a subordinate clause, e.g. *although, because, if, since, when,* etc.

Conjunctions consisting of two or more words are called COMPLEX *conjunctions* (e.g. *but that, in that, assuming (that), as if, in case, as soon as*).

Some that formally resemble other word classes are sometimes labelled accordingly, e.g.

adverbial conjunctions:
  immediately, now (e.g. *Immediately I hear, I'll let you know*)
verbal conjunctions:
  assuming (that), granted (that)
nominal conjunctions:
  e.g. *Every time/The moment he comes, I'll let you know*

2 Joining together, juxtaposition.

This is a very general term, but is sometimes narrowed down and contrasted with DISJUNCTION.

1976 R. HUDDLESTON The term 'conjoining' will sometimes be found in place of coordination; coordination with *and* and *or* are often distinguished as 'conjunction' and 'disjunction' respectively (with 'conjunction' here used in the logician's sense, rather than the traditional grammarian's, for whom it denotes a class of words).

conjunction group

(In Systemic-Functional Grammar.) A word group consisting of a conjunction and modifier(s), e.g. *and so*, *even if*, *not until*.

conjunctive

(*n.* & *adj.*) (A word or phrase) functioning as a conjunction or conjunct, or having some other similar function.

1985 M. A. K. HALLIDAY A conjunctive (that is, a conjunctive expression that is not structural but cohesive) such as *at that time*, *soon afterwards*, *till then*, *in that case*, *in that way*. Note also that some conjunctives, such as *meanwhile*, *otherwise*, *therefore*, *however*, *nevertheless*, are extending their use in modern spoken English so as to become structural conjunctions.

•• **conjunctive adjunct**: the same as CONJUNCT.

connected speech

Phonetics. Speech without pauses between words.

In normal speech several words are usually run together in a single TONE UNIT. This affects the pronunciation of speech sounds, and results in words being said differently from the way they would be said in isolation.

See ASSIMILATION, ELISION, JUNCTURE.

connective

(*n.* & *adj.*) (A word or other linguistic device) serving to link linguistic units. CONJUNCTIONS and/or CONJUNCTS are particularly classed as connectives. But copular and linking verbs are sometimes also included.

•• **connective device**: a way of binding a TEXT together; a term covering COHERENCE and COHESION.

• **connectivity**.

Compare COORDINATION.

connector

The same as CONJUNCT.

connotation

Semantics. An additional meaning that a word (or other linguistic unit) has by virtue of personal or cultural associations; in contrast to its DENOTATION.

Connotation is peripheral compared with 'dictionary meaning' and is considerably dependent on subjective judgement. For example, the connotations of *police* for some people may be 'reliable', 'helpful', 'protectors of law and order', 'the front line against crime', etc., but for others, the connotations may be 'harassment', 'breathalysers', 'arrests', 'water cannons', etc.

• **connotational, connotative**: pertaining to or involving connotation.

Connotative meaning is related to ATTITUDINAL meaning. See also AFFECTIVE, EMOTIVE.

consecutive

Expressing consequence or result.

•• **consecutive clause**: (an older name for) a RESULT *clause*.

consonant

1 (*a*) A speech sound with or without vibration of the vocal cords (voice) in which the escape of air is at least partly obstructed; contrasted with a VOWEL.

(*b*) A speech sound with or without voicing that functions marginally within a syllable.

The commonly accepted use of the term *consonant* is potentially ambiguous. Most consonants are defined in articulatory terms (as in 1 (*a*)), but also share the linguistic or phonological characteristic of being marginal to a syllable (1 (*b*)). Some speech sounds, however, overlap the two categories of vowel and consonant. Southern British /l/ and /r/ have vowel-like articulations, but are usually syllable-marginal; /m/ and /n/ can be either marginal (e.g. *man*) or syllabic (e.g. *frighten*); /w/ and /j/ (the initial sounds in *wet* and *yes*) are phonetically vowel-like but phonologically consonant-like and are classified as SEMI-VOWELS (or *semi-consonants*).

Because of these problems, it has sometimes been suggested that two separate pairs of words should be used, and the American linguist K. L. Pike (b. 1912) proposed retaining *consonant* and *vowel* for the sounds defined in phonological terms (i.e. in terms of their position in a syllable), and introduced *contoid* and *vocoid* for sounds as defined by acoustic or articulatory criteria.

2 (Traditionally) a letter representing a sound as defined in 1 (*a*)

The use of the word *consonant* to describe letters of the alphabet is better avoided, because of the discrepancy between symbol and sound; the letter *y*, for example, cannot satisfactorily be classified as only either consonant or vowel. *Consonant letter* may be considered more acceptable.

There are twenty-two consonant phonemes in standard English (RP): 6 PLOSIVES; 9 FRICATIVES; 2 AFFRICATES; 3 NASALS; 1 LATERAL; and 1 FRICTIONLESS CONTINUANT. [See chart p. 444]

•• **lost consonant**: see SILENT LETTER.

See also DOUBLE CONSONANT, SEMI-VOWEL, VOWEL.

consonantal

Phonetics. Like a consonant.

The term may be used in a straightforward non-technical sense, usually meaning 'like a consonant (i.e. contoid) in articulation'. In DISTINCTIVE FEATURE theory, *consonantal* versus *non-consonantal* is a binary phonetic opposition, consonantal sounds being produced with a major obstruction in the vocal tract, and non-consonantal sounds without it.

consonant cluster

Phonology. A series of consonants, occurring at the beginning or end of a syllable and pronounced together without any intervening vowels. Also called *consonant sequence*.

English has some quite complicated consonant clusters. Initial clusters can have up to three consonants, if the cluster begins with *s* (e.g. *spr*ead, *spl*endid, *str*eet, *squ*int /skwɪnt/, *sk*ewer /'skjuːə/).

Two-consonant clusters are much more usual, but only some combinations can occur. Native speakers who usually have no problem with an initial plosive followed by /r/ or /l/ (as in *proud, bread, true, drew, cream, grew, plate, blue, clue, glad*) might have difficulty in pronouncing *tlew* or *dlad* (which are also initial plosive + *l*), because /tl/ and /dl/ are not members of the English system. Other initial clusters are heard in *beauty* /'bjuːtɪ/, *quite* /kwaɪt/, *shred* /ʃred/, *through* /θruː/, *view* /vjuː/.

Final clusters can contain as many as four consonants, because of inflectional endings, e.g. *texts* /teksts/, *twelfths* /twelfθs/, *glimpsed* /glɪmpst/.

Thus phonologically we can represent English syllable structure as (CCC) V(CCCC).

constative See PERFORMATIVE.

constituency

The relationship of a unit to another unit of which it forms a part. See CONSTITUENT.

constituent

A unit forming part of a larger structure.

A very general term, which can cover clauses, phrases, words, and morphemes as parts of larger units.

See IMMEDIATE CONSTITUENT.

constriction See CLOSURE.

construct

(In older grammar.)

(*v.*) Combine (a word) grammatically; form into a construction (with another word).

e.g. *rely* is constructed with *on*. The term is not much used in modern grammar.
(*n*.) A rare term variously used for 'collocation' or 'construction'.

construction

1 A sentence, or smaller element of one, that is constructed from other CONSTITUENTS.
The term is a very general one, usually not precisely defined. For example *have got* can be described as a construction in contrasting it with the one-word verb *have* (e.g. *We have got/We have a problem here*).
2 The process of making constructions as defined in (1).

1755 S. JOHNSON *Construction* . . the putting of words, duly chosen, together in such a manner as is proper to convey a complete sense.

Not much in current use.
• **constructional**.

construe

(In traditional grammar.)
1 The same as the verb CONSTRUCT.

1902 *New English Dictionary* All the verbs and adjectives which are or have been construed with *of*.

2 Analyse the syntax of (a sentence).
Neither meaning is common in modern grammar.

contact clause

A relative clause joined to its noun phrase without any connecting word.

1970 B. M. H. STRANG Contact-clauses are ancient structures of independent origin, not just relatives with pronouns left out . . . At the beginning of II [the period 1570–1770] . . they were still extensively used where the 'relative' had subject function, as in Shakespeare's *I see a man here needs not liue by shifts*. This is ambiguous . . . . There was good reason for confining the structure to object relations, where there is no ambiguity (as in Defoe, *the same trade she had followed in Ireland*): since the 18c this limitation has been customary.

(*a*) Normally: a defining clause in which the absent relative pronoun would function as an object of the verb or of a deferred preposition.

the woman + *I love*
a crisis + *we could have done without*
a problem + *you know about*

Similar contact clauses are possible where the relative word expresses time, cause, or manner:

The moment (that/when) *I saw it*, I knew it was mine

The reason (that/why) *I asked* was that I needed to know
This is the way (that) *you should do it*

But where the absent relative word would express place, the use is possibly non-standard:

This is the exact place (where) *Latimer and Ridley were burnt*

(*b*) More rarely, since subject relative pronouns are not now normally omissible (see the 1970 quotation): a relative clause in which the absent relative pronoun would function as the subject.

Who was it + *said 'Inside every fat man there's a thin one trying to get out'*?
There's someone at the door + *says you know him.*

Compare APO KOINOU.

The term was introduced by Otto Jespersen (1927).

contamination

The process by which two more or less synonymous linguistic forms are blended by accident or through confusion so as to produce a new form.

The process occurs randomly in speech owing to hesitation between two equivalent forms. Examples of such words that appear to have arisen spontaneously (though they would now be used, if at all, self-consciously) include *insinuendo* and *portentious.*

The term was originated in English by the translator of Herman Paul's works on language.

1988 V. ADAMS We should perhaps distinguish . . between the contamination which arises because words are imperfectly known, or unfamiliar, and that resulting from 'slips of the tongue'. The dialectal examples appear to represent the former; it is the latter kind of contamination which Paul is defining.

Compare BLEND.

content clause

A clause, usually a noun clause, introduced by *that.*

That he can have simply disappeared is unbelievable
I don't believe *(that) the police made a thorough search*

The term was introduced by Otto Jespersen and is not now in common use. A popular modern label is *THAT*-CLAUSE.

In one modern analysis which rejects the popular division of subordinate clauses into noun (nominal), adjectival (relative), and adverbial clauses, *content clause* includes not only noun clauses but such 'completing' clauses as

He's covered his tracks so well *that the police cannot trace him*

content disjunct

An adverbial that comments on the content of an utterance, in contrast to a STYLE DISJUNCT. Also called *attitudinal disjunct.*

Regrettably, nobody bothered to tell them.

content word

A word with a statable meaning. Also called LEXICAL or FULL word; contrasted with a FORM word.

Content words include most open-class words—nouns, verbs, adjectives, and adverbs—but the distinction between content and form words is blurred rather than rigid.

Compare OPEN (2).

context

1 The words or sentences surrounding any piece of written (or spoken) text. See CONTEXT-FREE, CONTEXT-SENSITIVE.

2 **context of situation** (or **extralinguistic context**): the whole situation in which an utterance is made, i.e. who is addressing whom, whether formally or informally, why, for what purpose, when, where, etc.

The context of situation can be an important factor in interpreting MEANING. Although some 'texts' are complete in themselves, others rely heavily on the extralinguistic situation for the interpretation of pronouns, adverbials (e.g. *here, there, now, then, yesterday*), and tenses. *Context of situation* was first used widely in English by the social anthropologist Bronisław Malinowski.

• **contextual**.

Compare TEXTUAL.

context-free

(In generative theory.) Applicable in any context.

A *context-free rule* is one that is theoretically applicable in all contexts. It would be of the simple type 'Rewrite X as Y', with no exclusions or variants. A *context-free grammar* would be a grammar containing rules only of this simple type. Thus a context-free rule for forming noun plurals in English could be on the lines of 'rewrite the singular as singular + s' (but see CONTEXT-SENSITIVE).

context-sensitive

(In generative theory.) Applicable only in certain specified contexts.

Context-sensitive contrasts with CONTEXT-FREE. Thus rules for English noun plurals need to be context-sensitive, so that plurals with *-es* and such irregular forms as *men*, *teeth*, etc. are accurately generated. A context-sensitive grammar is more complicated but more accurate.

contingency

An event or state of affairs dependent on another, uncertain event or occurrence.

Contingency is used in some analyses of the meaning of adverbials as an umbrella term for cause, condition, concession, purpose, reason, and result.

contingent clause

A clause explaining the circumstances under which the main clause is true. Some contingent clauses are ellipted and analysed as adjective clauses, e.g.

The fruit should be picked *(when) unripe*.

continuant

Phonetics. (n. & adj.) (A speech sound) made without a complete closure of the vocal organs.

All vowels are by this definition among the continuants, but use of the term is often restricted to the classification of sounds with a consonantal role. The continuants of English therefore include the fricatives, the lateral /l/, the semi-vowels and /r/—i.e., all the consonants except the plosives and affricates, which involve total closure. (The nasals may or may not be included.)

More narrowly, *continuant* excludes sounds made with friction (i.e. fricatives and affricates), in which case the term may cover much the same sounds as APPROXIMANT. See FRICTIONLESS CONTINUANT.

continuative

One of a small group of linguistic items that are used to carry a conversation on.

This is not a very general classification. The term characterizes the function of words such as *yes, no, oh, well, right*, etc., in signalling that a new 'move' in the conversation is beginning, either a response which shows that the listener is paying attention, or a 'move' to show that the same speaker intends to continue.

continuative relative clause

A non-defining relative clause that continues the narrative.

e.g.

Bob had told Edwin, who passed the news to Henry, who came and told me.

Compare RESTRICTIVE, SENTENTIAL RELATIVE.

continuous

The same as PROGRESSIVE.

contoid

Phonetics. A specialized term for a speech sound made with some obstruction; in contrast to a VOCOID.

The term was introduced by the American phonetician Kenneth Pike (b. 1912): see CONSONANT. *Contoid* excludes [l], [r], [w], and [j].

contract

Shorten (a word, syllable, etc.) by omitting or combining some elements.

1884 *New English Dictionary* *Ain't* . . . A contracted form of *are not* . . , used also for *am not*, in the popular dialect of London and elsewhere.

•• **contracted form**: the same as CONTRACTION (2).

contraction

1 The action of shortening a word, a syllable, etc. by omitting or combining some elements (especially a vowel or vowels).

2 A shortened form of a word that can be attached to another word (usually as an ENCLITIC); the two words together. Also called *abbreviated form*, *contracted form*, or *short form*.

Thus both *'m* and *I'm* are described as contractions. Other contractions in English are:

's, 're, 've, 'd, 'll, n't (=*is/has, are, have, had/would, will, not*)

Compare REDUCTION.

contradictory

Semantics. (*n. & adj.*) (A word or proposition) that is in a relationship to another word or proposition such that they cannot both be true or both be false.

This is a logical concept. Thus *My father is older than my mother* and *My father is younger than my mother* are *contradictories*. Word pairs such as *life/death* and *male/female* are sometimes called contradictories, but more usually COMPLEMENTARIES.

contrafactive

Designating (*a*) a verb followed by a complement clause, where the meaning of the verb makes it necessary that the proposition in the complement clause is contrary to fact, or (*b*) the proposition itself.

In

I wish (that) I knew the answer
I'll pretend (that) I know

the verbs *wish* and *pretend* are contrafactive verbs, and the propositions following them here are contrafactive propositions.

Compare COUNTERFACTUAL, NON-FACTIVE, NON-FACTUAL.

contrastive

1 Showing contrast: a general term applied to clauses, conjunctions, and conjuncts.

Contrastive clauses (also called *clauses of contrast*, and sometimes classified with concessive clauses) are introduced by *contrastive conjunctions* such as *whereas*, *while*, *whilst*, e.g.

I adore jazz, whereas my husband prefers classical music

Contrastive conjuncts include *by contrast*, *alternatively*, *rather*, *more accurately*.

2 *Phonetics*. Designating stress on a word or syllable that would normally be unstressed, in order to convey a contrastive meaning:

What ˋARE you doing? (i.e. you are doing something surprising)
What are ˋYOU doing? (i.e. never mind about the others)

3 **contrastive analysis, contrastive grammar, contrastive linguistics**: the study of two languages, especially for purposes of translation or foreign language teaching.

Contrastive analysis is synchronic. Compare COMPARATIVE LINGUISTICS.

• **contrastivity**.

conversational implicature See IMPLICATURE.

conversational principle See COOPERATIVE PRINCIPLE.

converse

Semantics. (*n. & adj.*) (Designating) one of a pair of relationally opposite words.

Converses are a particular type of ANTONYM, e.g.

buy/sell, husband/wife, learn/teach

conversion

The process by which a word belonging to one word class gets used as part of another word class without the addition of an affix. (Also called *reclassification* or *functional shift*.)

Words produced by conversion are mainly nouns, verbs, or adjectives. Conversion is a very old process in English, as the date-range of the instances given below shows:

Nouns from verbs:
a bounce (E16), a meet (M19), a retread (E20), a swim (M16; M18 in current sense)
Verbs from nouns:
to fingerprint (E20), to highlight (M20), to holiday (M19), to mob (M17), to necklace (E18; L20 in current sense)
Adjectives from nouns:
average (L18), chief (ME), commonplace (E17), cream (M19), damp (L16; E18 in current sense), game (plucky; E18).

An unusual recent conversion is the use of *plus*, already a preposition and noun, as a colloquial conjunction:

10% bonus offer until 31st December, plus you'll get a mystery present.

Minor types of conversion include conversions of closed class words (e.g. *the ins and outs*, *the whys and wherefores*); of affixes (e.g. *So you've got an ology, isms and wasms*); and even of whole phrases (e.g. *his prolier-than-thou protestations*).

A distinction is sometimes made between *full conversion*, as here, and *partial conversion*. In this a word takes on only some of the characteristics of its new word class. The use of adjectives in constructions like *the poor*, *the handicapped* are cited as examples of partial conversion, since they do not permit marking for plural or countability (**six poors*, **a handicapped*); but this analysis is disputed by other grammarians, who prefer to treat such usage simply as an adjective functioning as the head of a noun phrase.

co-occur

Of linguistic units: occur together acceptably.

This term and *co-occurrence* were introduced by the American linguist Zellig Harris (1951).

co-occurrence

Acceptable occurrence of two linguistic units together.

Whether two units can co-occur depends on rules of grammar and vocabulary. For example, we say

a pound, *an* ecu

but not

**a/an* money

Have can occur with a past participle:

have done, have seen

but not with a past simple:

**have did*, **have saw*

Injured occurs with animate beings and their attributes:

its injured paw, my injured feelings

but not usually with inanimate things:

**an injured tank*

Such limitations are called *co-occurrence restrictions*, *relations*, or *rules*.

Compare SELECTIONAL restrictions.

cooperative principle

Pragmatics. An unspoken agreement to be truthful, relevant, and informative, especially in conversation. (Also called *conversational principle*.)

The concept comes from influential work by the philosopher H. P. Grice (1913–88). He suggested that in general speakers cooperate by following certain 'maxims of conversation', such as speaking the truth, giving enough information, avoiding irrelevancies, and so on. Of course speakers sometimes lie or deliberately mislead, but the cooperative principle is so strong that people usually try to make sense of what they hear.

See IMPLICATURE.

coordinate

(*v.*) (Pronounced /kəʊ'ɔːdɪneɪt/.) Join (linguistic units of equal status), commonly by means of **coordinating conjunctions** (also called COORDINATORS).

• **coordinating correlative**: see CORRELATIVE.

(*adj.*) (Pronounced /kəʊ'ɔːdɪnət/.) Of a linguistic unit: joined to another of equal status, as in **coordinate clause;** contrasted with SUBORDINATE.

See COORDINATION. Compare CONJOIN.

coordination

The joining together of two equal units, usually by means of a conjunction. Also called *conjoining*.

The units so joined may be anything from clauses (as in compound sentences) to single words (e.g. *knife and fork*, *poor but honest*, *double or quits*).

Some grammarians include under the term *coordination* structures that lack a conjunction (which could be supplied), called ASYNDETIC coordination. The more usual type of coordination, with a conjunction, is then termed SYNDETIC coordination.

Compare COMBINATORY *coordination*, CONJUNCTION, MULTIPLE *coordination*, PSEUDO-COORDINATION, SEGREGATORY COORDINATION, SUBORDINATION.

coordinator

A coordinating conjunction.

The main (or CENTRAL) coordinators are *and*, *but*, and *or*. They share certain characteristics with SUBORDINATORS and CONJUNCTS, but differ in some notable ways. In particular, coordinators disallow preceding conjunctions or conjuncts (**although and*: contrast *and although*, *but nevertheless*.)

See also MARGINAL *coordinator*.

copula

A verb that links subject and complement, especially the verb *be*.

The term *the copula* usually means the single verb *be*:

She is a pilot. They are pleased.

But any LINKING VERB can be described as a copula:

It *seemed* good at the time. Will it *turn* cold?

• **copular, copulative**: (of a verb) functioning as a copula.

core See COMMON *core*.

co-refer

Of words: have shared or identical reference.

1980 E. K. BROWN & J. E. MILLER The sentence [i.e. John thinks that he is intelligent] has two possible readings . . depending on whether John and he are used to refer to the same individual, or co-refer, or refer to different individuals.

co-reference

A relationship between two linguistic units such that they denote the same REFERENT in extralinguistic context.

Co-reference is often achieved through the use of pro-forms, but is to be distinguished from SUBSTITUTION. In the latter the pro-form may stand for another word or words, e.g.

I worked harder than I had ever *done*

With co-reference two expressions refer to the same person or thing in the world, e.g.

My cousin said *he* would help.

This may have effects on sentence grammar, e.g.

1973 R. QUIRK & S. GREENBAUM The passive transformation is blocked when there is co-reference between subject and object, *ie* when there are reflexive, reciprocal, or possessive pronouns in the noun phrase as object . . .
We could hardly see each other in the fog
~*Each other could hardly be seen in the fog

Co-reference is the phenomenon dealt with under *binding* in Government-Binding Theory.

• **co-referential. co-referentiality**.

coronal

Phonetics. Of a sound: made with the blade of the tongue raised.

This is a term in DISTINCTIVE FEATURE analysis of sounds. Different analysts use the term slightly differently, but a typical coronal sound in English is /ʃ/ as in *shoe*.

corpus (Plural **corpuses, corpora.**)

A collection of spoken and/or written TEXTS.

•• **corpus linguistics**: the study of language by means of corpora, now usually computerized ones.

The study of the English language has been transformed in recent years by the collecting of quantities of authentic texts into corpora on which grammatical and lexicographic analyses and descriptions of use can be based.

A pioneering project was the Survey of English Usage, begun in 1958, and containing a million words of running text (including transcribed spoken texts), all originally collected and analysed manually. Recent technological advances have revolutionized corpus compilation; several large electronic corpora are in existence or are being compiled. The Cobuild corpus at the University of Birmingham now exceeds 200 million words, while the British National Corpus, a government-sponsored collaborative project begun in 1991 and involving three publishers, the British Library, and the Universities of Oxford and Lancaster, aims at 100 million words.

correctness

Grammatical acceptability according to the rules.

Traditionally, grammar was thought to be concerned with PRESCRIPTIVE rules stating what is and is not 'correct' usage. Present-day linguists try to provide DESCRIPTIVE grammar, and tend to use the term *correctness* pejoratively. However, such a holier-than-thou attitude is somewhat disingenuous, since even the most permissive description must be based on some decisions about what to include.

Compare ACCEPTABILITY.

correlative

(*n.*) (One of) a pair of elements that join two similar parts of a sentence together.

(*adj.*) That is a correlative; made up of or joined by correlatives.

Coordinating correlatives include:

both . . . and
either . . . or
neither . . . nor
not only . . . but also

Subordinating correlatives include:

so/such . . . that
less/more . . . than
hardly . . . when
if . . . then

correspondence

A syntactic relationship between two structures containing similar lexical items, which is matched by a meaning relationship.

Systematic correspondences of this kind exist between a sentence containing an active verb and a direct object (e.g. *Everyone watched the eclipse*) and a passive sentence with the former object now the subject (e.g. *The eclipse was watched by everyone*).

count

Designating a noun that can be used with numerical values.

count noun (also called COUNTABLE noun) contrasts with UNCOUNT, *uncountable*, or *non-count* noun.

1973 R. QUIRK & S. GREENBAUM Abstract nouns may be count like *remark* or non-count like *warmth.*

Count nouns usually have different singular and plural forms (*book, books, child, children*), and when used in the singular must be preceded by a determiner: *a/my/this/one book*; not **I bought book.*

In the plural, count nouns have the potential for combining with certain determiners, some of them exclusive to the plural:

few, many, several, these

The binary division of COMMON nouns into count and uncount poses a few awkward problems. Some nouns belong to both categories while others do not neatly fit either. See MASS, PLURAL.

countable

(*n. & adj.*) (Designating) a noun with singular and plural forms. See COUNT.

A term introduced by Jespersen (1914).

• **countability**.

counterfactual

Of a conditional sentence: relating to a completely unreal or hypothetical situation (i.e. one entirely contrary to fact), in contrast to a real condition.

Thus

If he had been on that plane, he would have been killed

is counterfactual, since the meaning is 'He was not on the plane and he was not killed'.

Compare CONTRAFACTIVE. See also FACTUAL.

counter-intuitive

Contrary to what would be expected intuitively.

A term sometimes used in analysis. For example it is counter-intuitive to suppose that active verbs are derived from passive ones, or that affirmative sentences are derived from negative ones.

The term was introduced by Chomsky (1955).

courtesy subjunct See SUBJUNCT.

creativity

The ability of native speakers of a language to produce and to understand an infinite number of sentences of their language, many of which they have never produced or heard before.

creole

Sociolinguistics. A PIDGIN that has become a mother tongue.

Unlike a pidgin, a creole is the first language of a speech community and so has greater lexical and syntactic complexity. As with any language, there are usually several varieties, but they can usually be distinguished according to their closeness to the language on which the creole is based.

Most creoles have developed from contact between a European language (especially English, French, or Portuguese) and another (often African) language. But the process is not straightforward, and most English-based creoles and pidgins contain certain words of Portuguese origin, such as West African *palava* 'trouble' from Portuguese *palavra* 'word'.

English-based creoles are found in the Caribbean—the most widely spoken being Jamaican Creole—and in other ex-colonial territories.

Compare BASILECT.

creolize

Make into a creole.

•• **creolized language**: a CREOLE.

• **creolization**.

> 1980 R. A. HUDSON There is no research evidence of changes which have happened during creolisation which cannot be matched by changes to a pidgin without native speakers.

current See LINKING VERB.

current relevance

A concept often invoked in explaining the meaning of the present perfect in contrast to the past tense.

Thus

> Bob has gone to Edinburgh

implies 'and he is there (or on his way there) now'. Similarly

> My grandmother has lived in Oxford all her life

means that she is still alive. (Contrast *She lived in Oxford all her life.*) The concept, however, is not easy to pin down, and certainly past tenses can also have present effects or results (e.g. *I crashed the car yesterday—so I can't drive over to see you today*).

D

dangling modifier See HANGING PARTICIPLE.

dangling participle See HANGING PARTICIPLE.

dark *l* See LATERAL.

dative

1 (*n. & adj.*) (The case) expressing an indirect object or recipient.

In many inflected languages nouns and pronouns (and other words agreeing with them) have special forms to indicate a recipient meaning. The term is not really applicable in English, where the nearest comparable term is INDIRECT OBJECT.

2 (*n. & adj.*) (In Case Grammar.) (Expressing) the role taken by the noun (or noun phrase) referring to the person or other animate being 'affected' by the action expressed by the verb.

The referent may be the grammatical subject or object of traditional grammar. For example, in Case Grammar, *Tom* is a dative in both the following sentences:

Tom was attacked by the dog.
We forced Tom to listen.

In some formulations of Case Grammar it is said that in a phrase such as *Tom's chin*, Tom has a dative role (because the chin 'belongs to' Tom). The dative case was later renamed EXPERIENCER. The dative of Case Grammar is not equivalent to the traditional indirect object, which is analysed as *benefactive* or *recipient*.

de-adjectival

Of a word: derived from an adjective.

De-adjectival nouns are common, e.g. *falsehood* (ME) from *false*, *kindness* (ME) from *kind*, *subsidiarity* (M20) from *subsidiary*.

declaration

An utterance which by the mere virtue of being said brings about a result.

This is a specialist term from Speech-Act Theory for a particular type of illocutionary act, e.g.

I declare the meeting closed. You're fired.

In popular usage, verbs in such utterances would be included in the umbrella term PERFORMATIVES.

declarative

(*n.*)

1 A sentence in which the subject precedes the verb, typically used for making a statement.

Considerable confusion is caused by the use of words like *statement*, *question*, and *command* as both syntactic and semantic categories, and so some grammarians are careful to use separate terms. *Declarative* is often a formal syntactic category, contrasting with STATEMENT used as a functional category (though the reverse usage may be found). Sentences that are declarative in form may be used not only for making (pragmatic) statements, but also to ask questions (*You understand what you're doing?*) or to give orders (*You will report back to me tomorrow*).

Compare EXCLAMATIVE, IMPERATIVE, INTERROGATIVE.

2 (In Speech-Act Theory.) A verb that makes a declaration.

(*adj.*)

1 Of a sentence: that is formally a declarative.

•• **declarative question**: see QUESTION.

2 (In Speech-Act Theory.) Of or pertaining to a DECLARATION, or to the verb used in making one.

> 1990 D. VANDERVEKEN The primitive declarative verb is "declare", which names the illocutionary force of declaration. "Declare" . . also has an assertive use, but in its declarative use it exemplifies the characteristic features of the set in that the speaker purely and simply makes something the case by declaring it is so.

3 **declarative mood**: an alternative term for INDICATIVE *mood.*

declension

1 The variation of the form of a noun, adjective, or pronoun, to show different cases, such as nominative, accusative, dative, etc.; the class into which such words are put according to the exact form of this variation, usually called first, second, etc. declension.

2 The presentation of the various inflected forms of such a word.

The term is applicable to a language such as Latin, where nouns, pronouns, and adjectives are *declined* in this way, but not to English, where only six words (five PERSONAL PRONOUNS and *who*) show any case distinctions.

Compare GENITIVE.

declination

Phonetics. An intonation pattern starting fairly high, with a gradual dropping down of pitch during the utterance.

This is sometimes said to be the basic, 'unmarked' intonation pattern in English (or even in all languages).

decline

Inflect (a word) through different cases; give in set order the cases of; (of a word) inflect to show case.

See DECLENSION.

deep structure

(In Transformational-Generative Grammar.) The supposed abstract underlying organization of a sentence; in contrast to its SURFACE STRUCTURE.

In Transformational-Generative Grammar, an underlying abstract 'syntactic representation' is posited to explain the way in which actual sentences are interpreted. A classic example from the early theory was the type of ambiguous sentence such as

Visiting aunts can be boring

which was said to have two different deep structures depending on whether it means 'going to visit one's aunts can be boring' or 'aunts who visit (*or* when visiting) can be boring'. Each of these two interpretations was supposed to have its own syntactic representation in deep structure. Similarly, two sentences with identical surface structures such as

John is eager to please

and

John is easy to please

have different deep structures. Conversely, an active and passive pair of sentences, with different surface structures, e.g.

The dog bit the man
The man was bitten by the dog

are said to have the same deep structure.

In detailed theory, the deep structure is itself derived from some more primitive base. It is then transformed, by the application of more rules, into the surface structure, which in turn by the application of rules becomes the sentence as we recognize it. Both deep and surface structure are represented in a somewhat symbolic way. For example, *was bitten* might be shown as

bite [Past Passive]

More loosely, deep and surface structure are used as though they are terms in a simple binary opposition—the deep structure being a sort of unmarked underlying meaning, and the surface structure being the actual sentence we see.

defective

Of a verb: incomplete, lacking a complete set of forms.

The MODAL verbs are described in older grammar as defective, since they have only one form each (or at most two, if the pairs *will* and *would*, *can* and *could*, etc., are treated as single paradigms) and lack imperative and non-finite forms. Another defective verb is *beware*, which is used only as an imperative or a *to*-infinitive:

Beware of the dog
I warned him to beware of the dog

Compare PERIPHRASIS.

deferred preposition

A preposition that, instead of preceding its complement, comes later in the clause. (Also called *stranded preposition.*)

See PREPOSITION.

defining

Of modification or a modifier: that identifies or restricts the meaning of the modified head. Also called *restrictive*. Contrasted with NON-DEFINING (or NON-RESTRICTIVE).

Various kinds of linguistic unit, including adjectives and different kinds of postmodification, can have a defining or restrictive role. For example, in *my blind friend*, the adjective may well be understood to give uniquely defining reference, identifying one particular friend. But in *my blind mother*, the adjective is non-defining, merely adding some information about my mother.

Similarly, in

The man *wearing military uniform* is my uncle

and

The man *with all the gold braid* is my uncle

the non-finite clause and the prepositional phrase define *The man*, whereas the postmodification in the following is non-defining:

The British troops, *wearing bright red uniforms*, were an easy target
The Duke, *resplendent in his uniform*, led his army to victory

The terms *defining* and *non-defining* are, however, primarily applied to finite relative clauses.

• • **defining relative clause**: a finite clause that postmodifies a noun phrase and restricts its meaning:

News is what a chap [what sort of chap?] *who doesn't care much about anything* wants to read.
All the news [all of it?] *that's fit to print*.
There are only two posh papers on a Sunday—the one [which one?] *you're reading* and this one.

A defining relative clause is not separated from its noun phrase by a comma, and may sometimes, as in the third example, be a CONTACT CLAUSE.

Also called *identifying relative clause*, *restrictive relative clause*.

Compare ADNOMINAL, APPOSITIVE CLAUSE.

definite

Of a linguistic form: having or indicating identifiable particular or exclusive reference.

Contrasted with INDEFINITE.

The tense labels *past definite* and *past indefinite* are rarely used in English grammar, but in fact the past simple does usually imply some definite point or period in the past, when the present perfect has a more indefinite reference.

I have spoken to my bank manager. In fact, I *spoke* to him again yesterday.

•• **definite article**: the determiner *the*, which is typically used with a noun phrase whose referent has either just been mentioned (or implied) or is assumed to be familiar or uniquely identifiable in some way.

I had to call a taxi. *The* driver couldn't find *the* house
The sun's out at last
I heard it on *the* radio

Determiners other than *the* can make a noun phrase definite, including the demonstratives (*this*, *that*, etc.) and the possessives (*my*, *your*, etc.). All can be collectively labelled *definite determiners* or *definite identifiers*. Proper names are inherently definite (at least in context), and so are personal pronouns (*he*, *she*, etc.), in contrast to indefinite *somebody*, *something*.

Compare SPECIFIC.

definite frequency: see INDEFINITE.

• **definiteness**.

degree

1 Each of the steps on the threefold scale by which gradable adjectives and adverbs are compared; this scale as a feature of an adjective or adverb.

The three degrees are positive, comparative, and superlative:

good, *better*, *best*; *soon*, *sooner*, *soonest*.

Compare GRADABLE.

2 Greater or lesser intensity, as one category of adverb meaning.

•• **degree adverb, degree adverbial, adverb of degree,** or **adverbial of degree**: (usually) an adverb or adverbial expressing a meaning of greater or lesser intensity.

Examples:

much, *quite*, *so*, *too*, *very*, etc.

Degree is one of the traditional categories into which ADVERBS are divided, in contrast to manner, time, and place, though not all grammarians use the term. Some modern grammarians prefer to use INTENSIFIER more or less as a synonym. Others make subtle distinctions between *degree adverb*, *intensifier*, and *emphasizer*, sometimes differentiating them, sometimes making either *degree adverb(ial)* or *intensifier* the superordinate term.

deictic

(*n. & adj.*) (Designating or expressed by) a word that 'points', i.e. that has the function of relating the utterance to its extralinguistic context (time, place, etc.).

The four demonstrative determiners and pronouns are the prime deictics, *this* and *these* pointing to what is here or now, while *that* and *those* point to there or then. Other words commonly included in this category are *here* and *there*, *now* and *then*, *today*, *yesterday*, and *tomorrow*, and the personal pronouns (*I, we, you*, etc.). Tense too (present versus past) is a deictic category.

In some grammatical models, the category of *deictic* is extended to cover what are determiners in other analyses.

deixis

The process of indicating time and place in relation to the utterance, or the features of the language collectively that do this.

> 1977 J. LYONS The term 'deixis' . . is now used in linguistics to refer to the function of personal and demonstrative pronouns, of tense and of a variety of other grammatical and lexical features which relate utterances to the spatiotemporal co-ordinates of the act of utterance.

deletion

1 (In Transformational-Generative Grammar.) The process by which a constituent of some underlying structure is omitted from the surface structure.

Deletion of the word *you* from some deep structure is said to explain how it is we understand the meaning of usually subjectless imperatives (*Wait* here!). A more complicated deletion is said to be responsible for a sentence such as *I intend to go home*, derivable according to one theory from 'I intend SOMETHING' + 'I go home'. The missing item must be recoverable for deletion to be recognized.

2 Another word for ELLIPSIS.

delexical

Of a verb: having little meaning in itself.

Perhaps to give more end-focus or end-weight to a sentence English sometimes uses a verb + object noun where a plain intransitive verb could be used. For example instead of saying *I looked*, you can say *I had a look*; instead of *I'll think about it*, you can say *I'll give it some thought*. Verbs particularly used in this way include *do, have, give, make,* and *take*, and when so used they retain little of their usual meaning, and the main meaning is carried by the object noun.

delicacy

The scale determining the degree of detail in a grammatical analysis.

Delicacy is one of the three (or four) SCALES of Systemic Grammar. At the primary degree of delicacy, clauses can be classified as free (roughly, main or superordinate) or bound (roughly, subordinate). At the next degree of delicacy they can be classified to show how distant they are from the main clause, i.e., whether they are immediately subordinate, or subordinate to another subordinate clause.

Compare DEPTH, EXPONENCE, RANK.

demonstrative

(*n. & adj.*) (Being) a member of the set of pronouns and determiners (*this, these, that, those*) used in referring to things or people in relationship to the speaker or writer in space or time.

•• **demonstrative pronoun**: a demonstrative functioning as a pronoun.

e.g.

This doesn't suit me, but I really like *those*

demonstrative adjective: (the traditional term for) a demonstrative functioning as a determiner.

denominal

Of a word: derived from a noun.

Denominal nouns (nouns derived from other nouns) include many words formed by adding a suffix: *booklet* (M19), *childhood* (OE), *gangster* (L19), *lectureship* (M17), *lioness* (ME), *mileage* (M18), *spoonful* (ME), *teenager* (M20), *villager* (L16).

Denominal adjectives include words with pseudo-participles, such as *red-eyed* (M17), along with words formed from a noun + suffix, e.g. *childish* (OE), *hopeless* (M16), *friendly* (OE).

Denominal verbs include those formed with prefixes and suffixes such as *be-* and *-ize*, e.g. *behead* (OE), *dynamize* (M19).

An older term for this was *denominative*.

denotation

Semantics.

1 Relationship between a word (or other linguistic unit) and its referent.

2 The primary (often literal) meaning of a word, in contrast to its CONNOTATION.

Denotation relates to the naming function of words, and so to their generally definable 'dictionary' meanings. Thus the denotation of *(the) police* is 'the civil force of a State, responsible for maintaining public order'.

• **denotative**.

Denotative meaning is also called *cognitive* or *referential meaning*.

dental

Phonetics. (*n. & adj.*) (A consonant) made with the tongue coming in contact with the teeth.

The English dental consonants are the voiceless fricative /θ/ as in *thick* and *thin* and the voiced fricative /ð/ as in *this, them.*

In some languages the [t] and [d] sounds are dental. In standard English RP these are normally alveolar, but they may sometimes have dental articulation through ASSIMILATION, particularly when followed by /θ/ or /ð/.

The diacritic for marking an abnormal dental articulation is [̪]. Thus *not thin, had then* might be shown phonetically as /nɒt̪ θɪn/, /hæd̪ ðen/.

deontic

Of or relating to duty and obligation as ethical concepts. Deontic modality is sometimes called *intrinsic* or *root modality.*

The term is applied to those uses and meanings of modal verbs that are intended to impose an obligation or grant permission or otherwise influence behaviour, e.g.

> You must obey your parents
> You may go now
> You shouldn't mislead me

In a sense the deontic modals actually do something (e.g. they order, advise, permit, etc.), and can be regarded as a special type of PERFORMATIVE.

In the analysis of modal meaning and use, deontic modality is usually contrasted with EPISTEMIC modality. But some analysts make a three-way contrast, the third term being ALETHIC. Other analysts make a different three-way contrast: between *deontic*, *epistemic*, and DYNAMIC.

dependency

The fact of being dependent.

•• **dependency relation**: a relation between units where one unit is described as dependent on the other.

dependency grammar: a grammar that describes dependency relations.

dependent

(*n. & adj.*) (A linguistic unit) that is subordinate to some other linguistic unit. Contrasted with INDEPENDENT.

Some grammarians describe words other than the HEAD in an adjective, adverb, or noun phrase, as *dependent elements* in the phrase.

Compare EMBEDDED.

•• **dependent clause**: another term for SUBORDINATE *clause.*

depth

In Systemic Grammar, a scale measuring the degree of complexity of the analysis.

Depth is sometimes handled as part of the scale of DELICACY.

derivation

1 *Morphology*. The process of forming a new word by adding an affix to an existing word; contrasted with COMPOUNDING, and also with INFLECTION.
Examples:

alleviation (from *alleviate*)
interference (from *interfere*)
sub-editor (from *editor*)
unhelpful (from *helpful*)

Roughly speaking, derivation produces a new word (e.g. *driver* from *drive*), whereas an inflectional suffix produces another form of the same word (e.g. *driven, driving, drives*).

2 (In Transformational Grammar.) The process by which a structure of one kind is formed (or *derived*) from another structure through the application of the appropriate rules.

e.g. in early theory a passive sentence is said to be derived from an active one; in later theory both active and passive sentences are derived, through a series of layers, from a much more abstract representation.

derivational

Morphology. Pertaining to, used in, or due to derivation.

1940 C. C. FRIES Most of these derivational forms are, in Present-day English, chiefly vocabulary or word-formation matters and . . we have limited our study to grammatical structure and have excluded vocabulary.

•• **derivational suffix**: see SUFFIX.

derivative

Morphology.

1 (*n. & adj.*). (A word) formed from another word by a process of derivation.

2 (*adj.*) Of an affix: used in derivation.

1975 D. J. ALLERTON & M. A. FRENCH Derivative suffixes seem to fall into three main categories with regard to stress: (i) The majority of suffixes are unstressed and leave the stressing of the stem unchanged . . . (ii) Some suffixes are unstressed but shift the stressing of the stem . . . (iii) Suffixes which are usually stressed.

• **derivatively**.

derive

1 *Morphology*. Of a word etc.: descend or be formed by a process of word formation (from an earlier or more basic element in the same or another language). Also, form (a word etc.) by a process of word formation.

Thus current *denim*, noun, is derived from 17th century *serge de Nim*, from French *serge de Nimes* ('serge made in the town of Nimes'); *atonement* derives from the prepositional phrase *at one* + the suffix *-ment*.

1975 D. J. ALLERTON & M. A. FRENCH While it is quite common for English affixes to derive the same word-class from different stems, it is less common for the same affix to have words of different classes as derivatives.

2 **be derived**: (in Transformational-Generative Grammar) (of a structure) be formed from another, lower, structure; see DERIVATION.

descriptive

1 Describing the structure of a language at a given time, avoiding comparisons with other languages or other historical phases, and free from social valuations. Contrasted with PRESCRIPTIVE.

• • **descriptive linguistics**: describing the structure of a particular language at a particular time, in contrast with HISTORICAL or COMPARATIVE *linguistics*. Also called SYNCHRONIC linguistics. See LINGUISTICS.

Many modern grammarians try to describe language as it is used and try to avoid laying down idealized, unrealistic rules. They seek to practise *descriptive linguistics* and to produce *descriptive grammars*, in contrast to the more PRESCRIPTIVE aims of usage books.

See also DESCRIPTIVISM.

2 *Semantics*. (In some classifications of meaning.) Similar to COGNITIVE or REFERENTIAL.

Descriptive meaning, *descriptive statements*, etc. thus contrast with ATTITUDINAL meaning and utterances.

Compare DENOTATIVE, IDEATIONAL.

3 (In Systemic Grammar.) Designating verbs of non-directed verbal action.

1985 G. D. MORLEY A basic distinction is drawn between goal-directed action, labelled 'effective', e.g. *wash, hit, throw*, and non-directed action, known as descriptive, e.g. *march, rest, garden, walk* . . . The terms 'effective' and 'descriptive' thus form a sub-system of the feature 'extensive'.

4 **descriptive adjective**: an adjective that describes, in contrast to a LIMITING ADJECTIVE.

This is an old-fashioned term, since modern grammarians assign most so-called 'limiting adjectives' to a separate DETERMINER class.

5 **descriptive genitive**: a genitive construction in which the GENITIVE word has a describing function (also called CLASSIFYING GENITIVE).

In *a master's degree*, *a doll's house*, *ladies' shoes*, the genitive does not really indicate possession, but rather has an adjectival meaning (c.f. *a doctoral degree*, *a miniature house*).

6 **descriptive relative clause**: another term for *non-defining* or *non-restrictive relative clause*.

descriptivism

An approach to grammatical analysis, characterized by a concern with describing a language objectively. The term is often applied to the American

STRUCTURALIST school, which preceded Transformational-Generative Grammar.

• **descriptivist**.

descriptor

Linguistics. A word or expression used to describe or identify.

1985 R. QUIRK et al. Proper nouns often combine with descriptive words which we will call DESCRIPTORS, and which also begin with a capital letter, to make composite names like *Senator Morse*, *Dallas Road*.

destination See SOURCE (2).

determination

The function of determining what kind of reference a noun phrase has; as opposed to a function such as MODIFICATION.

This function is generally performed by a DETERMINER, but a genitive is also possible:

the/this/my/Albert's pet

1985 R. QUIRK et al. Determination. This term may be used for the function of words and (sometimes) phrases which, in general, determine what *kind of reference* a noun phrase has: for example, whether it is definite (like *the*) or indefinite (like *a/an*), partitive (like *some*) or universal (like *all*).

determinative

(*n.* & *adj.*) (A word or phrase) having the function of DETERMINATION: (*a*) more broadly, denoting any determinative word or phrase; (*b*) more narrowly, denoting a member of a (mainly) closed class of determinative words.

By definition (*a*), determinatives include not only closed class words such as *a, the, this, some, every, many*, etc., but also genitive phrases used in the same way (e.g. *President Clinton's* speech, *the Queen's* walkabout.) Definition (*b*) would exclude genitive phrases. See DETERMINER.

determiner

1 (Generally.) A member of a mainly closed class of words that precede nouns (or, strictly speaking, noun phrase heads) and limit the meaning in some way.

Determiners are sometimes called LIMITING ADJECTIVES in traditional grammar. However, they not only differ from the class of adjectives by meaning, but also must normally precede ordinary adjectives in noun phrase structure. Further, among determiners themselves there are co-occurrence restrictions and fairly strict rules of word order. On the grounds of these restrictions, some grammarians divide them into:

predeterminers, e.g *all, both, half, double, one-third, such, what!*
central determiners, e.g. *a, an, the, this,* etc., *my* etc., *every, each, no,* etc.
postdeterminers, e.g. *few, many, much, little,* and the CARDINAL and ORDINAL numbers.

The class is not entirely closed, as variants are possible on some items, e.g. *a (good) few, a (very) little, (a great) many*; furthermore, numbers are an open-ended set (e.g. *seventh, three times,* etc.).

Most determiners are restricted by number-related meaning as to the category of noun they can occur with, e.g.

many/few apples (count plural)

but

much/little food (uncount)

Determiners are also classified on grounds of meaning into such categories as DEMONSTRATIVES, QUANTIFIERS, etc. Some grammars favour a basically twofold classification: one model groups all determiners and numbers as either IDENTIFIERS or QUANTIFIERS (with *definite* and *indefinite* cutting across these groupings); another divides them, very differently, into GENERAL and SPECIFIC determiners. The multiplicity of labels resulting from these different approaches can be confusing. See SPECIFIC.

2 Any word or phrase in determinative function (for example, a genitive phrase).

The term *determiner* is widely used in sense 1, i.e. as a class label for a fairly closed class of words. But as a similar function can be performed by a wider set of elements (e.g. genitive phrases), some grammarians use the two terms *determiner* and *determinative* to make the distinction between the narrower and wider uses. Unfortunately there is disagreement as to which is used for which.

See also POSTDETERMINER (2).

deverbal

(*n. & adj.*) (A word) derived from a verb.

Examples of deverbal nouns:

dismissal (E19), *driver* (LME), *payee* (M18), *shrinkage* (E19), *starvation* (L18).

Examples of deverbal adjectives:

arguable (E17), *clingy* (E18), *innovative* (E17), *tiresome* (E16).

A less usual, probably obsolescent, term is *deverbative*.

Compare VERBAL NOUN.

deviant

Differing from the normal, dubious, ill-formed.

The term is favoured by some linguists as a way to avoid saying that a particular item is incorrect or downright wrong. Deviant sentences are often marked with an asterisk (*) and distinguished from dubious structures marked with a question mark (?), e.g.

*Looking out of the window, a fire-engine screamed past
?Having lived abroad so long, my outlook has changed

Terms such as *deviant* and *ill-formed* have a special status in some Generative Grammar, where grammatically correct sentences may be dubbed deviant if they break strict *selectional restrictions*. This has the effect of making some acceptable though perhaps metaphorical utterances deviant. For example *How sweet the moonlight sleeps upon this bank* would presumably be deviant, since the verb *sleep* normally requires an animate subject.

Compare GRAMMATICAL.

• **deviance**.

devoice

Phonetics. Articulate (a speech sound) with less VOICE than is usual.

• **devoiced. devoicing.**

English voiced sounds are often partly devoiced under the influence of surrounding sounds. Thus the voiced plosives /b/, /d/, and /g/ are normally devoiced or may even be completely voiceless in word-final position. Similarly, voiced fricatives tend to be partly devoiced except when occurring between voiced sounds; and /l/, /r/, /w/, and /j/ are usually devoiced when following initial voiceless sounds, as in *please, tray, twice, queue* /kjuː/.

Devoicing can be indicated by a small circle under the sound [̥] as in *end* /end̥/.

Compare VOICELESS.

diachronic

Linguistics. Concerned with the historical development of language; as opposed to SYNCHRONIC.

Study of the changes in pronunciation, grammar, or vocabulary between Anglo-Saxon times and the present day can be described as *diachronic phonology*, *diachronic linguistics* (also called HISTORICAL LINGUISTICS), etc.

The term was coined by F. de Saussure (1857–1913) in his *Cours de linguistique generale* (1916). *Diachronistic* has also been used.

Compare COMPARATIVE (2).

• **diachronically. diachrony** (somewhat rare): diachronic method or treatment.

diacritic

Phonetics. A mark indicating a modification to the usual value of a phonetic symbol.

Diacritics are employed in the International Phonetic Alphabet to indicate devoicing, as in /pl̥eɪ/ *play*; nasalization, as in /mæ̃n/, a possible pronunciation of *man*; and dental articulation, as in /eɪt̪θ/ *eighth*.

It is more usual to use them in a narrow transcription in order to indicate the exact sound, than in a broad or phonemic transcription, in which one symbol stands for each phoneme. (An exception to this is the nasalization symbol /˜/, which is used in the ordinary transcription of languages which distinguish nasalized and non-nasalized vowel phonemes, such as French.)

• • **diacritical mark**: the same as DIACRITIC. [See table p. 445]

diaeresis

The mark (as in *naïve, Chloë*) over a vowel to indicate that it is sounded separately. It resembles the German *umlaut*, but the latter is used to distinguish similar but distinct vowel phonemes (e.g. *ü* and *u*, etc.).

dialect

A variety of a language that is distinct from other varieties in grammar, vocabulary, and ACCENT.

Dialects may be regional, or based on class differences, (when they are usually called *social* or *class dialects*), or a mixture of the two. Although dialects are usually recognizable from the speaker's accent, the term primarily implies differences of grammar, e.g.

I likes it
I ain't done it
I didn't have no breakfast
It needs washed
We got off of the train
Look at them cows

and vocabulary, e.g. *while* meaning 'until' (*Wait while the lights are green* was allegedly a level-crossing notice confusing to southerners); *learn* meaning 'teach' (as in the book-title *Lern yerself Scouse*); *happen* meaning 'perhaps'; etc.

The term *dialect* tends to imply deviation from some standard educated norm, but linguists regard the standard variety as just another dialect. When it comes to global varieties of English, the term *dialect* is not used, and terms such as *American English* or *Indian English* are appropriate.

Compare STANDARD ENGLISH.

• • **dialect geography**: another term for DIALECTOLOGY.

• **dialectal**: of or pertaining to dialect or dialects.

dialectology

The study of dialects, particularly regional ones.

dictionary meaning See MEANING.

diglossia

Sociolinguistics. A situation in which two or more varieties of the same language are used by the same speakers under different conditions.

1964 E. PALMER [translating A. Martinet] Linguists have proposed the term 'diglossia' to designate a situation where a community uses . . both a more colloquial idiom of less prestige and another of more learned and refined status.

The term is particularly appropriate when applied to those languages that have distinct 'high' and 'low' varieties, like, for example, Arabic, which has a 'classical' and several colloquial forms. However, it has recently been suggested that a weakened form of the theory could be applied to English, where there are not only many 'high' and 'low' vocabulary equivalents (e.g. *purchase*, *buy*; *larceny*, *theft*; *sufficient*, *enough*) but where such a theory could account for some alternatives in grammatical usage, for example *whom* versus *who*.

The term is modelled on French *diglossie*.

• **diglossic**.

Compare ACROLECT, CODE-SWITCHING.

digraph

1 A group of two letters representing one sound, as *ph* in *phone* or *ey* in *key*.

2 Two letters that are physically joined together in a writing or printing system as in *æon*, *œdema*. Also called a LIGATURE.

diminutive

(*adj.*) Of a derivative word: denoting something small (literally or metaphorically) of the class which the base word denotes. Of a suffix: forming diminutive words.

(*n.*) A diminutive word or suffix.

Some diminutives are objective (e.g. *manikin, piglet*), but many are used as a mark of informality (e.g. *bunny, comfy, sweetie*), or to show affection (e.g. *auntie*), or to belittle (e.g. *starlet*). Proper names often have diminutive forms (e.g. *Teddy (Edward), Jimmy (James), Lizzie, Bessie, Betsy, Betty (Elizabeth)*).

Compare HYPOCORISTIC.

diphthong

1 *Phonetics*. A vowel that changes its quality within the same single syllable. (Also called *gliding vowel*.)

The English diphthongs in modern standard RP are:

Three that glide towards an /ɪ/ sound from different starting points:

/eɪ/ as in *day, late, rain, weigh, they, great*
/aɪ/ as in *time, cry, high, height, die, dye, aisle, eider*
/ɔɪ/ as in *boy, voice*

Two that glide towards /ʊ/:

/əʊ/ as in *so, road, toe, soul, know*
/aʊ/ as in *house, now*

Three that glide towards /ə/:

/eə/ as in *care, air, wear, their, there*
/ʊə/ as in *pure, during, tourist*
/ɪə/ as in *deer, dear, here, weird, idea*

A diphthong gliding to a closer sound (i.e. one ending in /ɪ/ or /ʊ/ in English) is called a CLOSING *diphthong*; a diphthong finishing at /ə/ is called a CENTRING *diphthong*.

Formerly a fourth centring diphthong was used, /ɔə/, which distinguished words such as *floor* /flɔə/ and *flaw* /flɔː/, but /ɔə/ has largely coalesced with /ɔː/ among standard speakers. A number of words formerly having /ʊə/, such as *moor* and *tour* are also often now said with /ɔː/: this has led to the proliferation of homophones (*moor, more, maw*; *tour, tore, taw, tor*; *poor, pour, pore, paw*; etc.).

Compare also MONOPHTHONG, TRIPHTHONG.

2 Two vowel letters representing

(*a*) a diphthongal sound, as in *rain* /reɪn/ or *toe* /təʊ/;

(*b*) (more fully **improper diphthong**) a single vowel as in *heat* /hiːt/, *soup* /suːp/. Compare DIGRAPH.

• **diphthongal. diphthongize**: (cause to) form or become a diphthong (in sense 1). **diphthongization**: the process by which a single sound has become a diphthong through historical or dialect change (for examples, compare BREAKING).

direct condition

direct condition See CONDITION.

directive

(*n. & adj.*)

1 (A sentence or clause) giving a command or order.

For grammarians who bother to distinguish sentence types defined syntactically (by their form) from sentence types categorized semantically (by their discourse functions), *directive* is a functional category often corresponding to an IMPERATIVE form; but directives can take declarative form, e.g. *You will apologize immediately*.

2 (In some theories, with a wider meaning.) (Pertaining to) an utterance which suggests, requests, or warns that a course of action should be carried out.

direct object

The noun phrase (or noun clause) most clearly affected or acted upon by the action of a transitive verb.

A monotransitive verb is normally followed immediately by the direct object in a declarative statement. Thus the direct object is *some vegetables* in:

Rachel ate/bought/cooked/grew some vegetables.

The direct object (*D.O.*) is often simply called the *object*, unless there is likelihood of confusion with the INDIRECT OBJECT.

In Case Grammar the direct object of the 'surface grammar' can be assigned different roles, not only the *objective*.

direct question See DIRECT SPEECH.

direct speech

The reporting of speech by repeating the actual words used, without making any grammatical changes; an example of this.

'Is there anybody there?' said the listener is an example of direct speech, including a *direct question*. This contrasts with the INDIRECT SPEECH exemplified in *The listener asked if there was anybody there*.

See also FREE *direct speech*, REPORTED SPEECH.

discontinuity

The splitting of a construction by the insertion of a word or words, or a particular instance of this.

For example:

Have you *finished*? (verb phrase)
Look the word *up* (phrasal verb)
That's a *hard* act *to follow* (modification)
There's *a man* outside *who wants to see you* (noun phrase and its relative clause)
The time has come, the Walrus said, *to talk of many things* (noun phrase and modification).

Discontinuity is very common in sentences containing comparative clauses, e.g.

I spend *more money* on clothes *than I can really afford*

• **discontinuous**: (of a construction) split into two parts by the insertion of a word or words.

discord

Lack of concord. See AGREEMENT.

discourse

A connected stretch of language (especially spoken language) usually bigger than a sentence, and particularly viewed as interaction between speakers or between writer and reader.

1991 M. HOEY Discourse . . is used in two ways . . . Firstly it refers to all aspects of language organization . . (whether structural or not) that operate above the level of grammar. Then, more specifically, it refers to the level of description that concerns itself with the structure . . of (spoken) interaction.

Some users confine the term to spoken language, contrasting discourse with written text.

• • **discourse analysis**: the analysis of how spoken and/or written stretches of language are structured.

This can include looking at grammatical and semantic connections between sentences (just as syntax is the study of such connections within a sentence), and in this respect discourse analysis is much the same as text linguistics. But usually discourse analysis is particularly concerned with sociolinguistic aspects of language, such as the organization of 'turn-taking' in conversation.

> 1991 M. McCarthy Discourse analysis is concerned with the study of the relationship between language and the contexts in which it is used. It grew out of the work in different disciplines, in the 1960s and early 1970s, including linguistics, semiotics, psychology, anthropology and sociology. Discourse analysts study language in use; written texts of all kinds, and spoken data, from conversation to highly institutionalised forms of talk.

Compare COHERENCE, COHESION, TEXT LINGUISTICS. See also FIELD OF DISCOURSE.

discourse adjunct

The same as CONJUNCT.

discourse marker

A word or phrase that helps to signal the direction in which language, particularly in a conversation, is going.

The term is somewhat specialized and not easily defined, but may include not only (*a*) conjunctions (e.g. *and*, *but*, *or*, *because*, etc.) but also (*b*) words outside the main syntax, such as *oh*, *well*, *you see*, *I mean*, etc. Words and phrases in this second group are sometimes labelled FILLERS or *pragmatic particles*.

discrete

Separate, not on a cline or continuum.

The term is particularly applied in phonetics and phonology. For example, English /p/ and /b/ are discrete items, since they are phonemes that distinguish words such as *pin* and *bin*, and there is no midway phoneme that also differentiates meaning. Even though different speakers may articulate these phonemes in various ways, a speaker must intend one or the other, and a listener too must make a choice.

Physically however these sounds are on a continuum, as is shown by the blurring of the sound in *spin*, where the /p/ loses its characteristic aspiration, and where no opposition is possible between /spɪn/ and /sbɪn/.

disjunct

An adverbial that has a more detached role in clause or sentence structure than other adverbials.

Disjuncts contrast in some grammatical models with ADJUNCTS, CONJUNCTS, and SUBJUNCTS. In other grammars they are called *sentence adjuncts*, SENTENCE ADVERBIALS, or *sentence modifiers*, or they are included, together with other adverbials, under such labels.

They express the speaker's or writer's attitude to the content of the sentence (CONTENT DISJUNCTS), e.g. *tragically* in

Tragically, the rescue party arrived too late

or they claim that the statement is being made in a particular way (STYLE DISJUNCTS), as in

Honestly, nobody could have done any better.
To be frank, the whole thing was hopeless.

disjunction

Semantics. Choice between two possibilities, or an instance of this.

See DISJUNCTIVE and compare CONJUNCTION (2).

disjunctive

(*n. & adj.*) (Designating) a word, especially the conjunction *or*, that expresses alternatives.

The terms *disjunctive* and *disjunction* are taken from logic. Further distinctions can be made. In *exclusive disjunction* the choice is 'one-or-the-other-but-not-both', e.g.

(Either) the train is late, or it's been cancelled.
He died in (either) 1940 or 1941.

With *inclusive disjunction* both alternatives may be possible, e.g.

(Either) the train is late—or my watch is fast. (Perhaps both.)
They certainly visited us in 1939 or 1940. (Maybe both.)

With *disjunctive interrogatives*, the presence of *either* allows the possibility that both alternatives may be true:

Did you visit (either) Edinburgh or Glasgow?

But where only one of the alternatives can be true, the word *either* is not possible in a question:

*Have you either passed or failed?
*Is their eldest child either a boy or a girl?

The second type of disjunctive interrogative is less usual than the '*either* . . *or* (maybe both)' type, and is therefore 'marked'.

dislocation

The use of a pro-form in addition to a noun phrase, so that the noun phrase is unusually positioned.

•• **right dislocation**: dislocation in which the pro-form comes first and the noun phrase is therefore dislocated to the 'right' (as written), e.g.

He's a good cricketer, your young son.

This is also called *postponed identification*.

left dislocation: dislocation in which the noun phrase is moved to the start or 'left', and an extra pronoun is added later, e.g.

Your young son, he's a good cricketer.

This is also called *anticipated dislocation*.

Dislocation is particularly a feature of informal spoken English.

Compare BRANCHING.

distal

The same as *non-proximal*: see PROXIMAL.

distinctive feature

Phonology. A characteristic of a speech sound within the phonology of the language that distinguishes it from another speech sound.

In traditional phonetic theory, the PHONEME is the smallest unit that makes meaning contrast possible. The difference between /f/ and /θ/ is what distinguishes *fin* and *thin*. However, phonemes can obviously be analysed into smaller phonetic features. English /f/ is labio-dental, fricative, voiceless, etc., and all other phonemes contrast with it in at least one feature: /θ/ is dental, /v/ is voiced, and so on.

In more fully developed distinctive feature theory, the features (fricativeness, voice, etc.) are held to be the minimal units of phonology; the emphasis has shifted, even in the analysis of consonants, from articulatory to acoustic contrasts; and there is a tendency to analyse contrasts as far as possible in binary (+ or –) terms.

1968 N. CHOMSKY & M. HALLE We take 'distinctive features' to be the minimal elements of which phonetic, lexical, and phonological transcriptions are composed.

distribution

The set of contexts in which a linguistic unit characteristically occurs.

Every speech sound and every word or phrase is limited in some way as to the contexts in which it can occur, and the set of such contexts is its distribution. Thus the English phoneme /p/ can occur in initial consonant clusters such as /pl/ (e.g. *please*), /pr/ (e.g. *praise*), and /pj/ (e.g. *pew*), but not in /pf/ or /pw/. The distribution of the articles *a* and *an* is restricted to use with singular count nouns (e.g. *a knife* but not **a knives* or **a cutlery*). And so on.

In traditional grammar, parts of speech are overtly defined in notional terms (e.g. 'a noun is a naming word'). In practice assignment of a word to a particular word class depends much more on its possible distribution in the structure of a clause. The distribution of two units may overlap, giving *overlapping distribution*, or the two may be mutually exclusive (so, for example, *a* or *an* cannot occur with *the*).

•• **distributional**: of or involving distribution (esp. in *distributional analysis*; for *distributional equivalence*, see EQUIVALENCE).

Compare COMPLEMENTARY DISTRIBUTION.

distributive

(*n. & adj.*) (A word or phrase) that relates to individual members of a class separately, not jointly.

Words like *each* and *every* are distributive words. Phrases like *once a week* and *three times per year* are distributive expressions.

Distributive plural concord is common in expressions such as *The children all had such eager faces* (where clearly each child had only one face), but a *distributive singular* is often possible, e.g. *They all had such an eager expression.*

disyllabic

Having two syllables.

The term is commonly used in discussions on comparative and superlative forms, where some two-syllable words take inflection (e.g. *cleverer, cleverest*) and others use *more* and *most* (e.g. *more, most eager*).

Compare MONOSYLLABIC and POLYSYLLABIC.

• **disyllable**: a disyllabic word.

ditransitive

1 (*n. & adj.*) (A verb) having two objects. Also called *double transitive (verb).*

Verbs that can take a direct object plus an indirect object are ditransitive verbs, as in:

I gave my mother flowers
I gave flowers to my mother

2 **ditransitive complementation**: complementation of a verb by a direct and an indirect object.

This is sometimes said to include indirect objects followed by various types of clause, e.g.

She told me *(that) she was delighted*
I hadn't asked her *what she wanted*
She urged me *to take a holiday*

But grammarians disagree over how to describe verb complementation; one analyst's *ditransitives* (e.g. the last example above) are another's *complex transitives*.

Compare COMPLEX TRANSITIVE, INTRANSITIVE, MONOTRANSITIVE.

domain

1 *Sociolinguistics*. The situation or sphere of activity to which an utterance relates as it affects the language variety used.

1982 G. LEECH et al. *Domain*. This has to do with how language varies according to the activity in which it plays a part . . . We can thus refer to the domains of chemistry, law, religion, and so on.

As will be seen from this quotation, there is a certain overlap and blurring between actual situations (which might be laboratory, courtroom, church, home, journalism, etc.) and subject-matter (e.g. chemistry, law, religion, etc.).

Domain is sometimes said (as it is by these authors) to be a part of REGISTER, but a distinction may be made between *domain* as the sphere in which the language is used and *register* as the speech variety used, which partly depends on the subject matter.

2 The same as (*semantic*) FIELD.

1968 J. LYONS In recent years, there has been a good deal of work devoted to the investigation of lexical systems . . with particular reference to such *fields* (or *domains*) as kinship, colour, flora and fauna, weights and measures, military ranks, [etc.].

dominate

(In Transformational-Generative Grammar.) Of a node: have as constituent.

1980 E. K. BROWN & J. E. MILLER S [meaning Sentence] dominates everything in the sentence.

• **dominance**.

dorsal

Phonetics. Of a sound: made with the back of the tongue.

do-support

The use of a part of the auxiliary verb *do* as a dummy verb. (Also called *do-insertion*.)

The use of *do, does, did* in questions, negatives, tag-questions, etc. in the simple present and past tenses means that there is a regularity of structure between these and other verb phrases. Thus *Do you understand?* has the same pattern of auxiliary + subject + verb as *Are you listening?* or *Can you hear?*, and *They knew* (or *did know*), *didn't they?* is comparable to *He could explain, couldn't he?*

The term comes from Generative Grammar, where *do-support* is expounded as a set of rules.

double accusative

(In older grammar.) A structure with a direct and an indirect object (in which the verb would in more modern grammar be called DITRANSITIVE).

double articulation

The same as COARTICULATION.

double-barrelled question

An interrogative in which the same question is repeated, as in

Who is *who*?

This is an older term, not in general use.

double consonant

(An instance of) two identical consonants coming together.

(*a*) Double consonant letters generally represent a single sound in English (e.g. *batter, puppy, shallow*, etc.). The main exception is *cc* before *i* or *e*, pronounced /ks/ as in *accident* or *succeed.*

(*b*) Double consonant sounds (also called *geminates*) occur when the phonemes at the end of one morpheme and the start of an immediately adjacent morpheme are the same, i.e., across syllable and word boundaries. Examples are

unnatural, shell-like, part-time, hat trick, fish shop

Such double sounds are not pronounced completely separately, as they would be in isolation. They are more like one sound lengthened. The main exception is when one affricate succeeds another: both are fully articulated, e.g. in *which child?* /wɪtʃ tʃaɪld/. Sometimes what in careful speech is a double consonant sound is elided into one, e.g. in *Prime Minister* /praɪ'mɪnɪstə/ (contrast *prime mover*).

double genitive

A structure consisting of a noun phrase + *of* + a genitive noun phrase or possessive pronoun, as in:

a home of their own
some books of Jane's
that cousin of yours

This structure might appear superfluous, but in fact it has the advantage that it can combine indefinite and definite meaning, whereas the single genitive can only have definite meaning (e.g. *their own home*, *Jane's books*, *your cousin*).

Also called *post-genitive*.

double marking

The use of a redundant grammatical feature where the idea is adequately expressed already.

The term is usually used of slightly dubious structures, such as

I would have liked to have seen it

where a second perfect is unnecessary if the meaning is 'I would have liked to see it' or 'I would like to have seen it'.

Compare BLEND.

double negative

The occurrence of two negative words in a single clause or sentence.

It is usually incorrect in standard English to negate a clause more than once (e.g. **I never said nothing*; *I haven't got none*), since the first negative could be omitted or the second replaced by a non-assertive form (e.g. *anything, any*). But the argument, from logic, that 'two negatives make a positive' is unjustified, as many English dialects and many other languages do have reinforcing multiple negation.

In a two-clause sentence, a common double negative structure—heard even from standard speakers—is potentially confusing. *I wouldn't be surprised if they didn't come* may mean that the speaker expects them not to come (both negatives justified) or it may mean the same as *I wouldn't be surprised if they came*.

However, in many two-clause sentences two negatives are essential: *I didn't ask him not to go* (though I hoped he wouldn't). And occasionally a double negative occurs quite legitimately in a single clause, where in a sense the two negatives do indeed cancel each other out: *You can't not worry about it. Surely nobody has no friends* (= You have to worry. Everybody has some friends). Even here, though, like true negatives, the sentences take positive tags (*You can't not worry, can you? Nobody has no friends, do they?*)

double passive

A clause containing two verbs in the passive, the second an infinitive, as in:

*Receipts are not proposed to be issued
Certificates are expected to be despatched next week

Usage books sometimes warn against all such structures, but their acceptability in fact varies. Verbs (like *expect* in the second example) that are also possible in the pattern verb + object + passive infinitive (e.g. *We expect certificates to be despatched*) seem grammatical when they occur in the double passive construction. Verbs that do not fit this pattern with a single passive (e.g. **We propose receipts to be issued*) do not happily take a double passive either.

doublet

Either of a pair of words that have developed from the same original word, but are now somewhat different in form and may be used in different senses.

Examples of doublets that have arisen from a single parent form within English are:

human (M16), humane (LME)
metal (ME), mettle (M16)
mood (in Grammar) (M16), mode (LME)
patron (ME), pattern (M16)
shade (OE), shadow (OE).

Examples of doublets that arose through borrowing from other languages at different times:

faction (L15), fashion (ME)
hostel (ME), hotel (M17)
ration (M16), reason (ME).

See BY-FORM. Compare HETERONYM (2).

double transitive

The same as DITRANSITIVE.

dual

1 (*n. & adj.*) (A form) expressing two or a pair, in contrast to *singular* and *plural*.

In some languages *dual* is an important category, and there are inflected dual forms of nouns, verbs, etc. In English *both* and to a lesser extent *either* and *neither* are the only grammatical words indicating *dual number*.

2 **dual class membership**: membership of two classes; said e.g. of nouns that can be both count and non-count (e.g. a *cake*, some *cake*).

dual gender (term): a (word) that can apply equally to a male or a female (e.g. *parent, guest*); contrasted with single-gender terms such as *father, hostess*.

Compare BINARY, PLURALE TANTUM.

dummy

Describing a sentence element that has no intrinsic meaning but maintains grammatical structure.

•• **dummy *it***: used especially as subject in sentences about time and weather. This is also called EMPTY *it* or PROP *it*.

It's five o'clock and *it* is snowing again.

A similarly vague *it* also appears in various idiomatic phrases:

We've made it. Well, that's it—let's go.

Other common dummy elements are *there* in existential sentences and forms of the verb *do* as *dummy operators* in questions and negatives:

There's someone at the door. What *does* he want? He *didn't* say.

Dummy elements are also described as EMPTY words.

See also INTRODUCTORY and compare ANTICIPATORY.

duration

1 (*a*) *Phonetics*. The actual time taken in the articulation of a speech sound, acoustically measured.

(*b*) *Phonology*. The 'linguistic' length of a speech sound, as perceived by the listener. More usually dealt with as LENGTH.

Like many words not restricted to linguistic use, *duration* is sometimes employed rather loosely. But in the discussion of speech *duration* is often reserved for objective measurable length, and contrasted with the phonological term *length*.

2 Length of time, considered as part of the meaning of a word.

The term is particularly used in relation to verbs, prepositions, and adverbials. Progressive tenses are usually said to imply a temporary state or *limited duration*, e.g.

I am living in a hostel

Prepositions and adverbials expressing duration include *since*, *for*, *from . . to*, and so on. e.g.

I've been here *for a month*

durative

1 Of an activity or process signified by a verb: taking place over a period of time; contrasted with PUNCTUAL.

The term is particularly used in relation to dynamic verbs that describe activities and processes (often of limited duration) and which are thus used especially in progressive tenses:

It was raining
The children are watching TV
The situation has improved
I'm redecorating the kitchen

Compare CONCLUSIVE.

2 **durative aspect**: the same as PROGRESSIVE aspect.

dvandva

(*n. & adj.*) (Designating) a compound consisting of two elements, neither of which is equivalent to the whole, and which could otherwise be syntactically joined by *and*.

Examples are very rare in English, except in place names such as *Alsace-Lorraine*, *Schleswig-Holstein*, *Bosnia-Hercegovina* and company names such as *Cadbury Schweppes*.

The term is Sanskrit, representing a reduplication of the word for 'two'.

dynamic

Having a meaning that implies action or change.

1 Of (chiefly verbal) meaning: relating to actions, events, happenings, and processes:

We'*ve bought* a new car.
War *broke out* in 1939.
They'*re playing* our tune.
I'*ve worked* hard all my life.
The lights *have gone* green.

Verbal meaning can be seen as either dynamic or STATIVE. It is common practice to describe verbs that can be used in progressive tenses as *dynamic verbs*, in contrast to *stative verbs* that cannot. It is more accurate to talk of *dynamic* and *stative meaning*, since many so-called stative verbs can be used dynamically with a shift of meaning:

I *have* two sisters.	We'*re having* a party.
They *are* hard-working.	You *are being* so silly.
They *look* alike.	My prospects *are looking* good.

Compare ACTION VERB.

Less generally, the terms *dynamic* and *stative* are applied to other word classes. Nouns and adjectives, for example, are characteristically stative, referring to stable things and attributing stable qualities to them, but both may sometimes have temporary, dynamic meaning:

He is being *foolish*
You are being *a nuisance*

Adverbs, and particularly adjuncts, on the other hand, are often dynamic:

He behaves *foolishly*

2 **dynamic modal**: a modal verb that 'predicates something'.

In some analyses of MODAL verbs, *dynamic* is added as a third category of meaning, contrasting with DEONTIC and EPISTEMIC.

Dynamic modals make factual statements, e.g.

She *can* read a novel in an evening
A cat *will* lie in front of the fire for hours

Contrast

You *can* borrow my paper if you like (deontic)
Can this really be true? (epistemic)
Of course we *will* help you (deontic)

Unlike modals with deontic or epistemic meaning, dynamic modality therefore sometimes refers to actual past time:

She *could* already read and write when she was four
Even as a small child, she *would* sit reading for hours

E

echo utterance

An utterance that repeats all or part of what the previous speaker has said. Echo utterances can take various forms, but function either as questions:

(A: He's a strange man.) B: He's strange?
(A: Yes, he collects beetles.) B: *What* does he collect?

or as exclamations:

(A: Beetles. And then he sings to them.) B: He does *what*?
(A: He sings to them.) B: *Sings to them*! You're joking.

-ed form

(A way of referring to) either:

(*a*) The past tense form of any verb.

This use includes both regular past tenses, which are actually marked by *-ed* (e.g. *looked*) and irregular ones, which are not (e.g. *rang*, *saw*, *wrote*).

or (*b*) The past tense and past participle form of a verb.

Since regular verbs have the same ending for both parts of the verb (e.g. *looked*), this usage is not unreasonable. But the existence of two different meanings is confusing.

See *-EN* FORM. Compare PARTICIPIAL ADJECTIVE.

editorial *we*

The use of *we* by a single writer, perhaps to avoid the more egotistical-sounding *I*.

Compare ROYAL *WE*.

effected object

The same as RESULT *object*.

effective

(In Systemic-Functional Grammar.) Of a verb (or a clause containing such a verb): describing goal-directed action; contrasted with DESCRIPTIVE (3).

An effective clause may be active or passive, and must either contain an AGENT or imply one, e.g.

The intruders smashed the door down The door was smashed (by the intruders)

egressive

Phonetics. Of a speech sound: produced with air going out through the mouth or nose or both. Also applied to the airflow producing such a sound.

Most speech sounds in most languages, and all the English phonemes, are egressive.

Compare AIR-STREAM MECHANISM and contrast INGRESSIVE.

elaborated code See CODE.

elaboration See EXPANSION.

element

1 (In some grammatical descriptions.) Any of the functional parts into which clause and sentence structure is analysed.

According to a widely-accepted modern analysis, there are five possible elements of structure, namely Subject, Verb, Object, Complement, and Adverbial (abbreviated to S, V, O, C, and A).

In Systemic Grammar, the analysis is slightly different. Predicator replaces Verb, Object is included within Complement, Adjunct replaces Adverbial, and there is a fifth Z element used for nominal groups whose status is indeterminate between subject and complement. Indeterminate status is assigned, for example, to titles, e.g.

A Tale of Two Cities

and elements in which subject and object are fused, e.g.

Jack persuaded *Fiona* to come

2 (More generally.) Any functional part of a larger structural whole.

Affixes, for example, are *elements* in word formation; a noun phrase consists of an essential *element*, the head, plus such *elements* as determiners, modifiers, and so on. [See table p. 445]

See also DUMMY, *WH*-ELEMENT.

elide

1 *Phonetics.* Omit (a sound or syllable) by ELISION; (of a sound or syllable) be omitted by elision.

2 Omit (a word or words) by ellipsis; (of a word or words) be omitted by ellipsis. The preferred term is ELLIPT.

• **elided.**

elision

Phonology. The omission of a speech sound or syllable.

Two broad types of elision may be distinguished:

(*a*) elided word forms that are long-established, where the spelling frequently reflects the earlier, fuller pronunciation; and

(*b*) forms heard today in colloquial or rapid speech, but where unelided forms are also current.

Long-established elisions include the reduction of some consonant clusters initially:

*g*nome, *k*night, *w*rong

medially:

lis*t*en, whis*t*le, san*d*wich

and finally:

hym*n*, lam*b*

along with the loss of vowels and whole syllables, as in:

Glou*ce*ster, Sal*is*bury, We*dne*sday.

In present-day speech, consonants within clusters often undergo elision (e.g. fac*t*s, han*d*bag, twel*f*th), but elision of weak vowels is particularly frequent, with the result that whole syllables may be lost:

fact(o)ry, cam(e)ra, nat(u)ral, batch(e)lor, fam(i)ly, med(i)cine, p(o)lice, Febr(uar)y.

Elision also occurs at word boundaries in connected speech, as in:

Nex(t), please; as a matt(e)r o(f) fact; mix (a)n(d) match.

The common elision of /t/ and /d/ when surrounded by other consonants means that the distinction between past and present tenses is sometimes neutralized:

I wish(ed) to help /aɪ ˈwɪʃtə help/

Compare ASSIMILATION, COMPRESSION, CONTRACTION, HAPLOLOGY.

ellipsis

Omission of a word or words from speech or writing that can be recovered by the hearer or reader from contextual cues.

Words are often omitted from informal speech where they can be recovered from the situation (*exophoric ellipsis*):

(Are you) coming?
(Is there) anything I can do to help?

In more formal speech and writing, words are often grammatically recoverable from the text (*anaphoric* or *cataphoric ellipsis*), and in many cases it is normal to omit words in order to avoid repetition:

We're as anxious to help as you are [anxious to help].

Unless you particularly want to [buy tickets in advance], there's no need to buy tickets in advance.

A: Tom's written to *The Times*. B: Why [has he written to *The Times*]?

A: I don't know [why he has written to *The Times*]. He's always writing letters and [he is always] complaining about something.

There are however grammatical constraints as to what can be omitted. Thus a coreferential subject can be omitted in a coordinated clause:

I telephoned my aunt and ‸ told her the news

but it cannot be omitted when the link is between main (matrix) and subordinate clauses:

*I told my aunt the news when ‸ telephoned her

Strictly, ellipsis exists only when the missing words are exactly recoverable. But the term is normally extended to include such sentences as:

He wrote a better letter than I could have [written].

and often to looser examples of omission, such as:

You remember that man [who/whom/that] I introduced to you?

• **elliptical**: characterized by or exhibiting ellipsis.

Compare RECOVERABILITY, REDUNDANCY, REFERENCE, SUBSTITUTION.

See GAPPING.

ellipt

Omit (an element) by ellipsis.

1990 S. GREENBAUM & R. QUIRK In MEDIAL ellipsis medial elements are ellipted: Jill owns a Volvo and Fred (owns) a BMW.

embed

Include (a linguistic unit) within another. See EMBEDDING.

• **embedded.**

1975 T. F. MITCHELL Sometimes the embedded sentence is more apparent than others. The subject-verb-object pattern of *he'd done it* in *John thought he'd done it* seems immediately to mark it as sentential.

embedding

The inclusion of a linguistic unit in another linguistic unit.

(*a*) The term is used in Generative Grammar, where it contrasts with CONJOINING, much as SUBORDINATION and COORDINATION do in traditional grammar.

(*b*) Most mainstream grammarians use the term more narrowly, restricting it to a clause or phrase that has been RANKSHIFTED so that it forms part of an ELEMENT in the superordinate clause. This includes sentences embedded into noun phrases, such as a defining relative clause or a defining appositive clause, e.g.

The news *that Cassandra announced* was not encouraging
The news *that the city was doomed* was met with disbelief

and nominal clauses functioning as subject or object, e.g.

That she was gloomy was characteristic
They did not believe *that they were doomed*

It also includes clauses of comparison, e.g.

She was gloomier *than she normally was*

where the clause postmodifies the adjective (or adverb). It may even include postmodifying phrases which are interpretable as 'reduced' clauses, e.g.:

The woman *in white* (= who was in white)
Prophecies *of doom* (= that are of doom)

Other clauses often classified as subordinate (dependent) clauses, however, are usually excluded, particularly non-defining relative clauses and adverbial clauses.

1984 R. HUDDLESTON Thus in *Ed liked it, whereas Max thought it appalling* we shall want to say that *whereas Max thought it appalling* is subordinate . . [but] the subordinate clause is an immediate constituent of the sentence, and . . it is not embedded.

(*c*) The concept of embedding is also invoked in the analysis of phrases. Thus the noun phrase *the gift of prophecy* contains an embedded prepositional phrase (*of prophecy*), which itself embeds another noun phrase (*prophecy*).

Theoretically the process of embedding can lead to sentences or phrases of indefinite length.

Compare PUSHDOWN.

emotive

Semantics. Arousing feeling, not purely descriptive.

(*a*) In theories of meaning, *emotive* refers to the kinds of meaning subjectively attached to words by some users (both individuals and communities). It is roughly equivalent to AFFECTIVE or ATTITUDINAL in some of their uses, and contrasts with COGNITIVE and REFERENTIAL. See also CONATIVE, EXPRESSIVE.

(*b*) Any words relating to the emotions can be labelled *emotive*, and some grammarians label verbs such as *dread, hate, loathe, love*, etc. *emotive* verbs.

(*c*) Various grammatical devices can be used to produce emotive effects, such as EXCLAMATIONS, the use of *do* in imperatives (e.g. *Do listen!*), unusual stress on operators (e.g. *I AM listening*), or on other words (e.g. *That WAS stupid*), the use of EMPHASIZERS (e.g. *Really!* That was very risky *indeed*), or the use of non-correlative *so* and *such* (e.g. *So* risky. *Such* a stupid thing to do).

emphasis

Special importance or prominence attached to a certain part of a sentence.

The term is used in its general sense. Emphasis is often achieved by marked FOCUS; by unusual STRESS (for example, on an auxiliary verb); or by grammatical devices such as CLEFT or PSEUDO-CLEFT sentences, the use of *do* in declaratives (e.g. *I do apologize*) or in imperatives (*Do be sensible*).

emphasizer

1 An adverb that adds to the force of the clause or part of the clause to which it applies.

e.g. *really* in

I really think you might have telephoned.

Only some grammatical models subdivide adverbs in this sort of detail. One model that does has *emphasizers*, along with FOCUSING adverbs, INTENSIFIERS, and others, as a subcategory of SUBJUNCT.

Also called **emphasizing adverb**.

2 In some grammatical models, a sub-species of INTENSIFYING adjective that has a heightening, reinforcing effect (generally only in attributive use): *pure* nonsense, a *real* idiot.

Also called **emphasizing adjective**.

emphatic

Imparting or expressing emphasis.

•• **emphatic pronoun**: a reflexive pronoun when used for emphasis.

e.g. *himself* in

He admitted himself that the whole thing had been a mistake.

Compare REFLEXIVE.

empty

Having no (lexical) meaning; DUMMY.

•• **empty it**: the same as DUMMY *it*.

empty word: the same as FORM WORD.

enclitic

(*n. & adj.*) (A word) pronounced with very little emphasis, and usually shortened and forming part of the preceding word.

Common *enclitic forms* include *-n't*, parts of the verbs *be* and *have* (e.g. they*'re*, he*'s* (= 'is' or 'has'), we*'ve*) and *'s* meaning *us* (as in *Let's go*).

Compare CLITIC and see CONTRACTION (2).

end

(Situated or occurring in) the last part of a clause or sentence.

•• **end position**: the same as FINAL position.

end-focus

The placing of the most important information in a sentence at the end.

It is normal to introduce the THEME of a message (often 'given' information) at the beginning of a sentence and to impart the important NEW information later. Hence *end-focus* is a normal characteristic of sentence structure.

In writing, therefore, the end of the clause or sentence will normally be taken to be the focus, unless there is some unusual punctuation or other sign (e.g. italics or capital letters); and marked word order may be used to maintain this focus. In speech, the focus is realized by nuclear pitch, which typically falls on the last stressed syllable, e.g.

Write to the li`BRARian

(There may of course be more than one nuclear pitch in a sentence if it is a long one.) The focus can be moved without changing the word order by moving the nucleus, e.g.

`WRITE to the librarian

Compare END-WEIGHT, MARKED. See FOCUS.

ending

The final part of a word, constituting an inflection.

In English the term applies to plural forms of nouns (e.g. cat*s*, child*ren*), the various inflections of verbs (e.g. look*ing*, look*ed*), and the comparative and superlative of certain adjectives and adverbs (e.g. bett*er*, be*st*, soon*er*, soon*est*).

endocentric

1 Of a structure: syntactically equivalent to the word which is its head. (Also called *headed*.)

Some types of phrase contain a HEAD word and have the same formal function in their clause as the single head would. Thus the following phrases, with their head words underlined, are all endocentric:

too *dreadful* to contemplate (adjective phrase)
rather more *surprisingly* (adverb phrase)
she who must be obeyed (noun phrase)

2 Of a compound: grammatically and semantically equivalent to one or other of its parts.

Many compounds belong to the same word class as their head word and preserve its basic meaning, e.g.

fire-alarm (alarm giving warning of fire) } nouns
girlfriend (girl who is a friend) }
duty-free (free from duty) } adjectives
rock hard (hard as a rock) }

Adjectives that contain a participle may be included here, since participles are often used adjectivally:

everlasting
typewritten
new-laid

Compare EXOCENTRIC.

endophora

Reference within a text. Contrasted with EXOPHORA.

This is a general term, covering both ANAPHORA and CATAPHORA.

• **endophoric**: (*endophoric reference*, the same as ENDOPHORA).

1976 M. A. K. HALLIDAY & R. HASAN We shall find it useful in the discussion to have a special term for situational reference. This we are referring to as EXOPHORA, or EXOPHORIC reference; and we could contrast it with ENDOPHORIC as a general name for reference within the text . . . As a general rule, therefore, reference items may be exophoric or endophoric; and if endophoric they may be anaphoric or cataphoric.

end-weight

The structural principle by which longer units of information tend to come at the end of a sentence.

Since 'new' information may need more detailed explanation than 'given' information, END-FOCUS is often accompanied by end-weight, e.g.

The bread industry and nutritionists alike (THEME) / have been trying to get the message across *that a healthy balanced diet should include sufficient bread.*

The 'weighty' message here comes naturally at the end. Notice how the word *across* comes immediately after *message* and not at the very end of the sentence. Grammatically it could, but it would be feeble and decidedly lightweight.

-en form

(A way of referring to) the past participle of any verb.

The name is based on the fact that many irregular verbs take this ending in their past participle, e.g.

broken, chosen, driven, forgotten, taken.

but it includes all past participles, e.g.

looked, hated, brought, kept, set, known, gone, drunk.

Because it is restricted, unlike *-ed*, to the participle, it is a useful shorthand way of distinguishing past participles from past tense forms. See also *-ED* FORM.

English

The West Germanic language that first developed in England and southern Scotland and is now spoken throughout the British Isles and in the United States, Canada, Australia, New Zealand, and the West Indies, as well as by significant communities in southern Africa, south and south-east Asia, and elsewhere.

See BASIC ENGLISH, BBC ENGLISH, MIDDLE ENGLISH, MODERN ENGLISH, NUCLEAR ENGLISH, OLD ENGLISH, STANDARD ENGLISH.

enhancement See EXPANSION.

entail

Semantics. Of a sentence: to necessitate the truth of (another sentence).

• **entailment**: a relationship between two sentences such that if the first is true, the second must also be true.

(1) *I am an only child* entails (2) *I have no brothers or sisters*. In this particular case sentence (2) also entails sentence (1). But such mutual entailment does not always happen. For example, sentence (1) clearly entails (3) *I have no brothers*, but as I might have any number of sisters sentence (3) does not entail sentence (1).

Compare IMPLICATION, PRESUPPOSITION.

epicene

Denoting either sex without change of gender.

An outdated term, more usually replaced by some expression like *sex-neutral* or DUAL *gender*.

epistemic

Of modality: concerned with likelihood or the degree of certainty of something. Contrasted with DEONTIC.

Out of context, sentences containing modal verbs are sometimes ambiguous. The modality of *You must love your mother* is epistemic if the meaning is 'I deduce, from some information or observations, that you love her'. It is deontic if the meaning is 'I am telling you to love her'. Similarly the modality of *Tom may keep the money* is epistemic if the meaning is 'He is quite likely to keep it', but deontic if it is 'He is allowed to keep it'.

Epistemic modality is sometimes called *extrinsic modality*. Compare ALETHIC and see MODALITY.

epithet

An adjective that indicates a quality of the noun it refers to. Contrasted with CLASSIFIER.

Epithets can include objective adjectives (e.g. *green*, *rectangular*) and subjective adjectives (*amazing*, *stupid*) though clearly some adjectives can be either (*old*, *expensive*). This is not a very general classification for adjectives and seems to be the same as QUALITATIVE in other models.

equational

Another word for EQUATIVE (particularly as applied to the verb *be*).

equative

Denoting that one thing is equal to, or the same as, another.

The verb *be* is the primary equative verb (or verb with *equative function*). Such verbs are more usually described as COPULAR (or COPULATIVE).

There is also a small group of conjuncts with equative meaning, such as *equally, likewise, similarly, in the same way*.

equivalence

The term is sometimes used in the analysis of degrees of comparison, where the *as X as* comparison can be described as *equative*. But this relationship is often described as EQUIVALENCE.

• **equatively**.

equivalence

1 Equality as regards meaning.

Comparisons taking the form *as X as* (e.g. *She is as generous as her mother*) are *comparisons of equivalence*. These are contrasted with comparisons using *more X* (or a comparative form ending in *-er*) *than* or *less X than*, which are *comparisons of non-equivalence* (e.g. *She is more/less generous than her mother*).

Equivalence of meaning is also a feature of strict apposition, and may or may not be overtly signalled:

> *The Young Pretender*, that is to say *Charles Edward Stuart*, or *Bonnie Prince Charlie*, as he is more familiarly known . . .

2 Equality as regards function.

Functional equivalence, also called *distributional equivalence*, is another way of saying identical syntactical DISTRIBUTION.

A collection of linguistic units that are formally different may have equivalence of function in a particular context. For example, four different forms have equivalence of function (adverbial) in the following:

> I have put the family silver *away* (adverb)
> I've put it *in the bank* (prepositional phrase)
> I've put it *somewhere safe* (noun phrase)
> I've put it *where you suggested* (adverbial clause)

equivalent

(*adj.*) That is semantically and functionally the same.

(*n.*) A linguistic unit that is equivalent in function to another (which may be specified, as *X-equivalent*).

> 1932 C. T. ONIONS A word or group of words which replaces a Noun, an Adjective, or an Adverb is called an Equivalent (Noun-equivalent, Adjective-equivalent, or Adverb-equivalent).

In more modern grammar these terms are replaced by such terms as *nominal*, *nominal/noun group/phrase*, *nominal/noun clause*, and so on.

ergative

Designating a particular kind of verb with which the same noun can be used as the subject when the verb is intransitive and as the object when the verb is transitive; designating the relationship between these two uses.

> 1968 J. LYONS The term that is generally employed by linguists for the syntactic relationship that holds between [The stone moved] and [John moved the stone] is 'ergative': the subject of an intransitive verb 'becomes' the object of a corresponding

transitive verb, and a new *ergative* subject is introduced as the 'agent' (or 'cause') of the action referred to.

The term (derived from the Greek word *ergates* 'workman') is applied in languages such as Eskimo and Basque to the case used for the subject of the transitive verb in the second sentence above (i.e. *John moved the stone*).

In English the term is commonly applied to verbs. For example:

My shirt has *torn*. I've *torn* my shirt.
The door *opened*. Someone *opened* the door.
The meat is *cooking*. I'm *cooking* the meat.

erlebte rede

The same as FREE *indirect speech*.

eternal truths See PRESENT.

etymological

Of, pertaining to, based upon, etc., etymology.

•• **etymological fallacy**: the belief that the true meaning of a word is that implied by its etymology; as, for example, that *awful* really means 'full of awe, awe-inspiring'.

• **etymologically**.

1926 H. W. FOWLER *Saxonism* is a name for the attempt to raise the proportion borne by the originally & etymologically English words in our speech to those that come from alien sources.

etymology

1 The historically verifiable sources of the formation of a word and the development of its meaning; an account of these.

2 The branch of linguistic science concerned with etymologies (in sense 1).

Compare FOLK ETYMOLOGY.

etymon (Plural **etyma**.)

The word that gives rise to a derivative or a borrowed or later form.

euphemism

1 A mild or vague expression substituted for one thought to be crude or unpleasant (e.g. *facilities* for 'bathroom or shower-room plus lavatory', *pass away* for 'die', *dehire* for 'dismiss'.)

2 The avoidance of unpleasant words by means of such expressions.

1942 E. PARTRIDGE Euphemism may be obtained by using an extremely vague phrase, as in *commit a nuisance*.

• **euphemistic**.

euphony

A tendency to make a phonetic change for ease of pronunciation.

This is dated. The usual term now is ASSIMILATION.

eventive

Of the semantic role of a noun phrase: designating an event.

This is not a very general term, but it is possible to describe some subjects—often deverbal nouns—as eventive:

The examination is next week (*compare* They will examine the candidates next week)

Similarly, we find eventive objects:

You must do some work (*compare* You must work)
Can I give you some advice? (*compare* Can I advise you?)

event verb

The same as ACTION VERB.

exclamation

A word, phrase, or clause expressing some emotion.

(*a*) (In traditional grammar.) Used in the classification of sentence types.

1932 C. T. ONIONS Sentences may be divided into the four following classes according to their form or the kind of meaning they express: I. Statements. II. Requests, i.e., commands, wishes, concessions. III. Questions. IV. Exclamations.

In this type of classification, a mixture of form and meaning, the term is used in a fairly non-specialized way to cover any word or group of words expressing anger, pleasure, surprise, etc. Some of these may lack normal sentence structure, e.g.

Marvellous! You poor thing! How kind of you! What a scorcher! Woe is me!

Sometimes single word INTERJECTIONS are included, e.g. *Alas*.

(*b*) (In some popular grammatical models.) A formal and more limited category, restricted to sentences beginning *How* or *What* (but without the auxiliary-subject inversion typical of questions):

How difficult it all is
What a muddle we are in

But see EXCLAMATIVE.

(*c*) (In some modern grammatical analyses that distinguish sentence function from sentence form.) A semantic label indicating the function of the sentence. Contrasted with EXCLAMATIVE.

In this kind of analysis, a (semantic) exclamation could be realized by various forms, e.g. *What* DOES *he think he looks like*? (perhaps meaning 'How very oddly he is dressed').

This definition is roughly the opposite of that in (*b*).

exclamative

(*n.* & *adj.*) (A clause or sentence) that expresses exclamation.

A term used by some grammarians as a formal label (i.e. in the same way as EXCLAMATION (*b*)), in contrast to the semantic (i.e. meaning) labels EXCLAMATION or EXCLAMATORY (compare EXCLAMATION (*c*)).

1988 R. HUDDLESTON Exclamative clauses are marked by one or other of the exclamative words *how* and *what*: i. How tall Ed is! ii. What a tall lad Ed is!

exclamatory

Expressing or containing an exclamation.

The term may be used as a formal or functional label to describe an EXCLAMATION or an EXCLAMATIVE, either of which may be called an *exclamatory sentence*. Grammarians who are careful to distinguish form from function tend to confine the term to semantic use, i.e. as the adjective corresponding to EXCLAMATION (*c*).

• • **exclamatory question**: (usually) a sentence that is interrogative in form, but an exclamation in meaning, e.g.

Isn't it a lovely day!

exclusive

1 Serving to exclude an option; indicating that something is excluded. Contrasted with INCLUSIVE.

(*a*) Commonly describing the use of a first person plural pronoun when the addressee is excluded, as in *We'll call for you tomorrow.*

(*b*) Applied to the meaning of the word *or* in contexts indicating that one alternative rules out the other.

Are you going to have tea or coffee? would often be interpreted as exclusive: you are expected to have one but not both. On the other hand *Do you take milk or sugar?* probably has an INCLUSIVE meaning: you can have both.

See DISJUNCTIVE.

(*c*) Applied to other elements of language which indicate exclusivity in some way.

Some adverbs serve to exclude all but what is specified, e.g. *alone, exclusively, merely, solely.*

2 Denoting a linguistic element the use of which excludes the use of another element.

Many determiners are mutually exclusive (**a the mistake*).

existential

(*n. & adj.*) (A grammatically marked structure) typically used to express a proposition that someone or something exists.

•• **existential *there***: the unstressed word *there* used as a dummy subject and followed by the verb *be*.

This is used in the commonest type of existential sentence or clause (the *there-existential*), e.g.

> There is a lot to do
> Can there be life on other planets?
> There has been nothing in the papers about this

The usefulness of this structure is that a NEW subject can avoid the GIVEN position (**A lot is to do*, ?*Can life be on other planets?*, ?*Nothing about this has been in the papers.*) and be presented as the new information that it is.

There-existentials also occur with verbs other than *be*:

> There comes a time in everyone's life when . . .
> Once upon a time there lived a beautiful princess

***have*-existential**: (in some analyses) a sentence or clause containing the verb *have* and corresponding semantically to a *there-existential.*

Thus:

> I have a hole in my pocket
> My pocket has a hole in it

correspond to

> There is a hole in my pocket

while

> You have visitors waiting to see you

corresponds to

> There are visitors waiting to see you
> ?Visitors are waiting to see you

exocentric

1 Of a structure: containing no element that is functionally equivalent to the whole structure. (Also called *non-headed* or *unheaded.*)

Some phrases are always exocentric. Thus a prepositional phrase, consisting of a preposition and its complement (commonly a noun) normally functions either as an adverbial:

> The boy stood *on the burning deck*

or adjectivally in postmodification:

> Who was the man *in the iron mask*?

A basic English sentence (consisting of subject and predicate) is always exocentric, since neither part can stand for the whole:

> The boy / stood on the burning deck.

2 Of a compound: not grammatically and/or not semantically equivalent to either of its parts.

One kind of exocentric compound is the BAHUVRIHI type, which is functionally of the same word-class as the head member of the compound, but exhibits a significant shift of meaning (e.g. *fathead* and *head* are both nouns, but a *fathead* is not a kind of head). Other kinds have neither semantic nor grammatical equivalence (e.g. *pullover* (noun made up of verb + particle), *outcome* (noun made up of particle + verb), *commonplace* (adjective made up of adjective + noun)).

1973 V. ADAMS An exocentric compound is one whose function is not the same as that of its head element, for example *blockhead, highbrow, pickpocket*.

Compare ENDOCENTRIC.

exophora

Situational reference; a linguistic unit that derives part of its meaning from the extralinguistic situation. Contrasted with ENDOPHORA.

• **exophoric**: of the nature of or pertaining to exophora (*exophoric* ELLIPSIS; *exophoric reference*, the same as EXOPHORA).

The understanding of language in general requires a certain knowledge of the world, but some language is particularly dependent for its intelligibility on the hearer's knowledge of the extralinguistic situation. *There she is!* out of context conveys little meaning: *there* could refer to all sorts of places; *she* might not even refer to a female person or animal.

See ENDOPHORA.

expansion

(In Systemic-Functional Grammar.) One of the two major logico-semantic categories held to explain the relationship between two clauses; the other being PROJECTION.

Expansion is subdivided into *elaboration*, which restates or exemplifies the main clause (e.g. I don't like it: *I hate it*); EXTENSION, which adds information; and *enhancement*, which in some way qualifies the clause with some feature of time, place, condition, or cause (e.g. I don't like it, *because it makes me ill*).

experiencer

Semantics. A person affected by the action or state of the verb.

The term comes from Case Grammar, in which it replaces the earlier term DATIVE. A verb of involuntary perception such as *hear* or *see* can be said to have an experiencer subject (e.g. *I heard/saw a bird*) in contrast to verbs such as *listen* or *look*, which have more deliberately involved agentive subjects. Other verbs of perception sometimes have experiencer meaning (e.g. *I smell burning* and sometimes agentive (e.g. *Smell this meat—is it all right?*).

experiential

Involving or based on experience.

In Systemic-Functional Grammar, the English clause has three functional components of meaning: IDEATIONAL, INTERPERSONAL, and TEXTUAL. The ideational part is further analysed into two parts. One is the *experiential*, meaning 'based on experience', which contrasts with a LOGICAL component, meaning 'based on certain general logical relationships'.

expletive

(*n. & adj.*) (A word or phrase) serving to fill out an utterance; especially applied to a swearword.

Expletives can be recognized as formulaic expressions of a particular type, many of them grammatically irregular in some way. For example *Damn you!* seems to be imperative in form, but an imperative verb is not normally followed by *you.*

• • **expletive *it***: another, not very frequent, term for PROP *it* or DUMMY *it*.

The term indicates that this *it* fills a syntactic space (late Latin *expletivus* serving to fill out, from *ex-* out, *plere* fill).

explicit performative See PERFORMATIVE.

exponence

Linguistics. A relationship between a level of linguistic analysis and an actual example, or EXPONENT, of this level.

In Systemic Grammar, exponence is one of three (or four) scales, along with RANK, DELICACY, and possibly DEPTH.

exponent

Linguistics. An example, an actual realization, of some level of linguistic abstraction.

> 1964 R. H. ROBINS All categories are part of the descriptive apparatus or frame set up by the linguist to deal with the particular language, and the relation between them and the material of utterance (or, *mutatis mutandis*, of writing) is one of class and category to exponent or manifestation.

expressive

Semantics. Designating a type of feeling-based meaning.

There are many different ways of classifying meaning; *expressive* corresponds to some extent with AFFECTIVE or EMOTIVE.

Extended Standard Theory See STANDARD THEORY.

extension

1 (In Systemic-Functional Grammar.) One type of semantic relationship between two clauses.

Extension is a sub-category of EXPANSION and can be embodied in a coordinate or a subordinate clause. Extension adds something new to the meaning of the other clause, if only by simple addition. For example, the 'extending' clause might begin (*both* . . .) *and*, (*neither* . . .) *nor*, *but*, *while*, *whereas*, and so on. Or the extension might add by variation (e.g. *but not*, *except* (*that*), *rather than*), or perhaps by an alternative (e.g. (*either* . . .) *or*).

2 *Semantics*. The range of referents covered by a particular term.

The word is taken over from logic, and contrasts with INTENSION. The extension of the word *dog* would include all breeds—and mongrels—to which the word is applied. The intension of the word *dog* would cover those features and characteristics that distinguish this animal from others.

• **extensional.**

extensive

Extending to someone or something beyond the subject. Contrasted with INTENSIVE.

In Systemic-Functional Grammar, where the *direct object* of traditional grammar is analysed as part of complementation, the direct object may be termed the *extensive complement* to distinguish it from the INTENSIVE *complement* (the ordinary complement). An *extensive verb* is therefore a verb taking an object and an *extensive clause* is one containing such a verb and its complement (e.g. *They are playing our tune*).

external sandhi See SANDHI.

extralinguistic

1 Referring to anything in the world outside language, but which is relevant to the utterance.

The term is applied to the situation (sometimes called the *situational context*) against which language operates. Extralinguistic features of an utterance often include shared knowledge, without which the language might make little sense. For example, the meaning of pronouns often depends on the situation: *Did he take them? Are they there yet? Isn't that lovely?* Extralinguistic features can also include body language, tone of voice, etc., which are called PARALINGUISTIC features in other analyses.

Compare EXOPHORIC, METALINGUISTIC.

2 Outside the normal phonology of a particular language.

CLICKS, which are normal phonemes in Xhosa, are an occasional extralinguistic feature of English.

extraposition

A special type of POSTPONEMENT involving the use of a substitute item for the postponed linguistic element.

Extraposition is frequently found with nominal clauses as subjects, where the subject is moved (on the principle of end-focus and end-weight) to a position after the verb, and an anticipatory *it* is put in subject position, e.g.

It's no use crying over spilt milk (*compare* Crying over spilt milk is no use)

It's disappointing (that) you can't stay longer (*compare* That you can't stay longer is disappointing)

Some examples of this structure have in fact no corresponding non-extraposed equivalents; extraposition is obligatory in these cases, e.g. *It seems that they're not coming after all* (**That they're not coming after all seems*).

• **extrapose**: postpone and substitute another item for (an element).

An *extraposed subject* is also called a *postponed subject*. Extraposition of an object also occurs, and sometimes this is obligatory in the same way, e.g. *You owe it to me to explain* (**You owe to explain to me*).

Extraposition is not to be confused with POSTPOSITION.

extrinsic modality

The same as EPISTEMIC *modality*.

F

factitive

1 *Semantics*. Of a linguistic element, usually a verb: expressing the notion of making something or causing a result.

This is a linguist's rather than a grammarian's term. In a broad sense it can apply to any transitive verb with the meaning of creating a result; e.g. the classic *kill* (if X kills Y, the result is that Y is dead). Other factitive verbs have meanings like *create, make, produce* (e.g. *make a cake*), which are usually termed CAUSATIVE in more popular grammar.

The term is also sometimes used of those COMPLEX TRANSITIVE VERBS that have a resultative meaning (e.g. *Paint the town red, They made him president*).

2 (In Case Grammar.) Describing the case of the noun phrase which refers to the thing made or created. In *I've been making cakes*, *cakes* is in the factitive case, in contrast to the same word in *They've eaten the cakes*, where *the cakes* is in the OBJECTIVE case. In later case grammar, this is the *result case*.

See RESULTANT.

factive

Semantics.

1 (*n. & adj.*) (A verb or a following *that*-clause) relating to the assertion of fact.

A verb that asserts the truth of a following clause is a *factive verb* (or *factive*), e.g.

I *know* that you were overcharged
I *regret* that you were overcharged

These verbs contrast with **non-factive** verbs which leave the proposition open:

I *believe* he was overcharged

and CONTRAFACTIVE verbs

He *pretends* that he was overcharged
I *wish* he had been overcharged

The proposition that follows a factive verb is a *factive predicate* (or a *factive*). Factive predicates can also depend on other parts of speech, not only verbs, e.g.

It is a pity that he was overcharged
I am sorry that he was overcharged

2 (In older usage.) The same as FACTITIVE (1).

• **factivity**: the quality of being factive.

See FACTUAL.

factual

1 Of a clause: referring to actual events, states, etc.

e.g.

I knew that last year
I arrived late, so I unfortunately missed dinner (i.e. result)
You didn't need to ask
Why were you late?
Were you late? (i.e. a factual question, as it requires a factual answer)

This type of clause contrasts with **non-factual** clauses, as in:

I arrived late *so that I would miss dinner* (i.e. purpose, intention)

Imperatives, subjunctives, and the infinitive are said to be *non-factual moods*. Both factual and non-factual predications contrast with COUNTERFACTUAL predications, where the truth is the reverse of the predication, as in

You needn't have asked (i.e. you did ask)
I wish I knew (i.e. I don't know)
If only I had arrived earlier! (i.e. I did not)

Open (or real) conditions are sometimes described as factual, in contrast to hypothetical ones, which are variously described as counterfactual or non-factual.

2 Of a verb: followed by a *that*-clause containing a verb in the indicative mood.

Factual verbs, in this formal, syntactical definition, include both PUBLIC VERBS of speaking and PRIVATE VERBS of thinking. Although these verbs may be generally described as referring to fact, the term is used with a rather wider meaning than the more semantically defined FACTIVE, and the predication may not be factual in sense (1). Thus factual verbs may include *allege, argue, bet, claim, pretend, realize, suspect,* and other verbs that would hardly be described as factive. Factual verbs contrast with SUASIVE verbs.

• **factuality**: the quality of being factual.

fall

Phonetics.

(*n.*) In the intonation of a syllable or longer utterance, a nuclear pitch change from (relatively) high to (relatively) low; contrasted with a RISE.

Phoneticians distinguish various kinds of fall, such as the *high fall* [ˋ], starting near the normal high limit of the voice and the *low fall* [ˎ], with a lower start. As with high and low notes in music, the use of such terms in describing intonation is of course metaphorical.

(*v.*) Of pitch: to change from high to low; usually in the term FALLING *tone*.

falling

Phonetics.

1 **falling tone**: the same as FALL (*n.*).

2 Of a diphthong: having most of the length and stress in the first part of the glide.

> 1988 J. C. CATFORD In English diphthongs, the stress-pulse is a *decrescendo* one, starting rather strong and then fading away . . . A decrescendo diphthong like this is often called a 'falling' diphthong because of the fact that the stress falls away from a peak near the beginning of the diphthong.

Most English diphthongs are normally articulated in this way, and *falling diphthong* is the normal label. Compare RISING.

fall-rise

Phonetics. A tone in which the pitch falls and then rises again [ˬ].

This tone is frequently heard in RP English. It has various conversational functions, but often suggests reservation or only partial agreement ('yes, but . . .'):

> A: Did you enjoy the film? B: ˬYes.

See INTONATION and compare RISE-FALL.

false friend

A word that has the same or a similar form in two (or more) languages but different meanings in each.

This term is used in contrastive analysis and foreign language teaching. For example, the French adjective *sympathique* (like Italian *simpatico*) often means 'nice', 'pleasant', or 'likeable' and is therefore a false friend to English *sympathetic*. In the same way French *actuel* means 'present', not 'actual'.

Sometimes also called *false cognate* and *faux ami.*

feature

A distinctive characteristic of some part of the language or of a speech act.

This is a very general term: some people even talk of EXTRALINGUISTIC features, such as body language. The term is particularly used in relation to phonology and semantics.

(*a*) *Phonology*. The sound system of English can be analysed in various ways, with reference to acoustic, articulatory, or auditory features. A detailed theory is that of the DISTINCTIVE FEATURE.

(*b*) *Semantics*. Vocabulary items have also been subjected to a similar kind of features analysis, sometimes called COMPONENTIAL ANALYSIS. This is often of a BINARY kind. Thus the word *bachelor* might be analysed as [+ male] [+ adult] [+ human] [– married]. But such analysis does not work equally well in all areas of the lexicon.

See also MORPHOSYNTACTIC.

feminine

(*n. & adj.*) (A noun etc.) of the gender that mainly denotes female persons or animals. Contrasted with MASCULINE.

In some languages grammatical gender distinctions of masculine and feminine (and sometimes also neuter) apply to all nouns and related words. In English, however, grammatical gender distinctions are found only in 3rd person singular personal pronouns and determiners, where the feminine forms (*she, her, herself, hers*) contrast with the masculine ones (*he* etc.) and the non-personal *it* etc.

The suffix *-ess* (e.g. in *lioness, hostess*) is a feminine marker in nouns, but feminism has reduced its use, so that, for example, some women who would previously have been described as *actresses* may now prefer to be called *actors*.

Some male-female pairs of words are morphologically related (e.g. *hero/heroine, widow/widower*) but many are not (e.g. *boy/girl, duck/drake*).

Compare GENDER.

field

Semantics. A range or system of referents that have some aspect of meaning in common. Sometimes called DOMAIN.

The theory of semantic fields asserts that the meaning of a word depends partly on the other words it is related to in meaning. All such words together constitute a *semantic field* (or *lexical field*). In different languages the same field is often apportioned differently, with dissimilar sets of terms used for that field. Classic examples are the fields of colour and kinship. It is well-known that other languages divide the spectrum differently, but in English the meaning of, e.g., *blue* is limited by the existence of *green* and *purple* and the latter is itself further limited by the existence of, say, *mauve*. Similarly the meanings of *brother*, *cousin*, and so on form a network of connected family terms.

field of discourse

Sociolinguistics. The subject matter being talked or written about.

In some cases different subject matter may involve few differences other than those of vocabulary. But some fields of discourse are characterized by their own distinctive grammatical styles, e.g. legal language, football commentaries, advertisements, sermons.

Compare REGISTER.

filled pause

Linguistics. The use of a hesitation noise.

Hesitation noises, inadequately represented as *er*, *erm*, etc., are a common feature of speech; they are used to give the speaker time to think or to prevent another speaker from taking over.

filler

1 A word or words that can fill a particular functional SLOT in a SLOT-AND-FILLER analysis of clause structure.

Identifying functional slots and the linguistic items that fill them is one way of analysing clauses and sentences. Thus

Jack and Jill / went / up the hill

contains three slots, which in current terms are identifiable as Subject/Verb/Adverbial. The fillers in this particular case are, formally, two coordinated proper names, a single-word verb, and a prepositional phrase, but these slots could be filled by other noun phrases, verb phrases, and adverbial phrases without changing the nature of the slots.

2 A word, usually outside the syntax of an adjoining clause, that serves to fill what might otherwise be an unwanted pause in conversation. Also called a *pragmatic particle*.

e.g.

Oh, well, you know.

Compare DISCOURSE MARKER.

final

Designating or occurring in the position at the end of a linguistic element.

(*a*) Final position in a clause or sentence normally means what it says, and is a useful phrase for describing whether a particular type of word or phrase is in a usual or unusual position.

Final position (also called *end position*) is often contrasted with INITIAL (or FRONT) and MEDIAL (or MID) position, and is particularly used in relation to ellipsis and to the placing of adverbials. The most usual place for an adverbial is final position, as is shown by the fact that SVOA is a normal, unmarked word order. For a single adverb or adverb phrase, the term does not necessarily mean 'at the very end of its clause', but rather 'after the (object and) verb'. Thus both *without question* and *immediately* are in final position in *The money was repaid without question immediately.*

(*b*) *Phonology*. Distinctions of initial, medial, and final are relevant in phonology since many phonemes regularly have different phonetic values according to their position. See ALLOPHONE.

Some English phonemes can occur only in certain positions. /h/, /j/, and /w/ never occur at the end of a syllable; /ʒ/ occurs there only in words of French origin (e.g. *rouge*); while /ŋ/ can only occur at the end of a syllable (e.g. *singer*, *sing*).

finite

Having tense. Contrasted with NON-FINITE.

Verbs and verb phrases having tense, and clauses and sentences containing them, can be described as finite.

The third person singular *-s* form (e.g. *looks, sees*) is always finite, as is the past form (e.g. *looked, saw*), whereas the *-ing* form (e.g. *looking, seeing*) and the past participle (e.g. *looked, seen*) are non-finite. The base form (e.g. *look,* *see*) can be either. It is finite as a present tense (e.g. *I see*), in the imperative (e.g. *Look out!*) and as a subjunctive (e.g. *The boss insisted that I see him*).

Although we talk of finite verb phrases, it is in fact only the first word of a verb phrase that is finite, e.g.

We *have* been wondering
It *may* be being changed

- **finiteness**: the fact or quality of being finite.

finite state grammar

A theoretical model of grammar discussed by Noam Chomsky in his book *Syntactic Structures* (1957).

The model was a deliberately oversimple one that could not account for many features of real language. It was introduced in order to show the need for more elaborate theories.

Compare PHRASE (2), TRANSFORMATIONAL.

first language See LANGUAGE.

first person

(Denoting, or used in conjunction with a word indicating) the speaker or writer, in contrast to the addressee(s) or others.

First person pronouns and determiners are *I, me, myself, my, mine* in the singular; and *we, us, ourselves, our, ours* in the plural.

Uniquely among English verbs, *be* has a distinct first person singular form in the present (*am*).

Compare SECOND PERSON, THIRD PERSON.

Firthian

(*adj.*) Of, pertaining to, or characteristic of the British linguist John Rupert Firth (1890–1960).

(*n.*) A person who subscribes to Firth's theories.

J. R. Firth, Professor of General Linguistics in the University of London (1944–56), was influential in the development of linguistics in Britain. Characteristic of his approach is the concept of polysystemicism (i.e. that language is not one system, but a set of systems). Systemic grammar is a development of his ideas.

- **Firthianism**.

  1975 T. F. MITCHELL There appear to be three salient features of Firthianism . . : (1) insistence on the centrality of meaning in all its aspects (2) adoption of a basic inductive approach to language study (3) recognition of the priority of syntagmatic analysis.

Compare NEO-FIRTHIAN.

fixed

Not subject to variation.

•• **fixed phrase**: a phrase of which few if any variants are acceptable (also called *set expression*).

In some phrases no change either of an individual word or of the word order is likely. Examples of fixed phrases are:

knife and fork (*fork and knife)
pay attention to (*pay attention towards/for/at)
heir apparent (*apparent heir)
from bad to worse (*from good to better)
beneath contempt (*below/under/underneath contempt)
no good, no different (*no bad, no similar)
for the time being (*for the time that is)
to and fro (*to and from)
as it were

Compare COLLOCATE, FORMULA.

fixed stress: (*Phonology*) the regular occurrence of stress on the same syllable in each word of a language; contrasted with FREE *stress*.

English is not a fixed-stress language, and in this it contrasts with some languages where the stress is fairly predictable. For example, in Polish, polysyllabic words are usually stressed on the penultimate syllable. However, the stress in individual words in English is largely fixed, so that deviant stressing can lead to misunderstanding or incomprehension. Contrast *im'portant* and *'impotent*, or the *'Aldwych (Theatre)* with the *Old 'Vic*.

fixed word order: a characteristic of some languages, whereby a change in the word order can change the meaning of (a part of) a sentence.

English is, relatively speaking, a fixed word order language, since Subject-Verb-Object is normal and a change can significantly affect the meaning. *Putting the cart before the horse* is quite different from *putting the horse before the cart*!

See FREE *word order*.

flap

Phonetics. A consonant sound in which a flexible speech organ makes a momentary contact with a firmer surface.

This is a 'manner of articulation'. In British English the voiced frictionless continuant /r/ is sometimes replaced by an alveolar flap [ɾ], with the tip of the tongue articulating against the alveolar ridge. This sound is commonly used in American English where *t* or *d* occur between vowels, so that the *t* and *d* may sound identical, as in *latter* and *ladder*.

A flap—consisting of a single rapid tap—is mainly contrasted with a ROLL (or *trill*). Some phoneticians make detailed distinctions between a flap and a tap, but for English the terms are used interchangeably.

flection

The same as INFLECTION. (Now old-fashioned.)

- **flectional. flectionless.**

1862 G. P. MARSH An important advantage of a positional . . over a flectional syntax.

focus

(*n.*) The most important part of a sentence in terms of its information content.

In analysing a sentence in terms of the structure of its information content or 'message' (as opposed to its syntactic structure), it is common to divide the sentence into two, using such terms as GIVEN and NEW, THEME and RHEME, TOPIC and COMMENT. Focus is another label for the second term in such pairs.

The pairs are sometimes treated as more or less synonymous, and indeed in many cases the theme or topic will be something 'given' or already known, while the second part of the sentence, the rheme, will present new information on which to focus. The term *focus* is not, however, the same as *rheme*. It often refers to the end of the rheme (END-FOCUS), rather than the entire rheme, and in speaking it roughly correlates with the final nuclear stress, as in

The telephone's still out of ˬorder

But the focus will be shifted to a point earlier in the sentence if the end is predictable, e.g.

The PHONE's ringing

And it can be marked on any part of a sentence:

You should phone YOUR mother (i.e., not mine)
You should PHONE your mother (i.e., not write)
You SHOULD phone your mother (i.e., it's your duty)
YOU should phone your mother (i.e., don't expect someone else to)

When written (without capitals or other marking) this sentence would normally be interpreted as having the focus at the end (MOTHER), so special devices may be used to mark the focus at a different place. See CLEFT, MARKED.

(*v.*) Make the focus or (use to) place the focus on.

1985 R. QUIRK et al. The item selected for being focused is generally 'new' information.

1973 R. QUIRK & S. GREENBAUM *Alone* must normally follow the part on which it is focused, *eg: You can get a B grade for that answer ALONE.*

1973 R. QUIRK & S. GREENBAUM Intonation can also focus more narrowly on a particular word of a phrase, rather than phrase of a clause.

focusing adverb

An adverb that focuses on a particular part of the sentence. (Also called *focusing adjunct*.) A few focusing adverbials are prepositional phrases.

Typical focusing adverbs are *even*, *merely*, and *only*, as in

Even in old age, she was immensely active
I *merely asked* them the time.
Only you would say a thing like that

The classification of adverbs is notoriously difficult and there is considerable variety in the way it is done. The categories of *focusing adverbs* and INTENSIFYING *adverbs* correspond to all or part of the more traditional DEGREE category. The scope of the category of focusing adverbs (which are sometimes called *focusing* SUBJUNCTS) may be restricted further in more refined systems.

folk etymology

A popular modification of the form of a word, in order to render it apparently significant. (Also called *popular etymology*.)

Welsh rarebit, an alteration of *Welsh rabbit*, is an example of folk etymology. It seems to be based on the belief that nobody could describe melted cheese on toast as 'rabbit'. But *Welsh rabbit* is correct: an ironical name that grew up in the days when only the better-off would be likely to eat much meat. Other examples of folk etymology are the more familiar-sounding *sparrow grass* (for *asparagus*) and *forlorn hope*, now meaning 'a faint remaining hope', but based on the Dutch *verloren hoop*, actually meaning 'lost troop' (i.e. a storming party etc.). The *cock-* of *cockroach*, the *-wing* of *lapwing*, and the *-house* of *penthouse* all result from the attempt to make at least a part of an arbitrary word into a meaningful element (their earlier forms being *cacarootch*, *lappewinke*, and *pentice*).

foot timed

The same as STRESS-TIMED.

force: see ILLOCUTIONARY.

foreign plural

A plural form of a noun that retains the inflection of the foreign word of which it is a borrowing.

Many nouns taken from foreign languages retain their foreign plurals. This applies particularly to Greek and Latin loanwords, many of which retain *classical plurals*, e.g.

crisis, crises
phenomenon, phenomena
larva, larvae
erratum, errata

Some have alternative anglicized plurals

candelabrums/candelabra
formulas/formulae
corpuses/corpora

Some have both types of plural but with different meanings (e.g. *appendixes/appendices*).

Because of its exceptional divergence in form from all other patterns of English plural formation, the Latin and Greek *-a* plural has shown a tendency to be reinterpreted as a non-count form, or as a singular with its own *-s* plural. This tendency has progressed furthest in *agenda* and has met with varying degrees of acceptance in *candelabra, criteria, data, media*, and *phenomena*.

Examples of foreign plurals from other languages include *bureaux/bureaus* (French) and *cherubim/cherubs*, *kibbutzim/kibbutzes* (Hebrew).

form

The shape or external characteristics of a linguistic unit; contrasted with its meaning or FUNCTION.

1 One of the ways in which a word may be spelt or pronounced or inflected.

All lexical verbs (represented by an abstract LEXEME) have several forms. For example, *see* has five forms: *see, sees, seeing, saw, seen*, while several verbs have two alternative past tense forms (both spoken and written), e.g. *spelt, spelled*. Some nouns have two alternative plural forms, e.g. *indices/indexes*.

2 Part of the internal structure of a word; the same as MORPHEME (2).

Words are made up of *free* and *bound forms* (or morphemes). Sometimes a bound form indicates what part of speech the word is; for example, the suffix *-ness* normally indicates a noun. But many words are not identified in this way, and the part of speech of these is in fact often identified by the function they perform in the sentence. See FORM CLASS.

3 The internal structure of a linguistic unit.

Above word level, phrases can be analysed by their formal constituents. Thus, a noun phrase may consist of a single noun (e.g. *people*), but often contains a determiner and an adjective (e.g. *the best people*) and possibly post-modification too (*all these people you were telling me about*).

Form contrasts with FUNCTION here. 'Noun phrase' is a formal category, but a noun phrase can function not only as a subject or object, but also (less typically) adverbially (e.g. We had a storm *last night*). Conversely, a noun phrase may rather exceptionally not contain a noun or pronoun; e.g. in *The poor are always with us*, an adjective (*poor*) is the head of the noun phrase.

4 (In early Scale-and-Category Grammar.) One of the primary levels of grammatical analysis (alongside SUBSTANCE and SITUATION) comprising grammar (i.e., roughly, syntax) and lexis (vocabulary).

In later systemic theory the concept of form is modified:

> 1985 G. D. MORLEY The phrase 'level of form' is replaced by the term 'lexico-grammatical stratum'.

formal

1 Relating to form as opposed to function. Contrasted with FUNCTIONAL.

'Noun', 'noun phrase', 'verb phrase', etc. are formal categories, defined largely by their structural composition, irrespective of their meaning or of their function in a sentence.

2 Relating to form as opposed to meaning.

Traditional grammar often defines linguistic units in NOTIONAL terms, e.g. 'A noun is the name of a person, place, or thing'. Modern grammar, seeking more formal criteria, prefers to define parts of speech largely by their syntactic distribution.

3 Of language: characterized by a relatively impersonal attitude and adherence to certain social conventions. Contrasted with INFORMAL.

Formal speech and writing is characterized by more complicated grammatical structures and more unusual vocabulary than informal language and by the avoidance of short forms (e.g. *can't, won't*) and colloquialisms. *Patrons are requested to refrain from smoking* is more formal than *Please Don't Smoke*. Formal and informal are however at the ends of a continuum, and much language is not marked as either.

4 **formal *it***: (a traditional label for) DUMMY *it* or ANTICIPATORY *it*.

formant

Phonetics. One of several characteristic bands of resonance, a combination of which determines the distinctive sound quality of a vowel or other voiced and unconstricted sound.

A term in acoustic phonetics.

> 1971 P. LADEFOGED Roughly speaking, we can say that the sound of a vowel consists of the pitch on which it is said . . and the pitches of the two or three principal groups of overtones (which can be associated with the resonant frequencies of the vocal tract). These groups of overtones are called formants.

formation

The same as WORD FORMATION.

formative

1 (*n. & adj.*) (Designating) the smallest meaningful physical element used in the formation of words. The same as MORPHEME (2).

Like *morpheme*, the term *formative* can be used in a physical sense. Formatives are divided into *lexical formatives*, used in the formation of derived words (e.g. *-ful* as in *hurtful, spoonful*), and *inflectional formatives* (e.g. *-ing* as in *hurting, spooning*).

2 A physical form of an abstract morpheme. The same as ALLOMORPH.

In this use, formative is not synonymous with MORPHEME (1) but means one of the variants of such a morpheme. Thus the three regular plural endings of English (/s/, /z/, and /ɪz/) are all *formatives* of the single 'plural morpheme', which is purely an abstraction.

form class

Generally, another term for WORD CLASS.

formula

An instance of stereotyped language that usually allows few or no changes in form and may not conform to current grammatical usage.

Common formulae include:

> Thank you!
> How do you do? (*How does your mother do?)
> See you! No way! You don't say!
> Many happy returns!
> Please God . . .
> Yours sincerely

This category overlaps with FIXED PHRASE and with FOSSILIZED.

- **formulaic**: that is a formula, that uses formulae.

formulaic subjunctive See SUBJUNCTIVE.

form word

A word that primarily has formal or grammatical importance rather than meaning. Also called *empty word*, *function word*, *grammatical word*, *structural word*, or *structure word*.

In some traditional grammar, word classes (parts of speech) are divided into *form words* and CONTENT WORDS (or LEXICAL *words*). Form words are mainly closed class words that glue the content words together: auxiliary verbs, determiners, conjunctions, and prepositions. The distinction is out of favour today, since all form words normally carry some meaning.

Confusingly, in view of the fact that form and function are often contrasted, form words are sometimes called *function words*.

> 1940 C. C. FRIES By a *function word* I mean a word that has little or no meaning apart from the grammatical idea it expresses . . . Henry Sweet calls these words *form words* . . . I prefer the term *function word* because *form words* sounds so much like the expression *the forms of words*, which I use for *inflections*, that students are often confused.

Compare CLOSED, MINOR, STRONG.

fortis

Phonetics. Of a consonant sound or its articulation: made with relatively strong breath force; contrasted with LENIS.

In English the voiceless plosives and fricatives (/p/, /t/, /k/, /f/, /s/, etc.) tend to be made with stronger muscular effort and breath force than their voiced counterparts. Such consonants are therefore said to be *fortis consonants* and to be pronounced with a *fortis articulation*.

Fortis consonants are sometimes described as *strong consonants* (and lenis as *weak*) but this terminology leads to confusion and has largely been abandoned.

forwards anaphora

The same as CATAPHORA.

fossilized

Having a structure that is no longer productive.

Fossilized structures in English are exemplified in:

> Handsome is as handsome does (*Good is as good acts)
> Come what may (*Occur what will)
> Long may it last (*Short may it last)
> How come . . .? (*How happen . . .?)

Fossilized phrases include *fore and aft*, *kith and kin*, *to and fro*.

Fossilized word-formation processes include such items as the plural *-en*, still present in *children, brethren, oxen*, but not available, except as a joke, for new words.

This category overlaps with FIXED PHRASE and with FORMULA.

free

1 Designating a linguistic form that can be used in isolation.

• • **free form, free morpheme**: the smallest linguistic unit that can stand alone; contrasted with a BOUND form. See MINIMUM FREE FORM.

2 **free indirect speech**: a form of INDIRECT SPEECH that retains the sentence structure of direct speech (e.g. subject-auxiliary inversion in questions) and usually lacks a reporting verb, but is signalled by backshifted verbs and changes in pronouns and in time and place references.

It is popular in fiction and narrative writing for reporting both speech and thoughts. *Free direct speech* is also found, marked by an absence of quotation marks and also by present tenses that contrast with the past tenses of the rest of the narrative.

In the following extract the conversation is reported in a mixture of the writer's free indirect speech (shown in italics) and the old man's free direct speech:

> As our conversation continued the old man became whiter and whiter. The dust clung to his hat, his face, his moustache, his eyelashes and his hands. He did not complain.
> *Who was doing all the killing?*
> —Oh, the guerillas of course. If the guerillas see you and do not know you they kill you.
> *And what about the army?*
> —The army are all right, if you greet them and have papers.

Was there enough food?
—There is enough food, but there is no business.

(P. Marnham *So Far From God* (1985), p. 148.)

Compare INDIRECT SPEECH.

3 **free relative clause**: another term for NOMINAL RELATIVE CLAUSE.

4 Of stress in a particular language: occurring on a different syllable in different polysyllabic words and not according to predictable patterns.

Although individual words in English have their own fixed stress patterns, stress falls on a different syllable in different words, e.g.

ADvertising, unFORtunate, diploMAtic, misunderSTANDing, commandEER, PHOtograph, phoTOGrapher, photoGRAPHic, photogravURE

English is therefore a free stress language. Compare FIXED stress.

5 Of word order in a particular language: that can be varied without alteration to the basic meaning of the sentence.

Free word order tends to be a characteristic of highly inflected languages like Latin, where inflections, rather than word order itself, are the primary indicator of relationships between words and meaning. English, by contrast, has a fairly FIXED word order.

free variation

1 *Phonology*. The possibility of substituting one phoneme for another without causing any change of meaning.

Sounds which contrast with each other in such a way that meaning is affected (i.e., distinct phonemes) cannot normally be interchanged. But in some words two normally contrasting phonemes are both acceptable and are therefore said to be *in free variation*.

Among British speakers, a majority are said to prefer the word *ate* to be pronounced /et/ to rhyme with *met*; but a large minority favour the pronunciation /eɪt/ like *eight*. The two pronunciations are therefore in free variation, as are /ekə'nɒmɪks/ or /iːkə'nɒmiks/ for *economics*, and /'iːðə/ or /'aɪðə/ for *either*. Similarly, though less noticeably, a standard speaker may or may not release the final plosive in a word such as *heap*, or use a glottal stop in place of the /t/ in the word *witness*.

Free variation contrasts with allophonically conditioned variation: compare COMPLEMENTARY DISTRIBUTION.

Free variation also occurs with the stressing of some words (e.g. *adult*, *subsidence*), but strong feelings are aroused by variation between such pairs as *CONtroversy* and *conTROversy*.

2 Occasionally, on other linguistic levels, the possibility of substituting one normally contrastive item for another.

Free variation occurs in the spelling of some words (e.g. *realize/realise*, *judgement/judgment*, *jail/gaol*).

Free variation in the use of words is normally labelled SYNONYMY.

frequency

1 Rate of occurrence.

The term is used in its everyday sense to group together semantically related adverbs, such as *always, usually, often, sometimes, never*. These are popularly called *frequency adverbs/adjuncts*.

See INDEFINITE.

2 *Phonetics*. The rate of occurrence of vibration in the production of a speech sound.

The term is used in acoustic phonetics. A distinction is made between the fundamental frequency of a sound—the basic vibration of the vocal cords over their whole length—and the frequencies of the higher harmonics, which are multiples of the fundamental frequency.

3 *Linguistics*. The rate at which a particular word in the language is used.

How commonly or rarely words occur is of special practical interest to lexicographers and educationists. With modern technology it is possible to obtain word frequencies from vast collections of texts and to confirm or disprove estimates of *high frequency* and *low frequency* words. The word of highest frequency in English, according to several different corpora, is *the*.

frequentative

(*n. & adj.*) (A verb) expressing frequent repetition or intensity of action.

In some languages there are special ways of forming frequentatives from ordinary verbs, e.g. in Latin, *rogito* I ask repeatedly, formed on *rogo* I ask. In English the suffixes *-er* and *-le* have been used to produce a number of verbs of this kind, e.g. *flitter* from *flit* and *crackle* from *crack*; but this is not a regularly productive system.

The term is also used to describe other ways of expressing repeated verbal action, and is sometimes synonymous with ITERATIVE (1).

fricative

Phonetics. (*n. & adj.*) (A consonant sound) articulated by two speech organs coming so close together that there is audible friction.

A fricative (sometimes called *friction consonant*) may be voiceless or voiced. There are four pairs of voiceless and voiced fricatives in RP, plus the voiceless /h/. The pairs are:

labio-dental /f/ as in fan, physics, rough and /v/ as in van, of, nephew (with some speakers)
dental /θ/ as in think, author, path and /ð/ as in the, father, with
alveolar /s/ as in soon, pencil, hopes, loose and /z/ as in zoo, easy, was, lose
palato-alveolar /ʃ/ as in shop, sure, machine, wish and /ʒ/ as in usual, prestige

The English /h/ is classified as a voiceless glottal fricative. The Scottish pronounciation of the final sound in the word *loch* is a voiceless velar fricative [x], which is not a regular phoneme in the RP system.

Many other fricative sounds feature in other languages.

Fricative sounds, lacking complete closure, contrast especially with AFFRICATES and PLOSIVES.

- **fricativeness**: the quality of being fricative.

friction consonant

The same as FRICATIVE.

frictionless continuant

A continuant speech sound lacking friction.

A frictionless continuant is neither a fricative nor a stop. In a very broad use, the term could be applied to vowels. Among consonants, several phonemes in RP can be so labelled:

the nasals /m/, /n/, and /ŋ/
the lateral /l/
the semi-vowels /w/ and /j/

These all have their own articulatory labels; hence the term is most often applied to the southern British /r/ (which can be described as a *post-alveolar frictionless continuant*.)

See CONTINUANT.

front

1 *Phonetics*. (*a*) (*n.*) The forward part of the tongue (but not the tip).

(*b*) (*adj.*) Related to the front part of the mouth. (*v.*) Produce (a phoneme) further forward in the mouth than normal (or than in the past). See FRONTED.

Standard (RP) English distinguishes four *front vowels*, so called because they are articulated with the front part of the tongue higher than any other part. Going from HIGH to LOW, they are the vowels (however spelt) of *seat* /iː/, *sit* /ɪ/, *set* /e/, and *sat* /æ/. Compare CENTRAL and BACK vowels.

Specifying the front or back of the mouth is less necessary in defining consonants, since more specific places of articulation can be given, but bilabials and labio-dentals, for example, can be described as *front consonants*.

In DISTINCTIVE FEATURE theory *front* is replaced by ANTERIOR.

2 (*adj.*) Of the position of a word: at, or relatively close to, the beginning of a sentence. (*v.*) Place (a sentence element) at the beginning of the sentence. See FRONTING.

Front position is referred to especially in the description of adverbs. See INITIAL.

fronted

1 *Phonetics*. Of a sound: produced further forward in the mouth than normal.

Phonemes produced further forward in the mouth than normal, usually under the influence of neighbouring sounds, can be described as *fronted*. Thus the velar stops /k/ and /g/ are often fronted when followed by a front vowel, as in *keen* or *geese* (and, by contrast, retracted when followed by a back vowel as in *cart*, *guard*).

2 Of a sentence element: placed (unusually) at the beginning of a sentence. See FRONTING.

fronting

The unusual placing of a sentence element at the beginning of the sentence.

English sentences typically begin with a subject, but other functional elements—object, complement, adverbial, and even part of the verb phrase—can be placed at the beginning in order to mark the THEME, e.g.

Loud music I do not like.	(fronted object)
Horrible I call it.	(fronted complement)
After half an hour, we walked out.	(fronted adverbial)
Walk out we did.	(fronted verb)

FSP See FUNCTIONAL *sentence perspective*.

full

Complete; answering in every respect to a description.

1 Designating (a case of) apposition where all possible criteria apply: see APPOSITION.

2 Designating complete and permanent conversion from one word class to another: see CONVERSION.

3 **full sentence**: a sentence with at least a subject and a finite verb.

A full sentence means a traditional sentence, in contrast to a grammatically incomplete utterance, which may well be completely adequate in context, such as an ellipted reply *In the evening*, or a *minor sentence* such as *What?*, *Yes!*.

4 **full stop**: the punctuation mark ⟨.⟩ used at the end of a sentence or abbreviation. Also called *full point* and *period*.

5 **full verb**: another term for LEXICAL *verb*. Compare MAIN VERB.

6 **full word**: another term for CONTENT *word*.

function

1 The syntactic role that a linguistic unit takes within a 'higher' unit, such as a clause or a sentence; distinguished from its FORM.

The five ELEMENTS of clause structure, namely Subject, Verb, Object, Complement, and Adverbial, are defined by virtue of their functions. Although the function of verb is always realized by a verb phrase, there is no one-to-one correspondence between the other functional sentence elements and their possible formal realizations. Thus the function of subject (like object) is often realized by a noun phrase but could, for example, be realized by a verb phrase (e.g. *To err* is human), while on a lower level such a verb phrase might function as part of a noun phrase (e.g. *a tendency to err*). Similarly, a single word may sometimes function as a different part of speech from the usual one without undergoing *full* CONVERSION. See FUNCTIONAL *equivalence.*

2 The semantic (or discourse) role of a sentence.

Sentences themselves may be classified in terms of their semantic (or discourse) functions. A typical functional classification of sentence types is into STATEMENTS, QUESTIONS, EXCLAMATIONS, and DIRECTIVES. Again, this is contrasted with FORM. Thus *Isn't it a lovely day!* (formally a question) functions as an exclamation.

See COMMUNICATIVE *function.*

3 One of the social roles for which language is used.

This sense is similar to (2) but less concerned with sentence types. It has been suggested that there are four primary speech functions: offer, command, statement, and question; and that these cover all the purposes for which language is used, whether the intention is to elicit a verbal response or a non-verbal one.

In the teaching of English (or any other modern language) to foreign learners, the concept of functions has been extended to cover a wide variety of social uses, such as making suggestions, complaining, and sympathizing. These social roles are also termed *functions.*

functional

As regards function; of, in, etc. function.

•• **functional conversion**: the same as CONVERSION.

functional equivalence: comparability of function in a particular context.

In the phrases *the university press* and *this ancient press*, the words *university* and *ancient* show functional equivalence, since *university* (normally a noun) is here *functionally equivalent* to the adjective *ancient*. Compare CONVERSION.

functional grammar: a theory of grammar primarily concerned with the ways in which language functions pragmatically and socially, rather than formally. See SYSTEMIC GRAMMAR.

functional load: (*Phonology*) the amount of meaning that has to be carried by a particular linguistic feature or contrast.

The term is applied especially in phonology. In English, for example, a great many words are distinguished solely by the difference between /t/ and /d/ (e.g. *toe/doe*, *true/drew*, *latter/ladder*, *late/laid*). The /t/ and /d/ contrast therefore has a heavy *functional load.* By contrast, there are very few pairs of words distinguished only by the phonemic contrast between voiceless /θ/ (as in *think*) and voiced /ð/ (as in *they*), and hence the functional load of this distinction is light.

functional phonetics: an occasional name for PHONOLOGY.

functional sentence perspective: a theory of linguistic analysis developed by the Prague School, concerned with the amount of information conveyed by different parts of a sentence or utterance. The theory of THEME and RHEME derives from this. Abbreviated FSP. See COMMUNICATIVE DYNAMISM.

functional shift: the same as CONVERSION.

function word

The same as FORM word.

1940 C. C. FRIES By a function word I mean a word that has little or no meaning apart from the grammatical idea it expresses.

functor

The same as FORM WORD or FUNCTION WORD. (A rare term.)

fundamental frequency See FREQUENCY (2).

fused participle

A structure containing a gerund preceded by a noun or pronoun, rather than a more strictly 'correct' possessive.

e.g.

Forgive *me asking*, but . . .
Were you surprised at *my father arriving* early?

Prescriptive grammar requires

Forgive *my asking*
Were you surprised at *my father's arriving* early?

But the fused participle form is common in spoken, and even written, English.

The term was coined by H. W. Fowler (1906).

fused relative construction

The same as NOMINAL RELATIVE CLAUSE.

fusion

Linguistics. The merging of more than one linguistic element in one form.

English verb inflections are an example of fusion. Thus the *-s* ending (e.g. *looks*, *puts*) signifies 'third person' + 'singular' + 'present'.

- **fusional**: exhibiting or causing fusion.

1980 E. K. BROWN & J. E. MILLER We can . . have rules of the form:
MAN + pl ⟶ *men*.
We call rules like this 'fusional'. Fusional rules have two morphemes on the left-hand side of the rule matched to a single morph on the right-hand side.

future

(*n. & adj.*) (A tense or form) relating to an event or state yet to happen.

In traditional grammar, tenses formed with *shall* and *will* are called *future tenses*. More specifically, *shall* or *will* + the bare infinitive is the *future simple* (e.g. *We shall overcome*, *Whatever will be will be*); while *shall* or *will* + *be* + an *-ing* form is the *future continuous* or *future progressive* (e.g. *They'll be coming round the mountain*).

In more modern analyses, though such labels are often retained, it is pointed out that strictly English has no future tense as such, but employs various ways of talking about future time, including (besides the *will* and *shall* construction):

the present progressive	e.g. *I am seeing Robert tomorrow*
the present simple	e.g. *His plane arrives at 8.33 am*
	If the plane is late . . .
	When he arrives . . .
the '*going to*' future	e.g. *It's going to rain*

See TENSE, TIME.

future in the past

A tense that from a time in the past looks towards its own future.

Traditionally this label is given to a certain type of verb phrase containing the word *would* such as:

They did not realize then that by 1914 the two countries *would be* at war.

However other structures can be used to describe what was seen as future time when viewed from a past perspective, e.g.,

We *were* bitterly *to regret* our decision
I *was going* to tell you (when you interrupted me)

future perfect

A tense formed with *shall* or *will* + *have* + a past participle, expressing expected completion in the future.

In the traditional labelling of tenses the future perfect simple is exemplified by

I will have wasted the whole morning if they don't come soon

and the *future perfect continuous/progressive* by

I'll have been waiting three hours by one o'clock

Future perfect passive tenses are also possible:

The best things will all have been sold by the time you get there.

G

gapping

The phenomenon caused by medial ellipsis (in contrast to the more usual initial or final ellipsis).

Gapping occurs when there is medial ellipsis in the second (or later) part of a coordination, e.g.

I play golf and my brother ^ tennis
I went by train and my friend ^ by car

geminate

Phonetics & Phonology.

(*n.*) A sequence of two identical consonants pronounced with some separation, though not as two separate sounds.

(*adj.*) Designating (a consonant within) such a sequence.

Geminates occur only where two consonants belong to separate morphemes, i.e. straddling syllable or word boundaries.

Compare *hopping* /ˈhɒpɪŋ/ and *hop-picking* /ˈhɒp pɪkɪŋ/.

See DOUBLE CONSONANT.

• **gemination**: the formation of a double consonant.

gender

1 A classification of nouns, pronouns, and related words, partly according to natural distinctions of sex (or absence of sex).

2 The property of belonging to one of such classes.

In some languages *gender* is an important grammatical property of nouns and related words, marked by distinct forms. In French, for example, all nouns are either *masculine* (*son livre*, masculine = 'his book' or 'her book') or *feminine* (*sa plume*, feminine = 'his pen' or 'her pen'). In these languages natural gender is usually, though not entirely, marked by the matching grammatical gender. In some languages (e.g. Latin, German, and Old English) there is a third gender, neuter, which marks nouns denoting inanimate objects (although many such nouns belong to one or other of the other two genders).

In Modern English overt grammatical gender hardly exists, except in third person singular pronouns:

he/him/his/himself (*masculine*)
she/her/hers/herself (*feminine*)
it/its/itself (often called *non-personal* rather than *neuter*)

Even here there can be some mismatch between natural and grammatical gender. Inanimate countries, ships, cars, etc. may sometimes be referred to by masculine or feminine pronouns; a baby may be *it*; animals may be referred to by personal or non-personal pronouns.

Natural gender distinctions are made covertly in many words referring to males and females. Pairs of words occasionally show a derivational relationship (e.g. *hero/heroine*, *widow/widower*), but many male and female noun pairs show no morphological connection (e.g. *brother/sister*, *duck/drake*).

Compare COMMON, DUAL, FEMININE, MASCULINE, PERSONAL.

General American

An accent of English used in the United States that lacks the especially marked regional characteristics of the north-east (New England, New York State) and south-east (the 'Southern States').

Introduced by G. P. Krapp (1924), the term has been widely criticized as implying greater uniformity than really exists, but it remains useful as a general label.

> 1982 J. C. WELLS 'General American' . . is not a single unified accent. But as a concept referring to non-eastern non-southern accents, the label has its uses. It corresponds to the layman's perception of an American accent without marked regional characteristics. It is sometimes referred to as '*Network* English', being the variety most acceptable on the television networks covering the whole United States.

general determiner See SPECIFIC (*b*).

generalized phrase-structure grammar

A linguistic theory of grammar originating in the 1970s and offering a radical alternative to Transformational Grammar.

It does not recognize transformations, but uses 'metarules' and other devices that result in generalized rules which were not possible in earlier Phrase-Structure Grammar.

generate

Produce (grammatical language) by the application of rules that can be precisely formulated.

Most grammars that have ever been written 'generate' sentences in the sense that the grammaticality of both existing and potential sentences can be tested against the rules. The term in a stronger sense is derived from one of its uses in mathematics and was introduced into grammatical theory by Chomsky in *Syntactic Structures* (1957):

> 1968 J. LYONS This second, more or less mathematical, sense of the term 'generate' presupposes, for its applicability to grammar, a rigorous and precise specification of

the nature of the grammatical rules and their manner of operation; it presupposes the *formalisation* of grammatical theory.

• **generation**.

generative

Able to generate grammatical utterances.

• **generativist**: a person who employs the methods of generative grammar.

generative grammar

A set of rules capable of producing an infinite number of grammatical (and only grammatical) sentences of a language.

This theory of grammar was first introduced by the American linguist Noam Chomsky in *Syntactic Structures* (1957). It has been developed and changed by Chomsky and others in diverging ways. It has also been challenged, but its influence on linguistic thought has been, and remains, considerable.

The term is sometimes used as a synonym of *transformational* (or *transformational-generative*) *grammar*, but generative grammar does not necessarily contain transformational rules.

In later writings, Chomsky redefined generative grammar as a 'topic', concerned with how the human mind or brain acquires language.

> 1986 N. CHOMSKY The concerns of traditional and generative grammar are, in a certain sense, complementary: a good traditional or pedagogical grammar provides a full list of exceptions (irregular verbs, etc.), paradigms and examples of regular constructions, and observations at various levels of detail and generality about the form and meaning of expressions . . . Generative grammar, in contrast, is concerned primarily with the intelligence of the reader, the principles and procedures brought to bear to attain full knowledge of a language.

See CHOMSKY, GOVERNMENT-BINDING THEORY, STANDARD THEORY.

generative phonology

A theory about the sound system of language, developed as a major part of generative grammar.

Instead of treating phonetics as a separate, almost independent, layer of language, generative phonology seeks to show, for example, that stress patterns depend on a knowledge of syntax, and at word level to explain relationships difficult to account for in a strictly phoneme-based analysis.

> 1968 N. CHOMSKY & M. HALLE We have seen . . how a variety of stress contours and a complex interplay of stress level and vowel reduction are determined by a small number of transformational rules that apply in a cyclical manner, beginning with the smallest constituents of the surface structure and proceding systematically to larger and larger constituents . . . We turn our attention to the phonological rules that do not reapply in this cyclical fashion. Among these, the ones that concern us most directly are the rules of word phonology.

Generative phonology at word level provides underlying rules that account for regular correspondences between such apparently different phonemes as:

/k/ and /s/ (e.g. *electric/electricity*)

or

/eɪ/ and /æ/ (e.g. *mania/manic, inane/inanity*)

generative semantics

A theory of grammar developed as an alternative to the standard transformational-generative model.

Despite its name, it is not a theory of semantics, but a grammatical model in which semantics have a generative role.

1991 J. LYONS The term 'generative semantics' refers to an alternative version of transformational-generative grammar—one which differs from the standard version of *Aspects* in that the rules of the semantic component are said to be 'generative', rather than 'interpretive'.

generic

Relating or referring to a whole class; in contrast to SPECIFIC.

(*a*) In describing the uses of the articles, a useful distinction is drawn between *generic* and *specific* which cuts across the distinction between *definite* and *indefinite*. In English, *the* + a singular count noun can have definite but generic meaning, as in

The dodo is extinct
Who invented the wheel?

The + certain adjectives is also used with generic (and plural) meaning

The poor are always with us
I don't understand the Chinese

Some kinds of indefinite meaning are often labelled generic, as in

Unexploded bombs are dangerous (plural count)
An unexploded bomb is dangerous (singular count)
Danger lurks everywhere! (uncount)

But see CLASSIFYING.

(*b*) Some personal pronouns are used with the generic meaning of 'people in general' or 'mankind':

One never can tell
Man seems to think *he* rules the planet
You can lead a horse to water
We still have many diseases to conquer

genitive

(*n. & adj.*) (Designating) the case of nouns and pronouns that indicates possession or close association. (Also called *Saxon genitive*.)

(*a*) The genitive is marked in nouns by the addition of *'s* to regular singular nouns and to plurals that lack *s* (e.g. *the boy's mother*, *the children's mother*). An apostrophe (') only (which of course has no spoken realization) is added to regular plurals (e.g. *the boys' mothers*).

The genitive thus contrasts with the COMMON case. Some grammarians, however, think that, although it is a relic of the earlier English case system, it should no longer be described in terms of case in modern English. On this view, the English genitive is not comparable to the inflected genitive case of Latin, but is a type of ENCLITIC, as evidenced by the fact that the *'s* ending can be added to a whole noun phrase, as in *the Prince of Wales's speech*, *the people in the flat below's radio*, *someone else's problem*. This phenomenon is sometimes called the *group genitive* or *group possessive*.

Possession is the most central meaning of the genitive form, but it may also indicate *descriptive* and *classifying* meanings, e.g. *a women's college*, *a stone's throw*, *a week's wages*, *London's West End*.

Meaning, rather than surface syntax, is distinguished by the terms *subjective genitive* and *objective genitive*. Thus the genitive has subjective meaning in

the enemy's plans (compare *The enemy* planned to attack)

The genitive has objective meaning in

the enemy's defeat (compare They defeated/will defeat *the enemy*)

The genitive seen in *See you at David's* or *I got it at the chemist's* is sometimes called the *local genitive*.

(*b*) The distinct genitive forms of the personal pronouns and determiners are called POSSESSIVE.

(*c*) When a possessive or similar relationship can be expressed by a postmodifying *of*-phrase (e.g. *the mother of the boys*), the term *of-genitive* can be used.

See also DOUBLE GENITIVE, INDEPENDENT *genitive*, PARTITIVE *genitive*.

• **genitival**: belonging to the genitive case (*rare*).

Germanic

(*n. & adj.*) (Designating) a family of related Indo-European languages that includes Danish, Dutch, English, German, Icelandic, Norwegian, and Swedish, or their presumed ancestor (*Common Germanic*).

gerund

The *-ing* form of the verb when used in a partly nounlike way, as in *No Smoking* (in contrast to the same form used as a PARTICIPLE, e.g. *Everyone was smoking*). (Sometimes called *verbal noun*.)

Both the term *gerund*, from Latin grammar, and the term *verbal noun* are out of favour among some modern grammarians, because the nounlike and verblike uses of the *-ing* form exist on a cline. For example, in *My smoking twenty cigarettes a day annoys them*, *smoking* is nounlike in having a determiner (*my*) and in being the head of a phrase (*my smoking twenty cigarettes a day*), which is the subject of the sentence; but it is verblike in taking an object and adverbial (*twenty cigarettes a day*), and it retains verbal meaning.

gerundive

If a word in *-ing* derived from a verb can inflect for plural and lacks verbal force, it is normally considered to be a noun and excluded from the class of gerund (e.g. *these delightful drawings*). Similarly, *-ing* words followed by an *of*-phrase are classified as nouns (e.g. *My drawing of the village*).

See *-ING* FORM.

gerundive

Relating to the gerund.

In Latin grammar, a *gerundive* (n.) is a form of the verb functioning as an adjective, and meaning 'that should or must be done'. There is no grammatical equivalent in English and the term is rarely used. Where it is, it seems to be synonymous with *gerund*.

> 1976 R. HUDDLESTON Complement clauses are subclassified according to the form of their V.Gp. All examples so far have been finite; others are infinitival ([50]) or 'gerundive' ([51]).
>
> [50] Everyone expects Jill to say something
> [51] I hate it raining so much.

***get*-passive**

A passive construction formed with the verb *get* instead of *be*.

The *get*-passive is sometimes considered informal, and its use may, for this reason, be discouraged. It can however be a useful way of making it clear that an action or event is meant, rather than a state. Contrast

> They got married
> The chair got broken

with the ambiguous

> They were married
> The chair was broken

As *get* has this more dynamic meaning, the *get*-passive is used for actions we do to ourselves (e.g. *get dressed*) and even when the action is done by someone else, a *get*-passive can imply that the referent of the subject was in some way responsible for what happened, or at any rate that there was a cause. Compare

> He got picked up by the police
> She got involved in an argument
> *The car got found abandoned

with

> He was picked up by the police
> She was involved in an argument
> The car was found abandoned

given

Semantics. (*n. & adj.*) (Designating) the already known, and therefore less important, information in an utterance; contrasted with the NEW.

Given and *new* are used in some analyses of the information structure of a sentence. The given information is information already supplied in the context. In speech, it usually receives little stress, while the important, new part of the message receives full stress.

Given and *new* may correlate with the syntactic distinctions SUBJECT and PREDICATE, as in *I'm at the end of my tether*, where *I* is the given in the situation, and is also the grammatical subject, and *(a)m at the end of my tether* is new and also the predicate. However, the concepts in these pairs are distinguishable. For example, if we are already talking about Tom, then in *Tom's at the end of his tether*, *Tom* would indeed be given. But if this were an answer to *Who did you say was at the end of his tether?* then *Tom* would be new and the rest given.

The terms *given* and *new* are often used interchangeably with THEME and RHEME or TOPIC and COMMENT, but distinctions are sometimes made whereby *given* and *new* are the terms for information structure and the others refer to a slightly different 'thematic' structure.

> 1985 M. A. K. HALLIDAY The Theme is what I, the speaker, choose to take as my point of departure. The Given is what you, the listener, already know about or have accessible to you. Theme + Rheme is speaker-oriented, while Given + New is listener-oriented.

glide

Phonetics. A gradually changing speech sound made in passing from one position of the speech organs to another.

(*a*) In some analyses of English the lateral /l/, the typical English /r/ (a frictionless continuant), and the semi-vowels /j/ and /w/ are labelled *glides.* Some analysts include /h/ and the glottal stop /ʔ/ as well. In some other descriptions, glides are called *gliding consonants.* Confusingly, /w/ and /j/ (often termed *semi-vowels*) are sometimes described as *vowel glides.* But note that some descriptions do not make use of the term *glide* at all.

(*b*) In the articulation of plosive sounds, there is sometimes an audible *on-glide* from the preceding sound, while after the final release stage there may be an *off-glide* to the following sound. These glides produce characteristic patterns when analysed acoustically.

Compare CONTINUANT, LIQUID.

gliding vowel

Phonetics. Another term for DIPHTHONG.

glossematics

Linguistics. A variety of structural linguistics introduced by the Danish scholar Louis Hjelmslev (1899–1965) in the 1930s and 1940s, concerned especially with developing an abstract theory of the distribution of minimal forms (*glossemes*) and their mutual relationships.

glosseme

Linguistics. In Glossematics, any feature in a language (e.g. of form, stress, order, etc.) that carries meaning and cannot be analysed into smaller meaningful units.

The term has not acquired general currency.

glottal

Phonetics. Of or produced by the glottis.

The /h/ sound of English is made in the glottis and is commonly classified as a *voiceless glottal fricative*. Some speakers use a voiced or partly voiced variant of this sound when it occurs between voiced sounds, e.g. in words such as *perhaps*, *ahoy*, *ahead*. Whispered speech is also produced with a considerably narrowed glottis.

•• **glottal reinforcement, glottal replacement**: see GLOTTAL STOP.

• **glottalic**: (used especially of the initiation of a sound) by means of the glottis. **glottalize**: articulate with closure of the glottis; pronounce with glottal reinforcement (or glottal replacement). **glottalization. glottalling**: see GLOTTAL STOP.

> 1982 J. C. WELLS Word-internally, a variety of realizations are again found for intervocalic /p, t, k/. If not glottalized, they are in Cockney usually aspirated.

glottal stop

Phonetics. The voiceless plosive sound made when air is released after a complete closure (with vocal cords held together) of the glottis.

(*a*) This sound, transcribed /ʔ/, is not a phoneme of English, but nevertheless may—exceptionally—be heard in standard RP.

(i) emphasizing a vowel at the beginning of a syllable, as in *You've got to act* [. . tə 'ʔækt]. Sometimes called HARD ATTACK.

(ii) optionally used to avoid a LINKING *R*, as in *overenthusiastic* [ˌəʊvəʔɪnθjuːzɪ'æstɪk]

(iii) reinforcing the voiceless plosives /p/, /t/, and /k/ at the end of syllables or words, especially when followed by a consonant, as in 'shor*t*, shar*p* sho*ck* treatment' [ʃɔːʔt ʃɑːʔp ʃɒʔk 'triːtmənt]. This is called *glottal reinforcement* and is sometimes also found with /tʃ/ (e.g. *butcher* ['bʊʔtʃə]) and /tr/ (e.g. *buttress* ['bʌʔtrɪs]).

(*b*) In some accents of English, the glottal stop is heard as an allophonic variant of /t/ at the end of syllables. It is a common feature of Cockney, where *better butter* might be pronounced /'beʔə 'bʌʔə/. But though the pronunciation is condemned by many speakers, it is widely heard in other accents, especially in informal RP, as in /kwaɪʔ raɪʔ/ for *quite right*. This is called *glottal replacement* or *glottalling*.

glottis

The opening between the vocal cords at the upper end of the windpipe.

goal

1 (Loosely.) The 'patient', 'recipient', or 'deep object' affected by the action of the verb.

In this usage, the term contrasts with ACTOR or AGENT in some analyses of sentences. Thus *dog* is the goal in both the following:

Man bites dog The dog was bitten by the man

The term may even include something that results from an action (*factitive* or *result* in other analyses), as in:

I've built *a path*
A path has been made

but is not normally used of mental processes.

See PHENOMENON.

2 (In later Case Grammar.) The 'target' of the verbal action, or the place to which something or someone moves.

In one model of Case Grammar, *goal* and *patient* were introduced with two contrasting semantic roles. Examples given included *I cut my foot with a rock* (*my foot* = goal) and *I cut my foot on a rock* (*my foot* = patient).

Goal, as a case, is used in its more everyday sense in relation to verbs of movement (e.g. *We reached harbour*) and here contrasts with SOURCE.

3 Used in semantic analyses of directional prepositions and adverbials.

Thus informally in *I always walk to the station* or *He threw the book at me*, *to* and *at* and the whole prepositional phrase can be related to *goal*.

In a more formal COMPONENTIAL analysis of spatial prepositions, *goal* may be one of the basic components of meaning:

1975 D. C. BENNETT Into: 'goal locative interior' . . off: 'goal locative *off*' . . onto: 'goal locative surface' and so on.

God's truth

(Designating) an extreme view of grammar which assumes that the 'rules' of grammar have an objective existence in the language, and that all good grammarians will therefore discover the same facts and propound the same description.

Invented by Fred W. Householder (1952). Compare HOCUS-POCUS.

govern

(In traditional grammar.) Especially of a verb or preposition: have (a noun or a pronoun and, where relevant, the case of the pronoun) depending on it.

The term is used in Latin and other inflected languages of verbs and prepositions which require the use of a particular case in a dependent noun or pronoun. Thus in *ab initio* 'from the beginning' *ab* requires the ablative case (realized by the ending *-o* in this noun), whereas in *ad infinitum*, literally 'to infinity' *ad* requires the accusative (realized by *-um*).

The term has sometimes been used to refer to similar syntactic connections in English grammar: for example, in *Jack built this house*, the verb *built* may be said to govern *house*. There seems little purpose in this, however. It is marginally more validly used in relation to verbs that require ('govern') a particular preposition, as in

They *deprived* him *of* his property
They *plied* him *with* drink

government

(Chiefly in traditional grammar.) The fact or property of governing some other element. Usually contrasted with CONCORD.

government-binding theory

A recent version of Transformational-Generative Grammar, developed primarily by Chomsky himself.

The term *government* (with the verb *govern*) is taken over from traditional grammar, where, for example, a verb is said to 'govern' its object, but the concept is extended to embrace other elements of language. *Binding* is a term taken from formal logic. In Government-Binding Theory, *binding* is particularly concerned with the relationship of pronouns to their (grammatical) antecedents and directly to their referents (in the outside world). To give a simple example: in both

When Tom arrived, he unpacked the case

and

When he arrived, Tom unpacked the case

Tom and *he* may be one and the same person, with the pronoun 'bound' to *Tom*. By contrast, in

He arrived and Tom unpacked the case

two people are involved and the pronoun is not bound.

gradable

Capable of being ranked on a scale.

The term is used in describing sense relationships between words, and is particularly applied to adjectives and adverbs.

Gradable adjectives and adverbs can take degrees of comparison

better, most northerly, soonest

and can be intensified:

very difficult, too quickly

They contrast with NON-GRADABLE or UNGRADABLE words, which normally cannot:

*more supreme, *very impossible, ?less unique, *most north, ?too occasionally, *less perfectly

Some determiners and pronouns are also gradable, e.g.

many/much: more, most
few: fewer, fewest
little: less, least

Compare ANTONYM.

• **gradability**.

gradation

The same as ABLAUT.

gradience

The quality of indeterminacy on a graduated scale connecting two linguistic elements.

Grammatical categories are not always clear-cut. Word classes, for example, have fuzzy boundaries. At one end of a scale are words that meet all the criteria for membership of a particular class. For example, some adjectives can be attributive and predicative, have comparative and superlative forms, and so on. (Confusingly, these are often called CENTRAL adjectives). Others are less adjective-like (e.g. *mere*), or even share characteristics with another class such as adverb (e.g. *afloat*). Again, the word *near*, even when used syntactically as a preposition, can be compared (e.g. *Stand nearer the table*). It is therefore said to show *gradience* between preposition and adjective or adverb. Similarly, while the conjunctions *and* and *or* are coordinators, and *if* and *because* are subordinators, other conjunctions have only some of their characteristics: *but*, unlike *and* and *or*, cannot link more than two clauses; *for* and *so that*, unlike true subordinators, cannot be preceded by another conjunction.

Gradience is also found in semantics (e.g. *cup . . . mug*, *blue . . . green*); and in phonetics (e.g. *voiced* versus *partly devoiced* versus *voiceless*).

Compare CLINE.

grammar

1 The entire system of a language, including its syntax, morphology, semantics, and phonology.

2 Popularly, the structural rules of a language, including those relating to syntax and possibly morphology, but excluding vocabulary (the semantic system) and phonology.

3 A book containing rules and examples of grammar (particularly in sense 2).

4 An individual's application of the rules, as in *This novel is full of bad grammar*.

Several adjectives are attached to the word *grammar*.

Traditional grammar can cover many periods. It is often used (as in this book) to mean the grammar of the eighteenth, nineteenth, and early twentieth centuries, which was often based on, or in the tradition of, Latin grammar. It is contrasted with the grammatical analysis that began in the nineteenth century but came to fruition in the twentieth, which depends on linguistic research into actual usage.

Pedagogical grammar (or *teaching grammar*) is variously used to mean grammar or a grammar for teachers or for learners.

Reference grammar may be restricted to meaning 'a large comprehensive work of grammar', but can be extended to include smaller books intended solely for reference purposes.

Theoretical grammar is concerned with language in general, rather than with an individual language.

Compare COMPARATIVE (2), DESCRIPTIVE (1), GENERALIZED PHRASE-STRUCTURE GRAMMAR, GENERATIVE GRAMMAR, GOVERNMENT-BINDING THEORY, PHRASE-*structure grammar*, PRESCRIPTIVE, RELATIONAL *grammar*, STRATIFICATIONAL GRAMMAR, STRUCTURALISM, SYSTEMIC GRAMMAR, TRANSFORMATIONAL GRAMMAR, WORD GRAMMAR.

grammatical

1 Relating to grammar, determined by grammar.

In this sense, *grammatical* is a formal term, relating to form rather than meaning, as in *grammatical category*, *grammatical collocation*, or *grammatical hierarchy*. *Grammatical concord* contrasts with NOTIONAL *concord*; *grammatical gender* with *natural* GENDER; *grammatical* SUBJECT with *logical* or *psychological subject*; *grammatical* (or *grammar*) *words* (also called FORM *words*) with CONTENT *words*; *grammatical* MEANING with *lexical meaning*, and *grammatical* MORPHEME with *lexical morpheme*. *Grammatical competence* means an ability to manipulate the syntactic rules and contrasts with *communicative competence*.

2 Conforming to the rules, particularly the syntactic rules; in contrast to *ungrammatical*.

(*a*) Popularly, sentences and other utterances are grammatical if they obey the rules of the standard language or, more narrowly, the PRESCRIPTIVE rules of usage books, and ungrammatical if they do not. Hence *I never said nothing to nobody* or *He were right angry*, though acceptable in some dialects, might be judged ungrammatical by some speakers of the standard dialect. But then so might *It's me*, or *It is essential that he checks in at Terminal 2*, which others would find acceptable.

(*b*) *Grammatical* is not synonymous with *meaningful*; a sentence may be grammatical even though it is nonsensical (e.g. *'Twas brillig, and the slithy toves Did gyre and gimble in the wabe*). Conversely, a sentence may be meaningful but ungrammatical (e.g. *Broked he the window?*). A sentence may also be grammatical despite being 'unacceptable' because it is (say) too long to be comprehensible. *Grammatical* in this sense does not entail any of the social value judgements of (*a*).

• **grammaticality**. **grammatically**.

1984 F. R. PALMER There is surely a sense in which knowing the grammar of a language means that you can speak it grammatically.

Compare ACCEPTABILITY, WELL-FORMED.

grammatical word

The same as FORM WORD.

graph

Linguistics. The smallest discrete unit of writing, especially a letter.

The term also includes marks of punctuation.

The term overlaps with ALLOGRAPH, which does not however include punctuation marks.

grapheme

Linguistics. The smallest meaningful contrastive unit in the writing system.

The grapheme is to writing what the PHONEME is to speech. A grapheme is an idealized abstraction which can be physically written or printed in various different ways. Thus the concept of the first letter of the alphabet (a, A, etc.) is a *grapheme*: each of the actual ways in which the letter may be written or printed is an ALLOGRAPH (sense 1).

• **graphemic**: of or pertaining to graphemes. **graphemically**.

graphemics

The study and analysis of graphemes.

graphology

Linguistics. The (study of the) writing system of a language.

The term is formed by analogy with PHONOLOGY. Graphology is concerned with graphemes, the smallest units of the writing system. It has nothing to do with the better-known meaning, the study of an individual's handwriting as a guide to character.

Great Vowel Shift

(In historical linguistics.) A series of phonological changes beginning around 1400 and affecting all dialects of English to some degree, by which long vowels were generally raised or diphthongized.

greengrocer's apostrophe

The use of an apostrophe in an ordinary plural, where it is incorrect.

e.g. *potato's 15p*.

Grimm's law See SOUND LAW.

group

1 A level of structure between clause and word.

This term is particularly used by Halliday and his followers. The main groups functioning as elements in clause structure are NOMINAL GROUPS, VERBAL *groups*, and ADVERBIAL *groups* (NG, VG, AG). In

The children were sitting on the grass

The children = NG; *were sitting* = VG; *on the grass* = AG.

The term *group* often, as here, corresponds to PHRASE in other analyses. It does not always do so, however, and Halliday himself uses both *preposition group* (roughly the COMPLEX *preposition* of other analyses) and PREPOSITIONAL PHRASE.

1985 M. A. K. HALLIDAY A phrase is different from a group in that, whereas a group is an expansion of a word, a phrase is a contraction of a clause.

See also WORD GROUP.

2 **group genitive, group possessive**: see GENITIVE.

3 *Phonetics*. See BREATH-GROUP, TONE GROUP.

group noun

A noun referring to a group of people, animals, or things, with particular grammatical characteristics.

This is usually synonymous with COLLECTIVE NOUN, as in *The committee has/have acted properly*. However, distinctions are sometimes made. In the *Oxford Advanced Learner's Dictionary of Current English* (1989) such nouns as *committee* are called *countable group nouns*; *group noun* is used for a word (often a proper name) similarly used with either a singular or a plural verb, but found only in the singular, as in *Whitehall believes/believe* . . .

H

habitual

Of a tense: denoting action that occurs regularly or repeatedly.

The term is often applied to one meaning of the simple present and the simple past tenses. Thus *I always catch the 8.15 train* or *He drives fast cars* are examples of the habitual present, in contrast to STATIVE meaning (e.g. *The house belongs to his son*) or the INSTANTANEOUS present (e.g. *I apologize*).

The habitual past (e.g. *He caught the same train for thirty years*) contrasts with the past used for a single event (e.g. *I caught the train just as it began to move*) or a past state (e.g. *The house belonged to her parents*).

half-close

Phonetics. Of a vowel: articulated in the second highest of the four levels of tongue position posited in the cardinal vowel system; between CLOSE and HALF-OPEN.

In RP the front vowel /ɪ/, as in *sit, symbol, pretty, build, women*, is slightly higher than half-close, as also is the vowel /ʊ/ heard in *put, woman, good*, and *could.*

The front vowel /e/, the vowel of *bed, head, many, friend*, and *bury*, lies somewhere between half-close and half-open, as also (in RP) does the back vowel /ɔː/ of *horse, saw, ought, all, door*.

half-open

Phonetics. Of a vowel: articulated with the tongue above the open (low) position, but lower than half-close according to the cardinal vowel system.

The English central vowel /ʌ/ of *son, sun, country, blood*, and *does* is articulated somewhere near a half-open position. The front vowel /æ/, as in *cat, plait*, lies somewhere between half-open and fully open in RP.

Hallidayan

Designating the type of grammar developed by the British linguist Michael A. K. Halliday (b. 1925).

Halliday's first grammatical model, developed in the 1960s, was SCALE-AND-CATEGORY GRAMMAR, in which three scales (RANK, EXPONENCE, and DELICACY) interacted with four categories (UNIT, CLASS, STRUCTURE, and SYSTEM). As the theory developed, increasing importance was attached to the category of system—and the term SYSTEMIC GRAMMAR (or *Systemic-Functional Grammar*) was used for later models.

hanging participle

A participle that is not related grammatically to an intended noun phrase of which it would be the modifier; also called *dangling*, *misrelated*, *unattached* or *unrelated participle*, or *dangling modifier*.

A participle clause usually contains no subject, but grammatically, if it is placed near the subject of the main clause, it is 'understood' to refer to this. Failure to observe this 'rule' results in a hanging participle, or often, more accurately, a *misrelated participle*. In this, the participle is grammatically attached to the subject, though according to the intended meaning, it has a different referent (which may not actually be mentioned in the main clause). For example:

> Speaking to her on the phone the other day, her praise for her colleagues was unstinting (*Daily Telegraph* 9 June 1989)

The meaning may be clear enough; equally clear is that neither the lady herself nor her praise for her colleagues was speaking to her on the phone.

> Shrouded by leaves in summer, the coming of winter for a deciduous tree reveals the true shape of its woody skeleton (G. Durrell *The Amateur Naturalist* (1982), p. 105).

What is shrouded?

The same rule (that the participle refers to the subject) can also apply when the participle clause is introduced by a conjunction or preposition:

> When buying statuary, old or new, its impact on the garden will be strong (*Weekend Telegraph* 9 June 1990)
>
> Every afternoon, instead of dozing listlessly in their beds, or staring vacantly out of a window, there is organized entertainment (*Daily Telegraph* 13 May 1986).

The hanging participle is generally condemned as ungrammatical, rather than as a mere error of style. But it has long been widely used, most famously by Shakespeare in *Hamlet*:

> Sleeping in mine orchard, a serpent stung me.

The rule does not extend to participles that refer to the speaker's or writer's comments (e.g. *Strictly speaking*, Monday is not the first day of the week) nor to apparent participles that are accepted as prepositions or conjunctions (e.g. *following*, *provided* (*that* . . .).

Compare MISRELATED.

haplology

Phonology. The omission of a sound or sound sequence (especially a syllable) when followed by another similar sound or sequence, as when *fifth* is pronounced /fɪθ/ rather than /fɪfθ/, *library* as /ˈlaɪbərɪ/ or /ˈlaɪbrɪ/ rather than /ˈlaɪbrərɪ/, or *deteriorate* as /dɪˈtɪərɪeɪt/ rather than /dɪˈtɪərɪəˌreɪt/.

Though often criticized as careless, omission of this sort is a very general feature of language, and there are many established examples. The phenomenon is more often dealt with today under the more general concept of ELISION.

Compare SYNCOPE.

hard

(In popular usage.) Designating the letter *c* when it represents the sound /k/ and the letter *g* when it represents the sound /g/.

See SOFT.

hard attack

Phonetics. Articulating a syllable-initial vowel with a preceding GLOTTAL STOP.

This method of pronouncing a word or syllable that begins with a vowel is not usual in English, but is sometimes used for emphasis.

hard palate

Phonetics. The part of the roof of the mouth lying behind the ALVEOLAR ridge but in front of the soft palate (or VELUM).

The term is used in articulatory phonetics to classify consonant sounds. See PALATAL.

head

1 The word which is an obligatory member of certain kinds of phrase and which, standing alone, would have the same grammatical function as the whole phrase of which it is part. (Also called **headword**.)

In the noun phrase *the ankle-deep propaganda which I waded through*, the head is the noun *propaganda*. Similarly, in the adjectival phrase *very misleading indeed* and the adverb phrase *somewhat superficially* the heads are *misleading* and *superficially*.

In Transformational Grammar the verb phrase and prepositional phrase are also said to have *heads*.

2 *Phonetics.* The part of a TONE UNIT that begins with the first stressed syllable before the NUCLEUS and ends with the last syllable before it.

In

'How ˛awful!

the head is the single syllable *How*; in

I 'thought it was ˛awful

the head is *thought it was*.

If there is no stressed syllable before the nucleus, there is of course no head.

See PREHEAD.

headed

The same as ENDOCENTRIC.

Headed or endocentric phrases contrast with *non-headed*, *unheaded*, or EXOCENTRIC phrases. Compare:

I was *very anxious* (headed adjectival phrase)

with

I was *in two minds about it* (non-headed prepositional phrase)

But this contrast would not be drawn in Transformational Grammar (see HEAD (1)).

headlinese

The grammar of newspaper headlines.

Newspaper headlines often employ grammatical conventions that differ from the norm. Articles and other minor words are often omitted (e.g. *Man hit by gunman 'critical'*); present tenses are used for past events (e.g. *Designer weeps in court*); a *to*-infinitive stands for a future tense (e.g. *UN to search for solution*); nouns are heavily stacked in noun phrases (e.g. *BA launches cut-price tickets sale*).

Compare BLOCK LANGUAGE.

headword

1 The same as HEAD (1).

2 The word which stands at the head of a dictionary entry and which is defined within that entry.

height

Phonetics. The degree of elevation of the tongue towards the roof of the mouth, as one of several features determining the articulation of vowels.

In the cardinal vowel system, the height of the tongue is described in terms of four equidistant levels. When part of the tongue is raised as near to the roof of the mouth as possible without friction (which would make the sound a consonant) it is in a HIGH (or CLOSE) position, with resulting high or close vowels; when the whole tongue is lowered, LOW (or OPEN) vowels are produced. Between these two extremes are tongue heights called HALF-CLOSE and HALF-OPEN.

When the system was devised, the four vowel heights were to some extent based on the auditory judgement of listeners, but later, X-rays showed that the heights were also equidistant in articulatory terms. Acoustic phonetic studies have since shown a correlation between vowel height and the frequency of the first FORMANT.

See HIGH (2).

helping verb

A somewhat dated term for AUXILIARY.

hesitation noise

A sound (or sounds) not classified as a word, but used by speakers to keep conversation going.

Hesitation noises are somewhat inadequately indicated by such sequences as *er*, *erm*, *uh*, *um*, etc.

heterograph See HOMOPHONE.

heteronym

1 A word having a different meaning from another word that is identical to it in spelling and possibly also pronunciation. Contrasted with SYNONYM.

Word pairs such as *bill* (statement of charges) and *bill* (beak) or *lead* (cause to go) and *lead* (metal) are usually considered together in the first place on the basis of their similarity, and are thus more likely to be labelled HOMONYMS (*bill/bill*) or HOMOGRAPHS (*lead/lead*). The term *heteronym* emphasizes difference.

2 Each of a set of morphologically different words that all have the same meaning but are each used by speakers of different dialects or in different localities. Contrasted with BY-FORM, DOUBLET.

Such words would be synonyms if they were used in the same dialect (and may become so if one dialect—for example, Standard English—adopts more than one of them). Examples are: the numerous truce terms used by schoolchildren, each of which is characteristic of different areas of England (e.g. *pax*, *fainites*, *barley*, etc.); or the varying local terms for rubber-soled canvas shoes (e.g. *plimsolls*, *gym shoes*, *sand shoes*, *daps*, *gollies*, etc.).

3 A word that has a semantic relationship (e.g. of antonymy or as a translation equivalent) with another word but is unrelated to it in form. Contrasted with PARONYM.

Examples are *husband:wife*, *debtor:creditor* (in contrast to *widow:widower*, *mortgagor:mortgagee*).

Heteronym (from Greek *heteros* 'other' + *-onym* as in *synonym* etc.) is a fairly uncommon word and it is not clear which use prevails. Sense (2) is favoured by continental dialectologists.

• **heteronymy**.

heterophone

A word having a different sound from another which is spelt the same.

Since a certain similarity is the reason for considering two words together as some sort of pair (e.g. *lead* (cause to go) and *lead* (metal) or *row* (a line) and *row* (a quarrel)), an alternative term would be HOMOGRAPH, or—more loosely—HOMONYM. The term might however be usefully used of two forms of the same lexeme, where both spelling and meaning are the same but the pronunciation is different, e.g. the verb *read* /riːd/ and its past tense and past participle, pronounced /red/.

hiatus

(Chiefly in historical linguistics.) A break between two vowels coming together in different syllables, as in:

*coo*perate, Goy*ae*sque, guff*aw*ing, r*ea*lign

hierarchy

A system of RANKS or classes in which each one includes the one below it.

The term is used in its everyday sense and can be applied to various types of grammatical classification. Hierarchy is shown in the analysis of sentences into clauses, phrases, and words or into subject and predicate, and of words into morphemes.

The concept can also be used in semantics where a series such as *furniture, chair, armchair, wing armchair* illustrate a hierarchy from the general to the particular.

• **hierarchical**. **hierarchically**.

high

Phonetics.

1 Of a vowel: produced with (part of) the tongue raised relatively close to the roof of the mouth.

The term is used in the articulatory description of vowels. Thus /iː/ as in *heat* is a high (or CLOSE) front vowel, in contrast to LOW (or OPEN) /æ/ as in *hat*.

2 (In intonation.) Of pitch: produced by relatively rapid vibrations of the vocal cords, as in a 'high level pitch'.

•• **high-fall**: a tone which starts near the highest pitch of the individual speaker's voice and glides to the lowest.

high-rise: a tone in which the voice rises from a medium to a high pitch.

Compare LOW, NARROW.

high frequency See FREQUENCY (3).

historical linguistics

The study of language change; the same as DIACHRONIC linguistics.

See COMPARATIVE LINGUISTICS.

historic present

The present simple tense (or sometimes the present progressive) when used with past reference. (Also called *narrative present*.)

The device is often used to make narrative more vivid and immediate, as in these chapter headings:

The Fleets Approach   The Armada Takes Shape
(C. Martin and G. Parker *The Spanish Armada* (1988))

hocus-pocus

An attitude to grammar that implies that grammarians must impose their own rules on language.

The term was coined by Fred W. Householder (1952) to denote a view of the linguist's work at one extreme, in contrast to the GOD'S TRUTH view of language.

holophrase

A single word used instead of a phrase, or to express a combination of ideas. See HOLOPHRASIS.

holophrasis

The expression of a whole phrase or combination of ideas by one word.

• **holophrastic**.

The concept originated in nineteenth-century philology. Today it is especially applied to the early, one-word stage of child language acquisition:

> 1968 D. BOLINGER & D. A. SEARS The first learning stage is *holophrastic*: word and utterance are one, an undivided word representing a total context. Only later are words differentiated out of this larger whole . . . These whole chunks that we learn persist as coded units even after analysis into words has partially split them up. An extreme example is *How do you do?*. That it is functionally a single piece is proved by its condensation to *Howdy*.

homograph

A word that is spelt the same (Greek *homos* 'same') as another but has a different meaning.

If such a word has only the same spelling as another word, but differs both in meaning and pronunciation, then such a word is a partial HOMONYM, and *homograph* is the more precise term. Another term, emphasizing the different pronunciation (rather than the identical spelling) is HETEROPHONE.

Examples are:

> lead (cause to go): lead (metal)
> routed (defeated): routed (sent by a route)
> row (as noun, line; as verb, propel boat): row (quarrel, noun and verb)
> slaver (slave-trader): slaver (dribbling saliva)
> slough (swamp): slough (cast off skin)
> sow (bury seed): sow (female pig)
> wound (injury): wound (past of *wind*)

If in fact a pair of homographs are also pronounced the same, the more usual term is HOMONYM.

• **homographic. homography**.

homomorph

A word that is identical to another in (written and spoken) form and shares the same meaning, but belongs to a different word class.

> 1985 R. QUIRK et al. There is no standard term for words which also share the same morphological form (eg: *red* as a noun and *red* as an adjective, *meeting* as a noun and *meeting* as a verb), but it seems appropriate to adopt the term HOMOMORPH for this purpose.

• **homomorphy**: the fact of being a homomorph or homomorphs.

Compare HOMONYM.

homonym

A word that has both the same pronunciation and the same spelling as another, but is etymologically unrelated to it.

Examples are:

> bill (statement of charges): bill (beak)
> fair (just): fair (sale, entertainment)
> pole (long slender rounded piece of wood or metal): pole (each of the two points in the celestial sphere about which the stars appear to revolve)
> pulse (throbbing): pulse (edible seeds)
> row (noun, a line): row (verb, propel boat)
> soil (earth): soil (make dirty)

Traditionally, homonyms of this type are treated as separate words and given distinct dictionary entries (e.g. 'pole 1' and 'pole 2') whereas more closely related meanings are treated as offshoots of the same word, which historically speaking they are (so 'Each of the two terminals of an electric cell or battery etc.' comes under 'pole 2'.)

Popularly, homonyms may or may not include pairs whose two words have the same meaning but do not belong to the same grammatical category (e.g. *red*, noun and adjective). See HOMOMORPH.

Loosely, *homonym* is sometimes used for a word that has either the same sound or the same spelling as another (but not both).

• **homonymic, homonymous**. **homonymy**.

Compare ANTONYM, SYNONYM.

homophone

A word that is pronounced the same as another.

The term is usually used of partial HOMONYMS, which are distinguished by both meaning and spelling. Another term, emphasizing the difference of spelling, is *heterograph.*

Examples are:

> feat: feet, no: know, none: nun, stare: stair

Some English pairs are homophones in some accents but not in others, for example:

saw: sore, pore: pour, wine: whine

If in fact the two words in a pair are both pronounced and spelt the same, the usual term is HOMONYM.

• **homophonic, homophonous. homophony.**

homophoric

Of reference: self-specifying.

Textual reference may be ANAPHORIC or CATAPHORIC. Extralinguistic reference, by contrast, is EXOPHORIC. The term *homophoric* particularly describes the kind of reference that can only be to one particular person or thing in the situation. Thus the use of *the* is homophoric in *That's the telephone!* in a normal domestic situation where there is only one telephone.

homorganic

Phonetics. Of two or more speech sounds: articulated in the same place.

English /p/, /b/, and /m/ are homorganic, all three being bilabial sounds. Similarly, /t/, /d/, and /n/ are homorganic, since they are all alveolar.

hybrid

Morphology. (*n. & adj.*) (A word) formed from words or morphemes derived from different languages.

Many affixes in common use in English word formation are ultimately of Latin or Greek origin (e.g. *a-, anti-, co-, ex-, in-, non-, post-, syn-; -al, -ation, -ic, -ist, -ive, -ize*), but they are so well established that they combine easily with words of Old English or any other origin. Examples are:

anticlockwise, disbelieve, interweave, refill
eatable, jingoism, starvation, talkative

There is a certain resistance to the mixture of affixes, so that *rational* forms *rationality* rather than ?*rationalness*, *cannibalize* forms *cannibalization* rather than **cannibalizement*, and *centralize* forms *decentralize* rather than **uncentralize*. Quite a number of hybrids, generally involving the longer combining forms, were criticized when first introduced, for example *appendicitis* (Latin *appendic-* + Greek *-itis*), *speedometer* (English *speed* + Greek *-(o)meter*), and *television* (Greek *tele-* + French/Latin *vision*).

hybrid speech

The same as FREE *indirect speech.*

hypercorrection

(An example of) the employment of a standard form in a context where it is not standard under the impression that this conforms to 'correct' (more prestigeful) usage. (Also called **hypercorrectness**.)

Typical instances of hypercorrection involve the use of pronouns:

That's a matter for John and *I* to decide
She mentioned some people *whom* she thought were cheating her

or the use of subjunctive *were* where *was* would be correct:

Even if that were true in 1950, it isn't now
We didn't know if he were old enough

or *as* when *like* would be correct:

He drinks as a fish

• **hypercorrect**: designating (use in this way of) a non-standard form.

1968 W. LABOV The lower middle class shows the sharpest shift towards *r*-pronunciation in formal styles, going even beyond the highest social group in this respect. This 'hypercorrect' behavior, or 'going one better', is quite characteristic of second-ranking groups in many communities.

hyperlect

A variety of language associated with the upper strata of society.
The term is intended to cover marked grammatical usage as well as marked accent:

1989 J. HONEY I will call any such special variety of language, associated not with the most highly educated but with those who are socially the most highly privileged, a HYPERLECT, remembering that this term can cover not just their accent (which in the case of contemporary Britain I have called 'marked RP') but may also refer to the complete range of accent, grammar, vocabulary, and idiom which constitute a social dialect.

Compare ACROLECT, BASILECT, PARALECT.

hypernym

Semantics. The superordinate term in a set of related words; contrasted with HYPONYM.
Animal is a hypernym of *tiger* and *kangaroo*.
See SUPERORDINATE.

hyperurbanism

The same as HYPERCORRECTION.

hyphen

(*n.*) The sign ⟨-⟩ used to join words semantically or syntactically (as in *sister-in-law*, *good-natured*), to indicate the division of a word at the end of a line; and to indicate a missing or implied element (as in *over*- and *underpayment*).
(*v.*) Join by a hyphen, write with a hyphen. (Also **hyphenate**.)

There is considerable variation and inconsistency in the use of hyphens in words and compounds. For example, one finds *coal field*, *coal-field*, or *coalfield*.

Basically hyphens are meant to aid comprehension. One useful convention is to separate with a hyphen vowels that could otherwise be run together in a word, as in *co-occur*, but this is by no means universally followed, even though such forms as *cooccur* are rather opaque. Another useful convention is to hyphenate words that would not normally be hyphenated in order to avoid ambiguity; a *spare room-heater* is not the same as a *spare-room heater*.

Hyphens are also useful for showing a close connection between words that might otherwise be understood as separate and equal (e.g. *a black-bearded pilot*) and are normal when a sequence of words is, unusually, used in attributive position before a noun:

> Black-bearded, lazy-voiced, dressed for the weather in a great plaid lumberjacket, the pilot roused my envy for his here-I-am, this-is-what-I-have, take-me-or-leave-me style. (J. Raban *Hunting Mr Heartbreak* (1990)).

• **hyphenated**. **hyphenation**.

hypocoristic

(*n. & adj.*) (Designating) a pet form of a word; (that is or has the nature of) a pet-name.

e.g.

> auntie, popsy, pussy

Compare DIMINUTIVE.

hyponym

Semantics. A word with a more specific meaning than, and therefore implying or able to be replaced by, another (more general or superordinate term, called the HYPERNYM).

Tiger and *kangaroo* are both hyponyms of *animal*; *knives* and *forks* of *cutlery*; *diamond* and *ruby* of *gemstone*. Words that are hyponyms of the same superordinate term are *co-hyponyms*.

• **hyponymy**.

Compare ANTONYM, SYNONYM.

hypotactic See HYPOTAXIS.

hypotaxis

The subordination of one linguistic unit to another in a relationship of inequality; contrasted with PARATAXIS.

• **hypotactic**: exhibiting, or in a relationship of, hypotaxis.

The relationship of a subordinate clause to a main (matrix) clause is hypotactic:

I'll believe it, when I see it (main + subordinate)
He wasn't too sure, he said (subordinate + main)

The term is however much wider than subordination. It extends to other types of dependency or embedding, involving not only clauses but phrases or single words. For example in *a light green shirt*, *light* is hypotactic to *green*. The shirt itself is not light. (Contrast *an expensive green shirt*, where the two adjectives are in a PARATACTIC relationship and could be overtly coordinated: *the shirt was expensive and green*.)

hypothetical condition See CONDITION.

I

icon

Linguistics. A linguistic form which has a characteristic in common with the thing it signifies.

• **iconic**: having the nature of or resembling an icon.

iconicity: the fact or quality of being an icon.

The term is derived from its more general use in semiotics, where it means a non-arbitrary sign. In languages which use hieroglyphic or ideographic writing systems some of the symbols are arguably icons.

In a language such as English, onomatopoeic words can be described as icons, though the form of such words is in fact conventional and different from that of foreign-language equivalents; for example, English *cock-a-doodle-do* contrasts with French *cocorico*, German *kikeriki*, and Dutch *kukeleku*.

It has also been argued that various sound combinations are iconic, and that they exhibit *secondary* or *weak iconicity*. For example the initial sequence *sn-* occurs in a number of words with unpleasant meanings, many of which are connected with the nose or mouth, e.g. *snap, snarl, sneer, sneeze, snicker, sniff, sniffle, snigger, snivel, snog, snore, snort, snot, snout, snub, snuff, snuffle*. Other words with disagreeable connotations are *sneak, snide, snob, snook, snoop, snooty*.

Compare ONOMATOPOEIA, PHONAESTHEME, SOUND SYMBOLISM.

ideational

Semantics. Concerned with objective meaning.

In Systemic-Functional Grammar the ideational function of a clause is the way it represents factual reality, experience, and this is contrasted both with its TEXTUAL function (how the text is organized as a message) and its INTERPERSONAL (social) function. The ideational function of a clause is said to contain an EXPERIENTIAL component, which is concerned with 'processes, participants, and circumstances', and a *logical* component, i.e. 'language as the expression of certain very general logical relations'.

Ideational is thus to some extent comparable with COGNITIVE, DESCRIPTIVE, or REFERENTIAL in other analyses.

identification See IDENTIFY.

identifier

A member of a particular class of determiners; contrasted with QUANTIFIER.

Many different classifications of determiners and related pronouns are made. The main division according to one model is into *identifiers* and *quantifiers*, with the contrast between definite and indefinite meaning cutting across both categories. Here *definite identifiers* include *the*, demonstratives (e.g. *this* etc.), *possessives* (e.g. *my* etc.), and some pronouns (e.g. *it, he, she, they*). *Indefinite identifiers* include *a*, *an*, *one*, and *a certain*.

identify

To specify exactly and uniquely.

Identify and the related words (*identified, identification,* and especially *identifying*) are used much in their everyday sense in Grammar. Thus the demonstratives (*this*, *that*, etc.) often have an identifying function, e.g. *This is my sister*.

The terms are also used in distinguishing different kinds of complements. For example, in

Henry's second wife was a beautiful young woman

or indeed

Ann Boleyn was a beautiful young woman

the complement is descriptive, but *non-identifying*. By contrast

His second wife was Ann Boleyn

is an identifying structure, in which *Ann Boleyn* is the identifier and *his second wife* is identified.

Similarly, the second term in an apposition may or may not have an identifying role. Compare

His second wife, Ann Boleyn, was beheaded (identifying apposition)
His second wife, a beautiful young woman, was beheaded (non-identifying apposition)

• **identifying relative clause**: the same as DEFINING *relative clause*.

Compare EQUATIVE. See also DISLOCATION.

ideogram

A written character symbolizing a word or phrase without indicating its pronunciation.

Ideograms are rather marginal to the English writing system, but include numerals and graphic symbols such as £ $ % & + –.

• **ideogrammatic**, **ideogrammic** (neither in frequent use).

Compare ABBREVIATION.

ideograph

The same as IDEOGRAM.

• **ideographic**. **ideography**.

idiolect

Linguistics. An individual's knowledge and command of the language; the speech habits of an individual person.

Speakers differ in their knowledge and use of the grammar and vocabulary of their language, so that in some ways everyone's idiolect is different.

The term is modelled on the word DIALECT, using the prefix *idio-* 'own, personal, distinct'.

• **idiolectal**. **idiolectally**.

idiom

1 A group of (more or less) fixed words having a meaning not deducible from those of the individual words.

e.g.

over the moon
under the weather
by the skin of one's teeth
up to one's eyes in work
for crying out loud
kick the bucket
paint the town red
throw a wobbly

give in
take up

fish out of water
had better
might as well
how goes it?

Some of these phrases allow no alteration except extremely facetiously (**over the stars*, **kick the pail*). Others allow some changes (*up to my/his/her/their*, etc. *eyes in work*).

2 A phrase that is fairly fixed (not necessarily with opaque meaning) but which shows or appears to show some grammatical irregularity.

e.g.

these sort of people
come to think of it

It is not unusual to find phrases such as *by car*, *on foot*, *in prison* (i.e. consisting of a preposition + a normally countable noun, but without an article) described as idioms, though this use of the base form where number is irrelevant is a regular feature of English (compare *bookcase*, *street guide*: not **bookscase*, **streets guide*).

In some cases there is no very clear distinction between *idiom*, COLLOCATION, and FIXED PHRASE.

The older meanings of *idiom* in English were (*a*) the form of speech peculiar to a nation or to a limited area; and (*b*) the specific character or property of a language, or the manner of expression natural or peculiar to it ('the idiom of the English tongue').

• **idiomatic**: marked by the use of idioms; peculiar to the usage of the language. **idiomaticity**.

1973 R. QUIRK & S. GREENBAUM Like phrasal verbs, prepositional verbs vary in their idiomaticity. Highly idiomatic combinations include *go into* . . , 'investigate', *come by* . . , 'obtain'.

if-clause

A subordinate clause introduced by *if*.

The term is particularly applied to clauses introduced by *if* expressing condition, but covers such clauses where the *if* is interpreted as concession etc. Loosely, the term may cover other subordinate clauses of condition (introduced by *unless*, *provided that*, *on condition that*, etc.).

ill-formed

Linguistics. Deviant.

The term is particularly used in Generative Grammar, and describes any structure that cannot be generated by the rules. It contrasts with WELL-FORMED. Compare UNGRAMMATICAL.

• **ill-formedness**.

illocution

Linguistics. An act effected by a speaker by the very fact of making an utterance, in that the stating, inquiring, requesting, commanding, or inviting itself constitutes an action; the communicative function of an utterance.

• **illocutionary**: (for *illocutionary force*, see the quotation of 1955).

The term derives from J. L. Austin's Speech-Act Theory, where it contrasts with LOCUTION (the act of making a referentially meaningful utterance) and PERLOCUTION (concerned with an addressee's response to the speaker's illocution). It is thus central to discussions of the social and interpersonal meaning of language behaviour, and has gained wider currency than the other two terms.

1955 J. L. AUSTIN I shall refer to the doctrine of the different types of function of language . . as the doctrine of 'illocutionary forces'.

1973 *Times Lit. Suppl.* The illocutionary act was the act performed by a speaker *in* saying something, such as the act of asking or answering a question.

An *illocutionary* PERFORMATIVE is a special type of illocution.

immediate constituent

One of the parts into which a linguistic unit is immediately divisible, by a process of *immediate constituent analysis*. Sometimes abbreviated *IC*.

The immediate constituents of a compound or complex sentence might be clauses, and each clause in turn might have noun phrase and verb phrase, or alternatively, subject and predicate, as constituents. In the sentence

The cost includes air travel by scheduled services

we might say that the immediate constituents are:

[The cost] [includes travel by scheduled air services]

and we might further break the predicate down into

[includes] [travel] [by scheduled air services]

What we would not do is say that there are any constituents such as **travel by* or **scheduled air*.

Immediate constituent analysis was an important practice in STRUCTURAL linguistics. Such analysis can continue until the *ultimate* constituents are reached; for example, the ultimate constituents of *services* are *service* + *-s*.

imperative

(*n. & adj.*) (A form or structure) that expresses a command, specifically

(*a*) The base form of the verb when used to express a request, command, order, exhortation, etc.; the *imperative mood*, e.g.

Listen!
Have fun!
Be sensible!

An imperative verb phrase can include *do* for emphasis (e.g. *Do listen!*, *Do be sensible*) or *don't* for a negative (*Don't forget*, *Don't be silly*). The subject (*you* 'understood') is usually omitted but may be expressed (*You listen to me*, *Don't you do that again*). As a verbal category, *imperative* contrasts with INDICATIVE and SUBJUNCTIVE mood.

(*b*) A complete clause in which the main verb is in the imperative mood.

Where a distinction is made between form and function in the analysis of sentence types, *imperative* is a formal category along with DECLARATIVE, INTERROGATIVE, and EXCLAMATIVE. Its discourse function is often, but not always, that of DIRECTIVE.

(*c*) (In popular grammar, in addition to (*a*) and (*b*)), a structure with *Let's* where *you* cannot be inserted, as in *Let's go*, *Don't let's worry*, *Let's not forget that* . . . .

• **imperatival**: pertaining to the imperative mood. **imperatively**: as or in the manner of an imperative.

Compare JUSSIVE.

imperfect

(*n. & adj.*) (In traditional grammar.) (A tense) denoting an action in progress, but not complete.

This term was derived from the classification of tenses in Latin grammar and was applied particularly in English to the *past imperfect* (e.g. *They were waiting*). It is now largely superseded by PROGRESSIVE or CONTINUOUS. But see IMPERFECTIVE.

imperfective

(*n. & adj.*) The same as PROGRESSIVE (1).

The term is used by some linguists, rather than *imperfect* (which has connotations of tense) to reinforce the concept of ASPECT as distinct from tense. Thus the *-ing* participle is said to be imperfective and contrasts in meaning with the *to*-infinitive (which is said to be PERFECTIVE). Compare PERFECT.

See IRREALIS.

impersonal

Used with or containing a formal subject, usually *it*.

The term is particularly applied to verbs such as *rain* or *snow* which express an action not attributable to any identifiable subject (e.g. *It is snowing*). More widely, it is applied to other structures which have *it* as subject:

> It is considered bad manners to eat peas with a knife
> It suddenly occurred to me that I'd forgotten to shut the windows
> It appears that nobody knew

Compare ANIMATE, GENERIC, NON-PERSONAL.

•• **impersonal *it***: *it* used with an impersonal verb or in other vague ways.

It is usually classified as a PERSONAL pronoun for syntactic reasons (e.g. it can substitute for a noun or refer to a 'thing'), but when used with an impersonal verb or in other vague ways *it* is variously labelled: in some traditional grammars it is called *formal it* or *unspecified it*.

Compare INTRODUCTORY *IT*.

implication

Semantics. What is implied by an utterance; the act of implying.

Linguists contrast *implication* with ENTAILMENT. Entailment is semantic: if the proposition in a sentence is true then various other propositions are also true. For example *My next-door neighbour brought me a birthday present yesterday* entails that I have a next-door neighbour. Implication is a less clear-cut relationship, based on a mixture of logic and shared knowledge. In normal circumstances an utterance will be understood to include other information not directly encoded in the words: in the above example there is an *implication* that yesterday was my birthday, but I could add *I don't know where she got the idea it was my birthday*.

implicature

The pragmatic implications of an utterance, possibly not mentioned in the words at all.

The term is taken from the philosopher H. P. Grice (1913–88), who developed the theory of the cooperative principle. On the basis that speaker and listener are cooperating and aiming to be relevant, a speaker may well imply something that he or she does not actually even refer to, confident that the listener will understand. Thus the *conversational implicature* of *Are you watching this programme?* might well be *This programme bores me. Can we turn the television off?* Compare PRESUPPOSITION.

improper diphthong See DIPHTHONG (2).

inanimate See ANIMATE.

inclusion

Semantics. The same as HYPONYMY.

inclusive

1 Designating a first person plural pronoun when it includes both speaker and addressee(s). Contrasted with EXCLUSIVE (1).

e.g.

Why don't we all go together? You'd enjoy it.

2 **inclusive disjunction**: see DISJUNCTIVE.

incomplete predication See VERB.

incomplete sentence

(In traditional grammar.) A sentence lacking elements that would normally be considered essential; an ELLIPTED sentence.

1968 J. LYONS 'Incomplete' or 'elliptical' sentences . . . One must distinguish between contextual completeness and grammatical completeness.

Thus an ellipted sentence may be incomplete in the sense that a word or words have been omitted, but may in fact be grammatically complete since the missing words can be readily supplied with no further context, e.g.

(Is there) *Anything I can do to help?*

On the other hand, an utterance such as *Whatever I can* is more seriously incomplete. It could be regarded as elliptical if it followed *What are you going to do to help?* but would be merely ungrammatical if it followed *Are you going to help?*

Sentences may be regarded as incomplete for a variety of reasons. For sentences that are unfinished by the speaker, see ANACOLUTHON. For sentences that are grammatically incomplete, but complete in context, see ELLIPSIS. For sentences that are not full sentences according to the rules of sentence structure, see MINOR *sentence type*.

indefinite

Of a word, tense, phrase, etc.: not having or indicating any particular reference. Contrasted with DEFINITE.

The word is particularly used in the terms INDEFINITE ARTICLE and INDEFINITE PRONOUN, but can also be used to describe certain other linguistic elements.

(*a*) Adverbs and adverbials of *indefinite frequency*, such as

generally, usually, always, repeatedly, occasionally, etc.

contrast with adverbs and adverbials of *definite frequency*

daily, twice a day, etc.

(*b*) In the use of tenses, a simple present perfect often refers to an indefinite time:

Have you ever read *Beowulf?*

in contrast to a simple past which characteristically implies a definite, even if unstated, time

(When) did you read *Beowulf?*

(*c*) An *indefinite noun phrase* has an indefinite pronoun as head or contains an indefinite determiner + noun:

Anyone with any sense would have realized
Some people have all the luck

(*d*) Most quantifiers are indefinite (e.g. *some*, *much*, etc.) but can be described as *indefinite quantifiers* by contrast with cardinal numbers which state definite quantities.

indefinite article

The common grammatical name for the determiner *a/an*.
See ARTICLE.

Although *a* and *an* are indefinite in meaning, reference may be general (any person or thing of that class or kind) or more particular (an actual example of the class or kind):

We don't expect letters—but send us a postcard (*general*: send us any postcard)
I've still got a card my grandfather sent from Kabul (*specific*: a particular card and no other)

Some grammarians distinguish these meanings by labelling the first meaning CLASSIFYING or GENERIC and the second SPECIFIC.

But many popular models ignore the distinction, often using terms like *general* or *specific* quite differently. One model includes the indefinite article, whatever its meaning, among *general determiners*, which are said to be 'general', 'indefinite' and 'without identifying meaning' and are contrasted with *specific determiners*, which include *the* but not *a*, *an*. Another model places *a*, *an* in a class of (*indefinite*) *identifiers* but treats *the* as a *definite identifier*.

indefinite partitive See PARTITIVE.

indefinite pronoun

A pronoun lacking the definiteness of reference inherent in personal, reflexive, possessive, and demonstrative pronouns.

Indefinite pronouns include compound pronouns (e.g. *everybody, something*, etc.) and *of*-pronouns of quantity (e.g. *all, either*, etc.).

The pronoun *one* is often indefinite in reference, as in:

one of these days
I haven't a car—I'd like one
One should drink in moderation

Corresponding determiners (*every*, *some*, *all*, etc.) have similar indefinite reference.

independent

1 Of a clause: that can stand on its own, in contrast to a DEPENDENT *clause*.

Independent clause is often synonymous with MAIN CLAUSE in contrast to SUBORDINATE *clause*. However, a distinction is sometimes to be made. For example, in

The more I think about it, the more I like the idea

neither clause is independent, though both are main clauses. On the other hand, where two main clauses are coordinated, with some ellipsis, e.g.

The plan is brilliant and the whole thing [is] excellent

some would argue that both clauses are independent, while others would again argue that the second clause is not.

Compare MATRIX.

2 **independent genitive**: a genitive apparently standing alone as a noun phrase because the head has been ellipted.

For example *Tom's* in

John's results were better than Tom's

A comparable possessive pronoun (e.g. *mine, yours*, etc.) is sometimes called an *independent possessive*.

Compare *local* GENITIVE.

independent relative clause: the same as NOMINAL RELATIVE CLAUSE.

indeterminacy

The quality of not being clearly limited and defined.

Clear-cut grammatical rules and unequivocal distinctions between 'correct' and 'incorrect' usage are not always attainable, since there are areas where native speakers disagree as to what is grammatically ACCEPTABLE. Indeterminacy is therefore a feature of grammar.

Grammarians also speak of *indeterminacy* in cases where two different grammatical analyses of an utterance offer solutions to its difficulties.

• **indeterminate**.

See MULTIPLE ANALYSIS. Compare GRADIENCE.

indicative

(*n. & adj.*) (A verbal form or mood or a sentence containing a verbal form) that denotes fact; contrasted with IMPERATIVE and SUBJUNCTIVE.

Traditional grammar follows Latin and similar models in making a threefold MOOD distinction. The paucity of verb inflections, however, makes such an analysis less appropriate for Modern English than for, say, Old English.

Compare DECLARATIVE, STATEMENT.

indirect condition See CONDITION.

indirect object

In clause structure, the noun phrase that is the 'recipient' of the DIRECT OBJECT of a ditransitive verb.

In *Can you tell me the time?* and *They bought her a bicycle*, *me* and *her* are indirect objects. An indirect object normally precedes the direct object in SVOO sentences, though there can be exceptions (e.g. *Give me it/Give it me*).

In traditional grammar, any noun phrase with a DATIVE function is an indirect object, whatever its position. Typically, when placed after the direct object the indirect object requires *to* or *for* (e.g. *They bought a bicycle for her, and gave it to her*). Modern grammars vary as to whether they classify such prepositional phrases as indirect objects or not.

There is also a different analysis when only the 'recipient' object of a potentially ditransitive verb is used. For example, in *He told his parents*, *his parents* can be analysed either as indirect object (compare *He told his parents the news*) or as direct object (compare *He informed his parents*).

Compare BENEFACTIVE, RECIPIENT.

indirect question

A question as reported in indirect speech.

Indirect questions are characterized by the use of unmarked SV word order, instead of auxiliary + subject inversion. Thus

> What do you want?
> [Object + *Aux* + *S* + *V*]

becomes

> I asked him what he wanted
> [Object + *S* + *V*]

Indirect yes/no questions are introduced by *if* or *whether*.

indirect speech

A way of reporting what someone has said, using an introductory reporting verb and a subordinate clause. Contrasted with DIRECT SPEECH.

In indirect speech the actual words of the original speaker are usually changed, in that the pronouns, time and place adverbials, and tenses are adjusted to the viewpoint of the person now reporting (see BACKSHIFT). Thus the first Duke of Wellington's advice to a new member of Parliament *Don't quote Latin, say what you have to say and then sit down* would in indirect speech become *He advised the member not to quote Latin, (but) to say what he had to say and then (to) sit down.*

The term *indirect speech* is often loosely used to cover the reporting of thoughts, using an introductory verb of thinking and a *that*-clause.

In general, *indirect speech* and REPORTED SPEECH are synonymous and interchangeable, but some people make a distinction. See REPORTED SPEECH.

Compare FREE *indirect speech.*

Indo-European

(*n. & adj.*) (Designating) the family of cognate languages (including English) spoken over the greater part of Europe and extending into Asia as far as northern India, or the hypothetical common ancestor of these languages (*Proto-Indo-European*).

infinite

The same as NON-FINITE. (Now old-fashioned.)

infinitival

Of or belonging to the infinitive.

•• **infinitival clause**: see NON-FINITE CLAUSE.

infinitival particle: the *to* used before an infinitive.

The *to* of the *to*-infinitive is not a preposition, as is shown by the fact that prepositions, including *to*, cannot be followed by an infinitive but require the *-ing* form of the verb (e.g. *They resorted to violence* or *to attacking him*, not **they resorted to attack him*). Nor does this *to* share the characteristics of any other word class. It therefore has a label of its own.

infinitive

(*n. & adj.*) The unmarked BASE form of a verb when used without any direct relationship to time, person, or number.

In English the simple infinitive is often preceded by *to*

> I wanted *to help*
> *To err* is human

but it is also used alone

> Don't *apologize*
> You can't *take* it with you

See BARE INFINITIVE, *TO*-INFINITIVE.

Some other non-finite verb phrases are analysed as complex infinitives.

> *To have made* a mistake is understandable (perfect infinitive)

It was upsetting *to be questioned* (passive infinitive)
I expected you *to be waiting* for me (progressive infinitive)

Other even more complex infinitives are also possible, e.g. *(to) have been doing*, *(to) have been done*, *(to) be being done*, *(to) have been being done* (rare).

Compare SPLIT INFINITIVE.

infinitive clause See NON-FINITE CLAUSE.

infix

An affix inserted within the main base of a word.

• **infixation**.

See AFFIX and compare TMESIS.

inflect

Morphology.

1 Change the form of (a word) in order to indicate differences of tense, number, gender, case, etc.

2 Of a word: change its form in order to indicate such differences.

English is a relatively uninflected language. Lexical verbs inflect for

3rd person singular in the present simple tense (*looks*, *sees*)
the past simple (*looked*, *saw*)
the present participle (*looking*, *seeing*)
the past participle (*looked*, *seen*)

Nouns inflect for

plural (*girls*)
possessive (*girl's*, *girls'*)

Some adjectives and adverbs inflect for

the comparative (*nicer*, *hotter*, *sooner*)
the superlative (*nicest*, *hottest*, *soonest*)

Inflection does not (except in certain uses of *-ing* forms) change the word-class to which a word belongs, in contrast to DERIVATION, which usually does.

inflection

1 The action of inflecting.

2 An inflected form of a word, or a suffix or other element used in order to inflect a word.

1874 H. SWEET Old English is the period of full inflections . . Middle English of levelled inflections . . and Modern English of lost inflections.

• **inflectional**: pertaining to or characterized by inflection (*inflectional formative*, see FORMATIVE (1); *inflectional suffix*, see SUFFIX).

The spelling *inflexion* is now old-fashioned.

informal

Of spoken or written style: characterized by simpler grammatical structure and more familiar vocabulary than FORMAL style.

The attitude of speakers (or readers) to their audience is often reflected in different levels of formality, with formal and informal style as two extremes, and a wide range of stylistically less marked language in between.

Compare ATTITUDINAL, COLLOQUIAL.

information content

Linguistics. The amount of information carried by a particular linguistic unit in a particular context.

The notion of information content is related to statistical probability. If a unit is totally predictable then, according to information theory, it is informationally redundant and its information content is nil. This is actually true of the *to* particle in most contexts (e.g. *What are you going . . . do?*).

Compare REDUNDANCY.

information question

The same as *WH*-QUESTION.

information structure

The way in which the words of a clause or sentence are arranged so that a particular part of the 'message' receives greatest attention.

The term is a general one, concerned with such contrasts as GIVEN and *new*, THEME and *rheme*, *topic* and COMMENT.

information unit

A TONE UNIT (or *tone group*) seen as a unit of information structure.

In speech the words, word order, and intonation combine in the structuring of whatever 'message' is being presented. The *tone unit* is defined phonologically; the term *information unit* emphasizes its role in the formation of meaning.

-ing form

The part of the verb that ends in *-ing*.

In traditional grammar, a distinction is made between the GERUND with its nounlike functions, e.g.

Seeing is *believing*

and the present PARTICIPLE used in the formation of progressive tenses or non-finite clauses, e.g.

We're *seeing* them tomorrow

Seeing them in that condition, I was greatly upset

But modern grammar considers nounlike and verblike uses to be on a cline and often prefers the neutral and comprehensive term *-ing form*.

The term is usefully extended to include borderline uses that lie between the participle and the adjective. In *running water, rising standards*, the *-ing* forms resemble adjectives in standing in attributive position, but their meaning is definitely verblike and the words would be analysed as participles if they were in predicative position (e.g. *Standards are rising*). This type must be distinguished from, on the one hand, the fully adjectival *-ing* form (e.g. *a very interesting book*; *the book is interesting*) and, on the other, the fully verbal *-ing* form (e.g. *That man is waving*; **Who is that waving man?*).

Compare PARTICIPIAL ADJECTIVE.

ingressive

Phonetics. Of a sound: made with inward-flowing air.

See CLICK.

inherent

Of an adjective: that directly characterizes the referent of the noun it is connected with; contrasted with **non-inherent**.

The meaning of most adjectives is not seriously affected by their position in a sentence, so that *rich* means the same in attributive position (e.g. *She's a rich woman*; *that rich woman*) as it does in predicative position (e.g. *That woman is rich*). Such adjectives are *inherent*.

By contrast some adjectives used in attributive position have a non-inherent meaning: *you poor darling* does not mean **Darling, you are poor*. Other examples of non-inherent adjectives are:

an old friend, pure invention, a complete idiot, a heavy sleeper, a real genius, a generative linguist.

Compare ATTRIBUTIVE, INTENSIFYING.

initial

Designating or occurring in the position at the front or start of a linguistic unit. (*Initial position* is also called *front position*.)

(*a*) *Initial position* in a clause or sentence is an expression used in analysing ELLIPSIS and also the position of adverbials. Since end-position in clause structure is normal for adverbials, initial position is marked. Contrast *I'm leaving tomorrow* with *Tomorrow, I'm leaving*, where *Tomorrow* is marked as topic. An ordinary manner adverb may become a disjunct when placed in initial position. Contrast *He spoke frankly* with *Frankly, we didn't believe him*.

Initial (front) position contrasts with FINAL (end) position and MEDIAL (mid) position.

(*b*) *Phonology*. In phonology, word- or syllable-initial position contrasts with MEDIAL and FINAL position, since the position of a phoneme conditions its pronunciation. See ALLOPHONE.

Among English phonemes, /h/ can only be syllable- (or word-) initial. The Scottish, Irish, and General American pronunciation of *wh-* in many words is actually the sequence /hw/, as in *when* /hwen/.

Compare CONSONANT CLUSTER.

initialism

The use of the initial letters of a name or expression as an abbreviation for it, each letter being pronounced separately, as in BBC, KO, RSPCA, RSVP. A type of ABBREVIATION.

Compare ACRONYM (2).

instantaneous

Of the simple present tense: referring to a brief action beginning and ending more or less at the moment of speech; contrasted with HABITUAL, STATE, and TIMELESS.

This is a rather restricted use of the simple present. It occurs in some sports commentaries (e.g. *Lineker takes the ball . . .*), in demonstrations (e.g. *I now add the gelatine to the hot water*), and in PERFORMATIVES (e.g. *I apologize*).

Compare PUNCTUAL.

institutionalized

Established as a norm.

The term is used in a fairly general way. It can be applied to (*a*) a distinct and acceptable variety of English such as Singaporean English or Indian English; (*b*) a set expression which appears in some way to deviate from the norm (e.g. *heir apparent* rather than **apparent heir*); and (*c*) a structure that would normally be unacceptable (e.g. a hanging participle) but which is prevalent and acceptable in certain contexts.

> 1985 R. QUIRK et al. In formal scientific writing, the construction has become institutionalized where the implied subject is to be identified with the *I*, *we*, and *you* of the writer(s) or reader(s):
>
> *When treating patients with language retardation and deviation of language development*, the therapy consists, in part, of discussions of the patient's problems with parents and teachers, with subsequent language teaching carried out by them.

instrumental

1 (*n. & adj.*) (A noun, phrase, case, or semantic role) that indicates the implement or other inanimate thing used in performing the action of a verb.

Old English had a few traces of an instrumental case (besides the four other cases marked by inflection).

The word is now particularly used in classifying the meaning of adverbials. Instrumental prepositional phrases used adverbially typically begin with *with*:

They attacked the police *with bricks*

Instrumental contrasts with AGENTIVE:

The officers were attacked *by the mob* with bricks

In Case Grammar, *instrumental* is extended to cover noun phrases with a semantic function of this sort, even though syntactically they are subjects or objects:

A brick injured the woman
They used *riot shields* to protect themselves

2 *Semantics*. Describing the function of language whereby the speaker gets the listener to do something.

Thus commands and requests can be said to have an instrumental function. It is similar to CONATIVE function.

3 *Phonetics*. Designating the study of any branch of phonetics using physical equipment.

1948b J. R. FIRTH The study of . . 'features' of syllables and words as wholes, or as systems rather than sequences, promises productive results, both by instrumental and perception techniques.

• **instrumentally**.

intensifier

An adverb that scales another element upwards or downwards in degrees of intensity.

Adverbs form such a broad class of words that there are many different ways of classifying them. In one current and very detailed classification *intensifiers* are technically described as a subcategory of SUBJUNCT, along with *emphasizers*, *focusing adverbs*, and others. *Intensifiers* in this analysis are exemplified in

We *thoroughly* disapprove and are *bitterly* disappointed
You worry *a lot*
I *hardly* know them
We were *kind of* wondering

intensify

Of a word: have a heightening or lowering effect on the meaning of (another word or a phrase).

• **intensification**: the action of intensifying; the fact of being intensified.

1973 R. QUIRK & S. GREENBAUM There are various ways of giving emotive intensification to a negative. For example, *by any means* and . . *a bit* are common alternatives to *at all* as non-assertive expressions of extent.

intensifying

Having a heightening or lowering effect.

The term is applied not only to adverbs (an *intensifying adverb* is usually called an INTENSIFIER) but also to adjectives. *Intensifying adjectives* (which include EMPHASIZERS) are exemplified in

pure joy, *downright* nonsense, a *firm* commitment, *utter* rubbish, *great* hopes

Compare ATTRIBUTIVE.

intension See EXTENSION (2).

intensity

Phonetics. The amount of energy used in the production of a speech sound. Intensity is a measurable physical phenomenon. The vibrating vocal cords set off patterns of air vibration, which can be objectively measured. Intensity is related to LOUDNESS, but is not the same.

intensive

Of elements in a structure: having a semantic relationship of 'sameness'; contrasted with EXTENSIVE.

•• **intensive verb**: an alternative term for a LINKING VERB, by which subject and complement are closely related (e.g. *He seems worried*).

Intensive also refers to the relationship between an object and an object complement

You've got *me worried*

and the identity existing between the two terms of an apposition

the Queen's daughter, the Princess Royal

interjection

A minor word-class whose members are outside normal clause structure, having no syntactical connection with other words, and generally having emotive meanings.

Examples:

aha, alas, eh?, mm, oops, sh!

Several interjections involve sounds that are not among the regular speech sounds of English, e.g. those represented in writing by *tut-tut*, which is actually a sequence of alveolar clicks, or *ugh*, in which *gh* represents a voiceless velar fricative /x/ (as in the Scottish pronunciation of *loch*). In these two particular cases a secondary pronunciation based on the spelling has arisen (/tʌt'tʌt/, /ʌg/).

Compare EXCLAMATION.

internal sandhi See SANDHI.

International Phonetic Alphabet

A system of written symbols designed to enable the speech sounds of any language to be consistently represented. Abbreviated *IPA*.

The alphabet was first published by the International Phonetic Association in 1889 and, adapted in various ways, is still in very wide use today. Some of the symbols are ordinary Roman letters, having the values that English speakers would expect: for example [p] and [b] for voiceless and voiced bilabial plosives. Other familiar letters have values that are found in other languages: e.g. [j] representing the initial sound of *Jaeger*, *you*, *yes*. Other symbols have been specially invented, such as [θ] and [ð] for the voiceless and voiced pronunciations of English *th* in *think* and *this*. The symbols are based on articulatory criteria, with additional DIACRITICS to indicate such features as nasalization, stress, and double articulations.

Compare BROAD TRANSCRIPTION, NARROW TRANSCRIPTION.

interpersonal

Semantics. Concerned with verbal exchanges between people.

In Systemic-Functional Grammar interpersonal meaning is contrasted with IDEATIONAL meaning and TEXTUAL meaning. The interpersonal meaning of a clause relates to its function as a connection between speaker and listener or writer and reader. Interpersonal meaning is concerned with social exchanges, the use of language to influence behaviour, and so on.

Interpersonal meaning thus shares features with ATTITUDINAL meaning. It is sometimes called *social meaning* or SITUATIONAL *meaning*.

interrogative

(*n. & adj.*) (A word or sentence) used to ask a question.

Interrogative and QUESTION are often used interchangeably. Where a distinction is made between the form and function of a sentence, *interrogative* may be reserved for the syntactic form, in which typically there is inversion of subject and auxiliary (e.g. *Is everybody ready?*, *What do you think?*).

Interrogative sentences normally, as here, function as questions, but may also function as STATEMENTS (e.g. *Haven't we all made mistakes at some stage?*), as DIRECTIVES (e.g. *Could you please make less noise*), and as EXCLAMATIONS (e.g. *Isn't it a lovely day!*).

• • **interrogative word**: an adverb, determiner, or pronoun (typically beginning with *wh-*), when used to introduce a question.

Interrogative adverbs include

> how (the main exception to the *wh-* rule), why, when, where, and the derived compounds (e.g. *wherever*)

Interrogative determiners and *pronouns* include

> what(ever), which(ever), who(ever), whom, whose

Compare RELATIVE PRONOUNS, *WH*-WORD.

intervocalic

Phonetics. Between two vowels.

The pronunciation of a consonant when it occurs between two vowels may differ from its pronunciation in other contexts. For example the voiced plosives (/b/, /d/, /g/) will probably be fully voiced in this position, but are not always so in other contexts.

• **intervocalically**.

1964 R. H. ROBINS Cockney speakers often have glottalized stops intervocalically.

intonation

Phonetics. The pitch variations and patterns in spoken language.

The total meaning of a spoken utterance derives not only from the actual words and patterns of STRESS, but also from the PITCH patterns used (the rises and falls in pitch). Intonation is concerned with the operation of such patterns over sequences of words (TONE UNITS). It is not easy to categorize, but a number of typical intonation patterns have been identified.

Intonation plays a part in speech not unlike punctuation in the written language. For example, it may indicate whether a sentence is a statement or a question:

You don't be`lieve me (statement)
You don't be͵lieve me (question)

or whether a relative clause is a defining or non-defining one (notice how the tone units correspond with the presence or absence of commas):

My sister who lives in New͵Zealand is a`teacher.
My͵sister, who lives in New͵Zealand, is a`teacher.

Intonation also has the important function of conveying attitude.

• **intonational**: relating to intonation. **intonationally**.

Compare FALL, RISE.

intransitive

Of a verb: not taking a direct object. See TRANSITIVE.

• **intransitively**. **intransitivity**.

intrinsic modality

Another term for DEONTIC *modality*.

introductory *it*

1 Another term for DUMMY *it*, EMPTY *it*, or *prop it*.

2 Another term to describe ANTICIPATORY or PREPARATORY *it*.

Like all these terms, *introductory it* is used by different people to cover (or exclude) different uses of the word *it*.

Compare IMPERSONAL *it*.

intrusive *r*

The pronunciation of an /r/ sound between two words or syllables in sequence, where the first ends in a vowel sound and the second begins with one, and where there is no *r* in the spelling.

Intrusive *r* is much criticized, but is quite commonly heard in standard RP and other NON-RHOTIC accents. It occurs after the vowels

/ə/ (e.g. *umbrella-r-organization*)
/ɜː/ (e.g. *a milieu-r-in which . .*)
/ɑː/ (e.g. *grandpa-r-is ill*)
/ɔː/ (e.g. *law-r-and order*)

These are among the vowels which are followed by LINKING *R* in words which have a written final *r*. In non-rhotic accents, *-r* has effectively become a SANDHI feature, applied in order to avoid the hiatus between a non-high final vowel and a following vowel, irrespective of whether this final vowel was originally followed by *r*. That this is so is shown by, among other things, the fact that intrusive *r* occurs (with some speakers) internally also, in such forms as /'drɔːrɪŋ/ (*drawing*), /mə'dʒentərɪʃ/ (*magenta-ish*), /ˌkæfkə'resk/ (*Kafkaesque*), etc.

Compare LINKING *R*.

invariable

(*n. & adj.*) (A word) that does not vary in form (i.e. by inflection); contrasting with VARIABLE (I).

The term is applied to words that have a single (uninflected) form (e.g. *but, for, over, sheep, politics*) in contrast to words that inflect.

inversion

The reversal of the usual word order.

The term is particularly used in relation to subjects and verbs. The unmarked word order subject + (auxiliary) + verb, e.g.

S	(Aux)	V	
I	*am*	*listening*	
I		*have*	*a complaint*
He		*understands*	

is changed to auxiliary (or single-word primary) + subject (+ verb); and if necessary DO is added, e.g.

Aux	S	(V)	
Are	*you*	*listening?*	
Have	*you*		*any complaints?*
Do	*you*	*have*	*any complaints?*
Does	*he*	*understand?*	

for direct questions.

Similar inversion occurs if a negative or near-negative adjunct is FRONTED:

Never *have I seen* such a ghastly sight

Only then *did I realize* how lucky I was

Inversion of subject and main verb, with no additional auxiliary, occurs in certain structures involving FRONTING and the placing of a subject in end position, e.g.

Here *comes the bride*
Just as surprising *was their refusal to help.*

IPA

The INTERNATIONAL PHONETIC ALPHABET; the International Phonetic Association.

irrealis

Unreality or extreme unlikelihood; potential mode. Contrasted with *realis.*

Irrealis relates to tense and aspect. For example, counterfactual CONDITIONAL clauses and so-called SUBJUNCTIVE tenses describe the totally impossible (e.g. *If I were you . . .*) or the extremely unlikely (*If I lived to be a hundred . . .*).

1984 R. HUDDLESTON Both subjunctive and irrealis are general terms used for moods characteristically associated with non-factuality; the difference is that the subjunctive is primarily used in subordinate clauses and tends to involve a wider range of types of non-factuality . . than irrealis, which primarily expresses counterfactuality or factual remoteness.

The *irrealis* versus *realis* distinction has also been used to describe the distinction of meaning between non-finite clauses with an *-ing* participle, and those with a *to*-infinitive.

1985 M. A. K. HALLIDAY The imperfective represents the real, or actual, mode of non-finiteness ('realis'), while the perfective represents the unreal, or potential, mode ('irrealis'). So for example

Reaching the monument, continue straight ahead (imperfective)
To reach the monument, continue straight ahead (perfective)

irregular

Not conforming to a rule.

The term is particularly applied to verbs that do not follow the general pattern of adding *-ed* to form past simple tense and past participle. Thus *see, saw, seen* or *put, put, put* are irregular verbs.

Among nouns, common irregular plurals include *men, women, children, mice, teeth*, and a number of FOREIGN PLURALS.

Irregular degrees of comparison include *good/well*, *better*, *best*; *bad/badly*, *worse*, *worst*; *little*, *less*, *least*; *much*, *more*, *most*.

For irregular sentence types, see MINOR *sentence*.

• **irregularly**.

Compare ANOMALOUS, DEFECTIVE.

isolating

The same as ANALYTIC (1).

it See ANTICIPATORY (1), DUMMY, EMPTY, EXPLETIVE, IMPERSONAL, INTRODUCTORY.

iteration

The same as RECURSION.

iterative

(*n. & adj.*)

1 (An expression) denoting repetition.

The term is used to describe verbal meaning, as in

I've been walking to work for the past three weeks

where the meaning is presumably a series of repeated actions. A possible alternative term here is HABITUAL.

Iterative meaning is also denoted by some types of coordination, as in

I wrote and wrote—but they didn't reply
They were running up and down the stairs

2 (A word) formed by repetition.

1961 F. G. CASSIDY In Standard English one finds three kinds of iteratives: the simple ones like *hush-hush* . . ; those with vowel gradation like *ding-dong* . . ; and the rhyming ones like *handy-dandy*.

In some cases, the meaning is iterative in sense (1), as in *knock-knock*, *tick-tock*.

This type of compound word is also called REDUPLICATIVE.

3 The same as FREQUENTATIVE.

J

junction

(In older grammar.) The action of combining into a single phrase words of different RANK.

> 1924 O. JESPERSEN If . . we compare the combination of *a furiously barking dog* . . with *the dog barks furiously* . . there is a fundamental difference between them which calls for separate terms for the two kinds of combination: we shall call the former kind *junction*, and the latter *nexus*.

The term is virtually limited to the writings of Jespersen.

juncture

Phonology. The transition between two words or syllables and the phonetic features that mark it.

> 1957 S. POTTER No less elusive than intonation are the related features of *juncture* and *pause*. Where precisely does one syllable end and another begin?

As listeners we normally distinguish the separate words of an utterance, but much speech is in fact a continuum without pauses between words. We are helped in mentally separating the words by the speaker's retention of certain phonological features that characterize the pronunciation of phonemes in different phonetic contexts. For example, *grey day* /greɪ deɪ/ potentially has *open juncture* between /eɪ/ and /d/, and *close juncture* between /d/ and /eɪ/; whereas *grade A* /greɪd eɪ/ has the reverse. These junctures entail subtle, but important, differences: for example, differences of vowel length (/eɪ/ is longer in *grey*, where it is word-final, than in *grade*); and differences in voicing, pitch, and stress (/d/ is more forceful in *day*, where it is initial, than it is in *grade*). Such differences may be lost in rapid speech.

Some phoneticians include LIAISON as part of juncture, and others distinguish them.

• **junctural**.

jussive

(*n. & adj.*) (A type of clause) that expresses a command.

Jussives include not only IMPERATIVES, as narrowly defined, but also related non-imperative clauses, including some in subjunctive mood:

> Be sensible
> You be quiet
> Everybody listen
> Let's forget it

Heaven help us
It is important that he keep this a secret

The term *jussive* is, however, used to some extent as a syntactic label, and in this use would not include commands expressed as straight declaratives, e.g.

You will do what I say

In popular grammars, where the term is not used, such structures would be dealt with under an expanded IMPERATIVE label and under SUBJUNCTIVE.

K

kernel

A basic unmarked form.

> 1984 R. HUDDLESTON A form which is maximally basic, one which does not belong to a marked term in any system, is called a *kernel* form.

The term is derived from early Generative Grammar, where the application of certain obligatory PHRASE-*structure* rules produced a set of basic structures called *kernel sentences.*

A *kernel sentence*, or, as some grammarians prefer to call it, a *kernel clause*, is a syntactically unmarked declarative clause, not 'optionally transformed' (i.e., by negation, inversion, passivization, etc.), and complete in itself (i.e. not ellipted, coordinated, or subordinated).

L

labial

Phonetics. (*n. & adj.*) (A speech sound) involving the active use of one or both lips.

The term is a rather general one. The lips are of course passively involved in all speech sounds, but the term *labial* is confined to those in which one or both lips actually contribute to the articulation.

English labial consonants are usually more specifically described as BILABIAL or LABIO-DENTAL. With respect to vowels, the position of the lips is usually described in terms of *lip-rounding* or *lip-spreading*.

labialize

Phonetics. Accompany (a speech sound) with lip-rounding, particularly where this is an unusual (and optional) feature.

The term is applied particularly where an articulation involves an unusual degree of lip-rounding which is not a requirement of the phonology. For example, speakers of standard RP English commonly labialize /r/ if the following vowel has some lip-rounding, e.g. in *rude* or *roar*; it is far less usual to labialize /r/ before unrounded vowels (e.g. in *rat*, *right*). The pronunciation of /r/ with no lip-rounding and with no articulation of the forward part of the tongue leads to the noticeable substitution of a /w/ sound. Other speech sounds may be labialized under the influence of adjacent sounds, as in /buːt̫ / where the /t/ of *boot* is influenced by the preceding vowel, or in /k̫waɪt/ (*quite*) where the /k/ is labialized by the semi-vowel /w/. The IPA diacritic for this is [̫].

- **labialization.**

labio-dental

Phonetics. (*n. & adj.*) (A consonant) articulated with the lower lip and the (upper) teeth.

English has two labio-dental phonemes, the voiceless and voiced pair of fricatives:

/f/ as in *fine, photograph, enough*
/v/ as in *vine, nephew, of*

Other phonemes sometimes have a labio-dental realization as a result of assimilation. For example, the bilabial stops /p/ and /b/ can become labio-dental under the influence of a following labio-dental sound (e.g. in *hopeful*, *obverse*).

labio-velar

Phonetics. (*n. & adj.*) (A speech sound) articulated with some lip-rounding and the tongue raised towards the velum.

The English sound /w/, as in *won, one, why, quick, suite,* is classified as a *labio-velar semivowel.*

lamina

The same as BLADE.

laminal

Phonetics. Of a consonant: made with the blade of the tongue.

This is very much a specialist term. Compare APICAL.

language

1 (In the most general sense.) The method of human communication, consisting of words, either spoken or written.

2 The variety of this used by a particular community or nation.

e.g. *the English language, the languages of the British Isles, Indo-European languages, Romance languages.*

More specifically called a *natural language* in contrast to an ARTIFICIAL LANGUAGE or a *computer language.*

•• **first language**: the earliest language that an individual learns to speak (also called *mother tongue*), and of which he or she is a *native speaker.*

3 The style of an utterance or text.

e.g. *bad language, graphic language, literary language, poetic language.* See also BLOCK LANGUAGE.

4 The variety of language used in a particular profession or specialized context.

e.g. *legal, religious, scientific language; the language of the law, the language of advertising.*

The language of an individual is an IDIOLECT; that of a region or community is its DIALECT; the individual's idealized knowledge of language is his or her COMPETENCE (contrasted with PERFORMANCE); the idealized abstract concept of the language shared by a particular group of people is the LANGUE (contrasted with the PAROLE); the study of language may be over a period of time (DIACHRONIC) or at one particular time (SYNCHRONIC).

language acquisition device

Psycholinguistics. The innate capacity of the human mind to learn language. Abbreviated *LAD.*

The theory—introduced by Chomsky in *Aspects of the Theory of Syntax* (1965)—is essentially generative. Children do not merely repeat phrases and sentences that they hear; they construct the grammar (of the language around them) for themselves and produce new utterances. They are able to do this because they are born with a LAD, a programme of 'language universals' and 'assumptions concerning the nature of language', which enables them to select and generalize. The theory has been considerably revised in recent years.

langue

Language as a system. Contrasted with PAROLE.

The terms *langue* and *parole* were introduced by the Swiss linguist Ferdinand de Saussure (see SAUSSUREAN) in order to separate two of the meanings of the word *language*. What exactly he meant has been the subject of argument, but, roughly speaking, *langue* is the language system of a particular language community, while *parole* is language behaviour, the way members of the community actually use the system.

Compare COMPETENCE, PERFORMANCE.

laryngeal

Phonetics. Relating to or produced by the LARYNX.

laryngealize

Produce (a speech sound) with the back parts of the vocal cords held together (as in a glottal stop) but with the front parts vibrating.

• **laryngealization.**

Laryngealization is a phonemic feature in some languages. In English it is not significant, but if present may produce a 'creaky' voice.

Compare PHARYNGEALIZE, VELARIZE.

larynx

The hollow muscular organ situated in the upper part of the trachaea (the windpipe).

The importance of the larynx in the study of phonetics is that it contains the VOCAL CORDS, between which is the GLOTTIS (hence the popular name for the larynx, the 'voice box').

lateral

Phonetics. (*n. & adj.*) (A speech sound) made with a partial closure, the air stream escaping on one or both sides of the closure.

Like the terms *stop* and *fricative*, *lateral* (from Latin *lateralis*, from *latus*, *lateris* side) describes the manner, not the place, of the articulation. In RP there is a single lateral phoneme, /l/, which is usually voiced and non-fricative. The tip of the tongue articulates with the centre of the alveolar ridge and air escapes at the side. Being a CONTINUANT, /l/ has some vowel-like qualities and is often syllabic (e.g. in *apple, final, camel*). It is, however, normally classified as a consonant.

There are two main allophones.

(*a*) *Clear l*: the front of the tongue, not merely the tip, is raised towards the hard palate, giving the sound a slight front-vowel quality. Clear *l* is heard before vowels (e.g. in *leave*, *let*, *fill it*) and /j/ (e.g. in *failure*, *million*).

(*b*) *Dark l*, phonetically [ɫ]: the tongue-tip again articulates with the teeth ridge, but the back of the tongue is somewhat raised, giving a back-vowel quality. Dark *l* is heard word-finally after a vowel (e.g. in *fill*, *oil*, *pale*); after a vowel and before a consonant (e.g. in *help*, *else*, *although*); and in syllabic l (see above).

A voiceless or partly devoiced allophone is heard after /p/, /t/, or /k/, as in *please*, *ghastly*, *clue*. The Welsh sound spelt *ll*, as in *Llangollen*, is a voiceless alveolar lateral fricative.

• **laterally**.

lateral plosion

Release of a stop consonant at the side of the tongue. (Also called *lateral release*).

When English /t/ or /d/ is followed by /l/, as in *cattle*, *muddle*, the alveolar stop can be released laterally instead of the usual way; that is, you can say these words without moving the tongue away from the /t/ or /d/ position. This is known as *lateral plosion*.

lax

Phonetics. Articulated with less effort than is normal; contrasted with TENSE.

Lax voice and *tense voice* are used by some phoneticians as middle terms among several others to describe different degrees of glottal stricture.

Lax and *tense* are among the BINARY contrasts held in one theory of phonology to be among the features of vowels.

Compare LENIS.

learned plural

An older and possibly obsolete term for FOREIGN PLURAL.

left-branching See BRANCHING.

left dislocation See DISLOCATION.

length

1 *Phonology*. The relative time taken in the articulation of different sounds or syllables, and a listener's perception of them. (Also sometimes called *duration*.)

Traditionally, English vowels have been categorized as *long* or *short*. The long vowels (in standard RP English) are

/iː/	green, heat, machine	/uː/	food, route, rude, blew
/ɑː/	cart, heart, father	/ɜː/	herd, heard, sir, nurse, worse
/ɔː/	cord, caught, saw, bought		

The short vowels are

/ɪ/	bit, wom*e*n, busy, hymn	/ʊ/	good, put
/e/	bed, head, any	/ʌ/	hunt, love, blood
/æ/	hat	/ə/	moth*er*, *a*way
/ɒ/	hot, wash, laurel		

The long vowels are traditionally shown, as here, with the length mark [ː]. But in fact the differences between similar long and short vowels, as for example /iː/ and /ɪ/ in *beat* and *bit*, or /uː/ and /ʊ/ in *food* and *good*, are differences of quality as well as length, so that it is now common, as here, to use different symbols. In some transcriptions length marks are dispensed with altogether.

An added complication is that length differences in actual articulation are conditioned by phonetic context. For example, the final voiceless /t/ in *beat* has a shortening effect on the preceding long vowel, whereas the short vowel in *bid*, being followed by a voiced consonant, is not shortened, with the result that the vowel lengths in these two words may be objectively the same. A distinction can therefore be made between 'linguistic' length, as the listener perceives it, and 'real world' length (or *duration*), as acoustically measured.

Compare QUANTITY.

2 *Phonetics.* The actual time taken in the articulation of a speech sound or syllable. This is more usually dealt with under DURATION (1).

lenis

Phonetics. Of a consonant sound: made with relatively weak breath force. Contrasted with FORTIS.

In English, voiced plosives and fricatives (e.g. /b/, /d/, /ð/) tend to be made with less muscular effort and less breath force than their voiceless counterparts. They are therefore sometimes called *lenis consonants* (Latin *lenis*, soft, easy).

Lenis consonants are sometimes described as *weak* consonants (and fortis consonants as *strong*) but this terminology leads to confusion and has largely been abandoned.

level

(*n.*) *Linguistics.* Each of the areas into which a language can be analysed, characterized by the distinct property (e.g. sound, meaning, function, etc.) which is the object of analysis.

A very general term for the different areas of language analysis, some of which are also described under such terms as HIERARCHY or RANK. At least three levels are generally recognized, grammar, phonology, and semantics, but some models of grammar recognize more. Different grammatical models relate the levels to each other in different ways. In Structural Linguistics, analysis proceeds from the lowest level, phonetics and phonology, to progressively higher levels dealing with morphemes, words, and syntax. In other analyses, the different levels are not considered to be so clearly separable.

(*adj.*) *Phonetics.* Of a tone: at a single pitch height.

In the analysis of intonation, contrasting pitch heights, e.g., high level [¯], low level [_], are posited. These are relative to the individual speaker, rather than absolute.

Compare PITCH.

lexeme

Semantics.

A word in the abstract sense, an individual distinct item of vocabulary, of which a number of actual forms may exist for use in different syntactic roles.

(*a*) In ordinary usage, *word* has more than one sense, so that it is possible to say 'The five words (i) *see*, *sees*, *seeing*, *saw*, and *seen* are different forms of the same word (ii)'. To distinguish these two meanings, *word* (i) is sometimes called a *form* or *word-form*, and *word* (ii) a lexeme.

Identical forms with unrelated meanings are treated as separate lexemes, just as they are treated as different vocabulary items in a dictionary. Thus *see* (the area under the authority of a bishop or archbishop) is a different lexeme from *see* (discern with the eyes). Compare HOMONYM.

Identical forms with different meanings, which in fact have a common origin, are similarly treated. Thus polysemous words such as *love* ((i) 'passion' and (ii) 'nil' (the relationship in this case being through the phrase *for love* ('without stakes', 'for nothing')) are also usually considered different lexemes, even though lexicographers may treat them under one headword in a dictionary. This is a disputable area, for while some 'words' are clearly distinguishable into several 'senses' (e.g. *crane* (i) 'bird', (ii) 'lifting device'), others can be subdivided with varying degrees of refinement (e.g. *company*, *high*, *office*, *thing*), and the assignment of any particular schema of subdivisions to separate lexemes would be open to challenge.

(*b*) *Lexeme* may be extended to include a sequence of words that constitute a single semantic item: e.g. *out of*, or *put off* 'postpone'.

(*c*) *Lexeme* is occasionally used in roughly the senses 'morpheme' and 'root', 'stem', but this is not now usual.

> 1958 C. F. HOCKETT The lexemes in the two-word sequence *twenty-eighth* are *twenty*, *eight* and *-th*.

•• **compound lexeme**: a compound word whose meaning cannot be deduced from its parts (also called *word lexeme*) or a similar combination of words.

e.g. *greenhouse*, *bucket shop*.

• **lexemic**: of or relating to lexemes. **lexemics**: the study of lexemes.

See also LEXICAL ITEM.

lexical

Relating to words as units of the vocabulary; of or pertaining to the lexis.

The term is extensively used in a fairly general sense, as in *lexical factor*, *lexical recurrence* (the repetition of a word), *lexical structure* (morphology) and so on. It is also used in more specific ways (see below).

•• **lexical ambiguity**: AMBIGUITY.

lexical entry: see LEXICON.

lexical field: see FIELD.

lexical formative: see FORMATIVE (1).

lexical meaning: see MEANING.

lexical morpheme: see MORPHEME.

lexical stress: see STRESS.

• **lexically**: as regards vocabulary.

lexical-functional grammar

A theory of grammar, developed in the 1980s, which—as its name implies—attaches more importance to the lexicon and less to purely syntactic rules than Transformational-Generative models.

lexical item

The same as LEXEME; a variant on it.

Although *lexeme* and *lexical item* are often used interchangeably, distinctions can be made.

(*a*) *Lexical item* may be preferred as the more general term, and *lexeme* reserved for the 'word' which has a group of variants (e.g. *see*, *saw*, *seen*, etc.).

> 1984 R. HUDDLESTON A lexical item . . may contain more than one lexeme or word: these are idioms such as *bury the hatchet* 'renounce a quarrel'.

(*b*) As the quotation above suggests, *lexical item* may also be the preferred term for a word, words, or a phrase whose meaning is not deducible from the meaning of its parts; e.g. *greenhouse*, *bucket shop* (normally dealt with as *compounds*), *bury the hatchet*, *show a clean pair of heels* (normally dealt with as *idioms*).

(*c*) *Lexical item* may also be used to mean a word form, such as an irregular inflectional form (of a lexeme) that would be expected to have a separate dictionary entry. For example, a dictionary, in addition to listing *buy*, might have an entry '*bought*: see *buy*'. *Bought* would therefore be a separate lexical item.

Lexical item is sometimes also called *lexical unit*.

lexicalize

Realize as a word (something that was previously, or could conceivably be, realized in an alternative way).

A notion or practice previously expressed by a syntactic phrase may be lexicalized for convenience and conciseness; e.g. *tailgating* (M20, US) for 'driving too closely behind the vehicle in front', or *doorstepping* (M20, UK) for 'canvassing support by going from door to door'.

• **lexicalization**.

lexical verb

A verb of a class that contains all verbs except MODAL and PRIMARY verbs. (Also called *full verb*.)

Lexical verbs constitute the majority of verbs. Formally they use auxiliary verbs in the formation of questions and negatives (e.g. *Do you understand?*, *I haven't forgotten*; not **Understand you?*, **I have forgotten not*).

Lexical verbs function as MAIN verbs, never as auxiliaries, and they can function alone. *Lexical* and *main* are often used interchangeably, though they can be distinguished. See MAIN.

lexical word

The same as CONTENT WORD.

lexicography

The art and practice of dictionary making.

lexicology

The study of the lexis.

lexicon

The complete set of vocabulary items in a language, especially considered as part of a theoretical description of the language.

In Generative Grammar, the lexicon is a theoretical component consisting of *lexical entries* which contain not only semantic information about each item, but also much more complete phonological and syntactical information than an ordinary dictionary would.

lexis

The stock of words in a language; the level of language consisting of vocabulary. Contrasted with GRAMMAR (or SYNTAX).

liaison

Phonology. The linking together of words in connected speech, particularly where this involves some unusual phonetic feature.

INTRUSIVE *R* and LINKING *R* are prime examples of liaison. Some speakers are so anxious to avoid an intrusive *r* that they use a vowel glide or a glottal stop even where a linking *r* would be natural, e.g. /mɔːən mɔː/ or /mɔːʔən mɔː/ for *more and more*.

A word-final consonant is not usually carried over to a word beginning with an accented vowel, and successive words usually keep their separate identities; but occasionally liaison is heard (e.g. /nɒt ə 'tɔːl/ for *not at all*.

Compare JUNCTURE.

ligature

A written or (especially) printed character combining two letters in one, e.g. ⟨æ⟩, ⟨œ⟩. Sometimes called DIGRAPH.

limiter

A name for a type of focusing adverb with a RESTRICTIVE meaning.

e.g.

just, only, merely, simply

as in

I just/only/merely/simply said it was expensive

limiter adjective

Another name for RESTRICTIVE adjective (which is currently the more usual term). See ATTRIBUTIVE.

limiting adjective

1 (In traditional grammar.) Another name for DETERMINER.

The term was used to distinguish this class of words from adjectives proper, which were more fully called *descriptive adjectives*. It is now old-fashioned.

2 The same as RESTRICTIVE ADJECTIVE. See ATTRIBUTIVE.

lingual

Phonetics. Involving the tongue; a term occasionally used in the classification of speech sounds.

A *lingual roll* is an [r] produced with the tongue making a series of rapid taps against the alveolar ridge. This is a possible though unusual pronunciation of an English /r/.

linguist

A student of or expert in linguistics.

This is nearly as old (E17) as the lay meaning 'someone fluent in several languages' (L16). To make a clearer distinction, the term **linguistician** (L19) has been suggested as the term for the student of linguistics, but it is not used by those involved in the profession.

linguistic

1 Relating to LINGUISTICS.

e.g. *linguistic analysis, linguistic theories.*

2 Relating to language.

e.g. *linguistic complexity, linguistic phenomena, linguistic properties, linguistic rules, linguistic unit*; *linguistic knowledge* is a speaker's knowledge of language (e.g. of the mother tongue), contrasted with NON-LINGUISTIC knowledge (e.g. about the properties of fire).

•• **linguistic determinism, linguistic relativity**: see SAPIR-WHORF HYPOTHESIS.

linguistic phonetics: an occasional name for PHONOLOGY.

linguistics

The scientific study of language.

Linguistics as an academic subject has burgeoned in recent years, so that surveys of the whole field are sometimes called *general linguistics*, while more specialized areas have specific labels.

•• **applied linguistics**: the practical application of linguistic studies to other areas, especially the teaching of foreign languages.

When contrasted with this, general linguistics is sometimes called *theoretical linguistics.*

computational linguistics: linguistics that uses computers in the gathering, analysis, and manipulation of linguistic data (including speech synthesis, machine translation, and aspects of artificial intelligence).

corpus linguistics: see CORPUS.

structural linguistics, structuralist linguistics: the study of language structure, particularly as practised by the STRUCTURALISTS in the 1940s and 1950s.

taxonomic linguistics: language study concerned particularly with classification; often used as a prejorative synonym for *structural linguistics.*

See also COMPARATIVE, CONSTRASTIVE, DESCRIPTIVE (1), DIACHRONIC, HISTORICAL, SYNCHRONIC; PSYCHOLINGUISTICS, SOCIOLINGUISTICS, TEXT LINGUISTICS.

linker

Another term for COORDINATOR.

Not a very general term.

linking adverb, linking adverbial

The same as CONJUNCT. See LINKING WORD.

linking *r*

The pronunciation of a written word-final *r* as /r/ when the next word begins with a vowel.

In standard RP a written word-final *r* is not pronounced before a pause or a following consonant sound. However it is usually pronounced when the following word begins with a vowel (as in *Here it is* or *far away*).

Compare INTRUSIVE *R*. See RHOTIC.

linking verb

A verb that semantically joins the rest of the predicate, particularly a noun complement or adjective complement, back to the subject. (Also called *copular verb*.)

The prime linking verb is *be*. But other verbs are used in a similar way, e.g.

She'*s become* a courier
He'*s looking* much better

Linking verbs are sometimes divided into verbs of *current* meaning

It *is/looks/remains* a mystery

and verbs of *resulting* meaning

It *became/proved* difficult

linking word

(An umbrella term for) any word that joins two linguistic units.

Linking words include

(*a*) coordinating conjunctions (e.g. *and*, *but*)
(*b*) subordinating conjunctions (e.g. *although*, *because*, *when*)
(*c*) conjuncts (e.g. *in addition*, *moreover*, *meanwhile*, *nevertheless*)

Conjuncts are sometimes called *linking adjuncts* or *linking adverb(ial)s*.

link verb

The same as LINKING VERB.

lip position

Phonetics. The configuration of the lips during the articulation of a speech sound.

The lips have an important role in the production of speech sounds. Even where they are not primary articulators (as they are with BILABIAL and LABIO-DENTAL consonants) their position affects sound quality.

Each English vowel has its own characteristic lip position, and these are variously described. One binary distinction is between *rounded* and *unrounded*. Other terms used are *spread*, *neutral*, *close-rounded*, and *open-rounded*. English /iː/ (as in *bead*) is usually said with *lip-spreading*; /ɑː/ (as in *hard*) is pronounced with the lips neutrally open; while /uː/ as in *boot* is a rounded vowel, said with *lip-rounding*.

Compare LABIALIZATION.

liquid

Phonetics. (*n. & adj.*) (Being or pertaining to) one of the consonants /l/ and /r/.

This is chiefly a traditional term. It is a borrowing of Latin *liquidae*, itself a translation of Greek *hugra*, applied to the four sounds /l/, /m/, /n/, /r/ on account of their flowing and easy sound. But the term would not now be applied to the 'nasal stops'.

Compare APPROXIMANT, LATERAL, FRICTIONLESS CONTINUANT.

load See FUNCTIONAL *load*.

loan-shift

A change in the meaning of a word resulting from the influence of a corresponding word in a foreign language.

Examples are *arrive*, which has borrowed the sense 'achieve success' (L19) from French *arriver* and *suspicion*, which has similarly borrowed the meaning 'slight trace or suggestion (of)' (E19) from French *soupçon*.

loan translation

An expression adopted by one language from another in a more or less literally translated form; a CALQUE.

One of the earliest is *gospel*, OE *godspell*, from *god* 'good' + *spell* 'news', translating Greek *euaggelion*. The expression *lose face* is a loan translation of Chinese *tiu lien*.

loanword

A word adopted or borrowed, usually with little modification, from another language.

Loanword is itself a loan-translation of German *lehnwort*.

See BORROWING.

local genitive See GENITIVE.

locative

(*n. & adj.*) (A word, phrase, case, etc.) expressing location.

In Case Grammar, the locative is one of the original six cases proposed. Some noun phrases which in traditional grammar are subject or object are reclassified as locatives, e.g. in

London can be very lonely
We pounded *the pavements*
Hyde Park was the venue for the concert

More generally, prepositions and adverbials of PLACE can be described as *locative* words and phrases, or as having a *locative* role.

locution

Semantics. An utterance, an act of speaking.

This is a term in Speech-Act Theory, and refers to the mere act of speaking, with no reference to the intention of the speaker (ILLOCUTION) or the resulting effect (PERLOCUTION).

1955 J. L. AUSTIN *Locution.* He said to me 'Shoot her!' meaning by 'shoot' shoot and referring by 'her' to her.
Illocution. He urged (or advised, ordered, etc.) me to shoot her.
Perlocution. He persuaded me to shoot her.

•• **locutionary**.

1955 J. L. AUSTIN The act of 'saying something' in this full normal sense I call . . the performance of a locutionary act.

logical component See IDEATIONAL.

logical gap See CONDITION.

logical subject See SUBJECT.

long vowel See LENGTH.

lost consonant See SILENT LETTER.

loudness

Phonetics. A perceptual category, along with PITCH, sound quality, and LENGTH, in terms of which speech sounds are heard.

Loudness is primarily related to INTENSITY, but the two are to be distinguished. *Intensity* is the speaker's physical effort used in producing a speech sound, and is objectively measurable. *Loudness* is a matter of the listener's perception, which is affected by other factors, such as the pitch of the voice and length.

Compare STRESS.

low

Phonetics.

1 Of a vowel: produced with the tongue raised only a small degree towards the roof of the mouth. Also called OPEN. Contrasted with HIGH (or CLOSE).

The sound /æ/ as in RP *hat* is a low front vowel, and /ɑː/ as in *hard* and *heart* is a low back vowel.

2 (In intonation.) Of pitch: produced by relatively slow vibrations of the vocal cords.

A *low fall* glides from a mid pitch to the lowest pitch of the speaker's voice, while a *low rise* extends from a low pitch to somewhere about the middle range.

low frequency See FREQUENCY (3).

M

main clause

A clause that is not subordinate to any other.

(*a*) *Main clause* is traditionally contrasted with SUBORDINATE *clause*. Thus in *I was ten when I got my scholarship*, the main clause is *I was ten* (and *when I got my scholarship* is a subordinate adverbial clause of time).

In grammatical models where *clause* is the prime unit of utterance, a simple sentence such as *I was ten at the time* is also, by this definition, a main clause.

(*b*) *Main clause* is often defined additionally as a clause that could stand alone as a complete sentence, but the important criterion for a main clause is that it is not in a subordinate relationship. The coordinated clauses of a compound sentence, such as *I was ten and I got the scholarship* are called main clauses. So also are some that are not INDEPENDENT.

(*c*) In some models, however, a main clause can actually contain a subordinate clause. By this definition the whole sentence *I was ten when I got my scholarship* is a main clause of the pattern SVCA, with the subordinate clause supplying the Adverbial element.

Compare INDEPENDENT, MATRIX, SUPERORDINATE.

main verb

1 A verb functioning as the head of a verb phrase.

This is basically a functional label, contrasting with the functional sense of AUXILIARY.

If there is more than one word in an unellipted verb phrase, the final word is the main verb (and the others are auxiliaries). e.g.

Have you been *waiting* long?
It may have been *forgotten*

If there is only one word then that is the main verb, e.g.

I *know* nothing about it
What *is* the matter?

In a catenative verb phrase, the first lexical verb is the main verb:

I have been *meaning* to telephone you

2 Loosely, a LEXICAL verb.

As the main verb function is often realized by a LEXICAL (or *full*) verb, the terms *main* and *lexical* are sometimes used as synonyms.

1990 *Collins Cobuild English Grammar* **Main verb** all verbs which are not auxiliaries. Also called lexical verb.

This definition ignores the fact that primary verbs (sometimes classified as auxiliaries, but never as lexical verbs) can function as main verbs.

See AUXILIARY, PRIMARY.

major

Being or belonging to a group that is of more importance than another. Contrasted with MINOR.

This is used somewhat loosely, and not as a particularly technical term.

• • **major part of speech**: one of the parts of speech (or word classes) that were traditionally considered to contain most of the meaning of an utterance, i.e. nouns, verbs, adjectives, and adverbs; the *minor* word classes (pronouns, prepositions, conjunctions, interjections) were considered to have little meaning in themselves apart from their function of connecting the others.

The terms are now somewhat out of favour.

Compare CONTENT WORD, OPEN (2).

major sentence type, major clause type

1 A structure containing at least a subject and a finite verb, though possibly an imperative is covered by the term even though with it a subject is optional (see 2).

2 (More specifically.) One of the commonest types of sentence (or clause) in a formal analysis, namely Declarative, Interrogative, Imperative, or Exclamative.

Sentences as defined in (1) or (2) are also called *full sentences* or *complete sentences*. In this type of analysis, sentences that fall outside these structures are variously classified as *minor*, *irregular*, or *fragmentary sentences*, or perhaps as *utterances*.

3 (In Systemic-Functional Grammar.) A clause with a verb in indicative or imperative mood. This definition would appear to exclude sentences containing a subjunctive verb; but as *minor clauses* are defined as 'clauses with no mood or transitivity structure', it may be that this definition is basically the same as (1) and (2).

mandative subjunctive

The main type of subjunctive in use today. See SUBJUNCTIVE.

manner

1 A semantic category of adverbs or adverbials, consisting of those that answer the question 'How?'

Adverbs are traditionally classified by meaning, a fairly standard basic division being into *adverbs of place* (answering the question 'Where?'), *time* ('When?'), *manner* ('How?') and *degree*.

Single-word *manner adverbs* are often derived from adjectives by adding *-ly* (e.g. *brightly*, *stupidly*, *quietly*) and are felt to be central or typical adverbs. In more detailed semantic analyses of *adverbials* the label *manner adverbial* is more narrowly applied: and manner adverbials are classified as a subclass of some larger grouping of adverbials, such as PROCESS adjuncts.

Some manner adverbials also occur as style disjuncts, e.g.

Seriously, do you mean that?

Clauses of manner usually imply comparison

He looked / *as if* he had seen a ghost
He speaks / *just as/like* his father always did

and so such clauses are sometimes subsumed under *comparison*.

2 **manner of articulation**: (*Phonetics*) the method by which a speech sound is made, described in terms of the degree and type of closure of the speech organs.

Manner of articulation, along with PLACE OF ARTICULATION, forms a major part of the framework used in describing the production of speech sounds, particularly consonants. Among the different types of closure involved are PLOSIVE, AFFRICATE, NASAL, LATERAL, and FRICATIVE.

marginal

1 Of a language feature or unit: not CENTRAL, less important, infrequent.

(*a*) Sometimes applied to any of the metaphorical meanings of a word, in contrast to the basic or central meaning.

(*b*) More usually, applied to members of a word class that only have some characteristics of that class, in contrast to CENTRAL members.

• • **marginal coordinator**: e.g. *for*, *nor*, *so* (='therefore'), *then* (='after that') and others.

marginal modal (verb): the same as SEMI-MODAL.

marginal preposition: a word that functions in many ways like a central preposition, but also has some characteristics of other word classes.

Many marginal prepositions share certain features with verbs or adjectives:

He's remarkable, *considering* his age
Given the provocation, the outcome was understandable
That must be *worth* a fortune
It was well written, *bar* a few trivial mistakes

Among the marginal prepositions are *less*, *minus*, *plus*, and *times*:

What's five *times* six?
He arrived *minus* a ticket

marginal subordinator: a word or phrase that can be followed by a subordinate clause, but overlaps with other word-classes or sentence elements

The moment (that) you said so, I remembered
Provided the train is on time, there should be no problem

2 Forming the boundary of a syllable.

The term is used in describing the typical phonological characteristics of consonants.

• **marginally**: so as to form the boundary of a syllable.

marked

Of a linguistic feature: distinguished in some way from the *unmarked*, more basic or central form to which it is related.

The concept was originated by the Russian linguist Nikolay Trubetzkoy (1890–1938) in relation to phonology.

See MARKEDNESS.

markedness

The condition, quality, or state of being marked.

The concept of markedness can be applied in many areas of language. Thus a simple declarative sentence (e.g. *I love Lucy*) is unmarked, whereas *I don't love Lucy* is marked for negation, and *Do you love Lucy?* is marked for interrogative. Similarly, *Lucy I love* has a marked word order (OSV) compared with the unmarked SVO that is usual in English.

With nouns and verbs, and other words that can be inflected, the base forms are said to be *unmarked* (e.g. *look*, *table*, *nice*) while inflected forms (*looked, looks, tables, nicer*), are marked for past, plural, comparative, and so on. Similarly the active voice is unmarked, the passive marked.

Markedness also applies in semantics, where features used in COMPONENTIAL ANALYSIS can be described in this way. Thus, *horse* is unmarked for sex, whereas *stallion* and *mare* are so marked. This type of marking is *semantic* marking. Other words exhibit *formal* marking (e.g. *host* versus *hostess* (marked for 'female'), *widow* versus *widower* (marked for 'male')). In a neutral context the unmarked term in a pair is used. Thus of the pair *old* versus *young*, *old* is the unmarked term (e.g. *How old is the baby?*).

In phonology, DISTINCTIVE FEATURES can be described in terms of markedness: [+] features are regarded as marked, and [–] features as unmarked.

• **marking**: causing something to be marked (see also DOUBLE MARKING).

masculine

(*n. & adj.*) (A noun etc.) of the gender that mainly denotes male persons and animals.

See FEMININE, and compare GENDER.

mass noun

1 An UNCOUNT noun (or *uncountable noun* or *non-count noun*). Contrasted with COUNT *noun*.

In general, *mass noun* is another synonym for one of the two main classes of common nouns, the other being *count* (or *countable*).

Many English nouns seem to belong to both classes:

beer, a beer; cloth, a cloth; ice, an ice; iron, an iron;
paper, a paper; war, a war.

One solution is to consider the words in such pairs to be two separate dictionary entries, but when the two meanings are close it seems preferable to talk of a single word with both mass and count usage. Hence terms like *mass usage*, *mass meaning*, and *mass interpretation* are sometimes introduced.

2 (More narrowly.) A particular type of uncount noun:

1990 *Collins Cobuild English Grammar* **Mass noun** (in this grammar), a noun which is usually an uncount noun, but which can be used as a count noun when it refers to quantities or types of something; EG . . . two sugars . . cough medicines.

This definition gives a very distinct meaning to the term *mass*, since it specifically includes many words that are used as countables. It excludes, however,

Some other nouns that can be uncount nouns when they refer to a thing in general, and count nouns when they refer to a particular instance of it (*ibid.*).

Examples of words that *Cobuild* classifies as both *uncount* and *count* (but *not mass*) are *victory* and *conflict*.

The term appears to have originated with Jespersen, who used *mass-word* in 1914.

Compare BOUNDED, NOUN, SINGULAR.

material process See PROCESS.

matrix clause

A superordinate clause minus its subordinate clause.

This is what in popular grammar is called the main clause of a complex sentence.

In grammatical descriptions that use the term *main* for the entire superordinate clause, *matrix* is used to distinguish the part from the whole:

I was ten (*matrix*) when I got my scholarship (*subordinate*)
Nobody had expected (*matrix*) that I would get one (*subordinate*)

meaning

What is meant by a word, phrase, clause, or longer text.

Many different types of meaning are distinguished, and different classifications are made:

Objective, factual, verifiable meaning is considered as DENOTATION or as COGNITIVE, DESCRIPTIVE (2), IDEATIONAL, PROPOSITIONAL, or REFERENTIAL MEANING.

Subjective, emotional, personal meaning is labelled CONNOTATION or AFFECTIVE, ATTITUDINAL, EMOTIVE, or EXPRESSIVE *meaning*.

Meaning as particularly involving social interaction is variously labelled INTERPERSONAL, SITUATIONAL, or SOCIAL *meaning*.

Meaning as derived from the context of the surrounding text is sometimes called TEXTUAL meaning. Compare CONTEXTUAL *meaning*.

The meaning of words as given in the dictionary is called *lexical meaning* (sometimes also *dictionary meaning* or *central meaning*).

The meaning inherent in the grammar (for example, the meaning of tenses, the relationship of subject and object, and the difference between declarative and interrogative) is called *grammatical meaning* or *structural meaning*.

Compare SPEECH ACT.

means

Method by which something is done: a term used in the semantic analysis of adverbials and prepositions.

Adjuncts of means are usually contrasted with adjuncts expressing INSTRUMENT or AGENT, for example:

We came *by train*
They got in *by breaking a window*

in contrast to

They got in *with a skeleton key* (instrument)
It was stolen *by a cat burglar* (agent)

medial

Designating or occurring in the middle of a linguistic unit. (*Medial position* is also called MID-POSITION.)

(*a*) The term is used in analysing ellipsis (see GAPPING), but particularly in describing the position of adverbs in the clause; it contrasts with INITIAL (or *front*) and FINAL (or *end*). Broadly, it indicates the position between the subject and the verb:

The train *soon* gathered speed
I *hardly* think so

With verb phrases made up of several elements several positions can be called *medial*. A common medial position is after the first auxiliary:

I have *definitely* made up my mind

but other medial positions (between the subject and the lexical verb) may be possible:

He *definitely* had intended to go (initial medial)
It could have been *intentionally* overlooked (end medial)

(*b*) *Phonology*. The realization of a phoneme when it occurs in the middle of a word may be different from its realization when initial or final. For example, the voiceless plosives /p/, /t/, and /k/, which have strong aspiration when initial in an accented syllable are only weakly aspirated when medial following an accented syllable, as in *rapid*.

• **medially**.

medial-branching See BRANCHING.

medium

The means by which something is communicated.
e.g. *spoken* versus *written* (or *graphic*) *medium*.

mental process See PROCESS.

mesolect

Sociolinguistics. A level of language showing greater speaker competence than a BASILECT but less than an ACROLECT.

As with the related terms, *mesolect* can be used of the language competence of native or non-native speakers.

1989 J. HONEY With every year that passes, fewer and fewer young children [in Britain] are introduced to the meanings of the old dialect words, and the accents of more and more of them move to at least an intermediate stage in the direction of RP, which is called the MESOLECT.

metafunction

(In Systemic-Functional Grammar.) Each of the three fundamental components of meaning that are posited for any language, namely the two main kinds, IDEATIONAL meaning and INTERPERSONAL meaning, plus a third metafunctional component, the TEXTUAL, which is said to 'breathe relevance' into the other two.

• **metafunctional**.

metalanguage

A (form of) language used to discuss language.

Many of the terms used in this book are examples of metalanguage.

• **metalinguistic. metalinguistics**.

metanalysis

Reinterpretation of the division between between words or syntactic units.

e.g. *adder* from OE *nædder* by the analysis in ME of *a naddre* as *an addre*.

The term was coined by Jespersen (1914).

• **metanalyse**: alter by metanalysis.

metaphor

The application of a name or descriptive term or phrase to an object or action to which it is imaginatively but not literally applicable (e.g. *a glaring error*, *a loud check*).

• **metaphorical**.

These are not specifically grammatical terms, but semanticists like to point out that language is much more metaphorical than is commonly realized. Even prepositions of time (e.g. *at* (the weekend), *on* (Sunday), *in* (the morning)) are metaphorical applications of prepositions of space; *high* and *low* applied to voice pitch are metaphorical uses.

In a *mixed metaphor*, incongruous and incompatible terms are used of the same object or event, e.g.

> If we want to take part in the new space frontier, we must get in on the ground floor. We have the key to the twenty-first century.

metathesis (Plural **metatheses.**)

The transposition of neighbouring sounds.

> 1989 J. HONEY His occasional lapse into a minor linguistic quirk called 'metathesis', the transposing of sounds in a word. On his election as party leader, he used . . the word 'em-nity' (enmity).

• **metathesize**: transpose (neighbouring sounds); undergo metathesis. **metathetic**.

This phonetic change has occurred quite often in the history of English; it accounts for such correspondences as *three* and *third* (ME *thridde*) and *work* and *wrought* (OE *worhte*), and for some doublets such as *curd* and *crud*. In current English it is found in such alternations as *wasp* and (dialect and humorous) *waps*, and *ask* and (dialect) *aks*. Spoonerisms (the inadvertent exchange of the initial sounds of neighbouring words) are a special kind of metathesis: e.g. 'Sir, you have tasted two whole worms; you have hissed all my mystery lectures and have been caught fighting a liar in the quad; you will leave Oxford by the town drain' (attributed to W.A. Spooner, 1844–1930).

metonym

A word or expression which is used as a substitute for another word or expression with which it is in a close semantic relationship.

e.g. *Whitehall* for 'the British civil service', *the Turf* for 'the racing world', *per head* for 'per person'.

• **metonymic**. **metonymy**.

mid

1 *Phonetics*. Of a vowel sound: produced with (part of) the tongue in a middle position between high and low.

2 The same as MEDIAL.

mid-branching See BRANCHING.

Middle English

The form of English used in Britain between circa 1150 and circa 1450; the stage in the development of the English language intermediate between Old English and Modern English. Abbreviated *ME*.

middle verb

One of a small group of apparently transitive verbs that do not normally occur in the passive.

The term is not in very general use, but is a way of classifying verbs such as *have* (in its possessive meaning: *We have a house* does not have a passive **A house is had by us*), *consist of*, *lack*, *possess*, *resemble*, and some other verbs in certain of their meanings, e.g.

> *You are suited by blue
> *I am not fitted by this jumper
> *Twenty is equalled by 4 times 5

The term is adapted from Greek grammar, which has a middle voice distinct from both active and passive.

mid-position

(Occupying) the position between subject and verb; (in) medial position.

Used chiefly with reference to the placing of an adverb, the term is a popular alternative to MEDIAL position. For example, frequency adverbs (e.g. *always, often, sometimes, never,* etc.) can be described as usually taking mid-position, e.g.

> We *always* watch the news.

or even as being *mid-position adverbs*; though they can in fact appear in other positions.

minimal

1 That is the least possible, especially in terms of distinctiveness.

> 1988 J. C. CATFORD Phonemes are *minimal* sequential units, because if you take a stretch of speech and chop it up into a sequence of phonological units, the *shortest* stretch of speech that functions as a contrastive unit in the buildup of the phonological forms of words is the phoneme.

•• **minimal free form**: see MINIMUM FREE FORM.

2 (More specifically.) That is, or is characterized by, a distinction based on only a single feature. See MINIMAL PAIR.

• **minimality**.

minimal pair

Phonology. A pair of words that differ only in respect of one meaningful sound contrast.

English *pin* and *bin*, *pin* and *pen*, *pin* and *pick* are all minimal pairs. The significance of the concept is that the difference of meaning between each member of a pair shows that each of the sounds involved (/p/ and /b/, /ɪ/ and /e/, /n/ and /k/) is a distinct PHONEME. Minimal pair tests are used as a discovery technique by linguists recording a new language, and in the teaching of foreign languages.

minimum free form

Linguistics. The smallest linguistic unit that can function on its own as a complete utterance. (Also called *minimal free form*.)

In a now classic definition, the American linguist Leonard Bloomfield, in 1926, defined the *word* as a minimum free form. This definition is open to criticism, as Bloomfield himself realized, since it would exclude *a*, *the*, *my*, and similar forms that are generally considered to be words, but which do not function alone. The definition largely holds, however.

There is a second objection, that in certain circumstances even BOUND forms can be used alone, e.g.

> 'Did you say disinterested or uninterested?'
> 'Dis'

But this can easily be discounted on the grounds that any speech segment, even a single sound, can be used alone as a citation form.

Minimum free forms contrast with BOUND *forms*.

Compare MORPHEME.

minor

Less important, occurring less frequently.

•• **minor word class**: see MAJOR.

minor sentence

An acceptable utterance that does not conform to normal sentence rules, in contrast to a regular or FULL sentence.

Various classifications and terms are used in the analysis of utterances that do not conform to the main rules of clause or sentence structure. Some grammarians use terms such as *irregular sentence* or NON-SENTENCE for such utterances.

Utterances classified under such headings occur with some frequency however and often follow their own patterns. These terms may therefore include:

wh-question lacking a verb and/or subject:
  How about a drink?
  Why not forget it?
Subordinate clause lacking matrix clause (often with exclamatory meaning):
  That you should be so lucky
  To think you were there all the time
  As if you didn't know

Adverbials (as commands):
Out with it!
'Proportional structures':
The quicker the better
Instructional language:
Flavour to taste
Not to be taken

Terms like *minor* or *irregular sentence* may also be extended to include BLOCK LANGUAGE, ELLIPTED sentences, FORMULAE, INTERJECTIONS, and the optative SUBJUNCTIVE.

misrelated

Not attached grammatically to the word or phrase intended by the meaning; either joined to the wrong word or phrase, or completely unattached.

Although terms such as *misrelated*, *dangling*, *hanging*, *unattached*, etc. are most commonly applied to participles, verbless phrases can also be misrelated:

A rock-climber of some note, there is a story, never denied, of how he tackled the treacherous Aonach Eagach ridge in Glencoe by moonlight, dressed in a dinner jacket.

Here the first noun phrase semantically refers back to someone mentioned in the previous sentence and forward to the *he* of this one, but grammatically it is entirely unconnected to anything in the sentence.

Now nine years old, one day out walking, . . the question of her parentage arose.

Here the adjective phrase refers back to someone previously mentioned, the person whose parentage is mentioned in this sentence. But grammatically the phrase is misrelated to *the question*. The sentence also contains an unattached participle (*walking*).

Compare HANGING PARTICIPLE.

mixed metaphor See METAPHOR.

mobile

Of a linguistic unit: capable of being moved to different positions in the clause or sentence.

This concept is of some importance in differentiating units according to the way they function. For example, adverbials tend to be more mobile than other clause constituents:

It will be winter *soon*
It will *soon* be winter
Soon it will be winter

• **mobility**.

1985 R. QUIRK et al. The adverbial nature of the particle in . . phrasal verbs . . is generally shown by its mobility, its ability to follow the noun phrase: . . They turned *down* the suggestion. They turned the suggestion *down*.

or POSTMODIFICATION, e.g.

Travels *with my Aunt*

or a mixture of both, e.g.

Our Man *in Havana*

2 (In Systemic-Functional Grammar.) The same as *premodification*.

In this model, words following the headword in a noun group (i.e. a noun phrase) are generally treated as QUALIFICATION.

3 *Phonetics*. Any change to a linguistic unit.

1933 L. BLOOMFIELD The typical actions of the vocal organs described in the last chapter may be viewed as a kind of basis, which may be modified in various ways. Such modifications are: the length of time through which a sound is continued; the loudness with which it is produced; the musical pitch of the voice during its production; the position of organs not immediately concerned in the characteristic action; the manner of moving the vocal organs from one characteristic position to another.

4 *Morphology*. A change within a lexical word.

Examples are:

man → men
get → got

Compare SUPPLETION.

modified RP

The same as NEAR-RP.

modifier

1 A dependent element, normally a word, that affects the meaning of another element, normally the headword.

A *modifier* in noun phrase structure may be a PREMODIFIER or a POSTMODIFIER.

Other sentence elements, apart from noun phrases, may be modified. Adverbs frequently function as modifiers of other words, e.g.

really + useful (adjective)
very + badly (adverb)
almost + everyone (pronoun)
right + at the end (prepositional phrase)

In traditional grammar, an adverb (or adverbial) in clause structure is sometimes said to be the modifier of the verb: hence the name adverbial.

Sentence modifier is a way of describing a DISJUNCT, an adverbial referring to a complete sentence or clause.

•• **modifier clause**: a clause that postmodifies a headword (e.g. a defining relative clause).

2 (In Systemic-Functional Grammar.) A word or words preceding the headword in a noun group. Contrasted with a QUALIFIER.

modify

Of a word or similar element: affect the meaning of (a headword or principal element). See MODIFICATION, MODIFIER.

modulation

1 Deontic modality; the expression of obligation by a modal verb.

The use of a specific term for this type of modality allows *epistemic modality* to be called simply MODALITY.

2 A paralinguistic feature conveying some attitude not necessarily implicit in the actual words spoken.

> 1977 J. LYONS By the modulation of an utterance is meant the superimposing upon the utterance of a particular attitudinal colouring, indicative of the speaker's involvement in what he is saying and his desire to impress or convince the hearer.

Compare ATTITUDINAL.

monolingual

(*adj.*)

1 Speaking only one language.

2 Written in one language.

(*n.*) A person who speaks only one language.

monomorphemic

Of a word: consisting of a single morpheme.

E.g. *able*, *no*, *interest*.

monophthong

Phonetics. A vowel in which there is no change in the position of the vocal organs during articulation. Contrasted with DIPHTHONG.

> 1885 E. SIEVERS The sound of the so-called long *a* in *make*, *paper*, &c., although once a monophthong, is now pronounced as a diphthong.

English monophthongs are usually referred to as PURE VOWELS.

• **monophthongal**.

monophthongization

The conversion of a diphthong or triphthong into a single pure vowel.

• **monophthongize**: convert into a monophthong; become a monophthong.

The sound /ɔə/ as a distinctive phoneme in English, once frequently heard in such words as *door* and *more*, has in most varieties of RP been monophthongized into the pure long vowel /ɔː/.

Another noticeable tendency is for the triphthongs /aɪə/, as in *tyre*, and /aʊə/, as in *tower*, to monophthongize; either or both may fall in with the existing monophthong /ɑː/, giving us new homonyms such as *tyre/tower* and *tar* or *pyre/power* and *par*.

monosyllabic

Of a word: having only one syllable. Contrasted with DISYLLABIC, TRISYLLABIC, and POLYSYLLABIC.

The term is used in describing the formation of the comparative and superlative of adjectives, since almost all monosyllabic adjectives that are gradable can be inflected for these (e.g. *kind, kinder, kindest*). Many disyllabic adjectives have the option of inflecting or taking PERIPHRASTIC forms, while virtually all adjectives of more than two syllables can take only periphrastic forms.

It is also used in describing the pronunciation of some disyllabic words in which a syllable may be lost by elision, such as *police* pronounced /pliːs/.

• **monosyllable**: a monosyllabic word.

monotransitive

Of a verb: taking one and only one object. Contrasted with DITRANSITIVE and INTRANSITIVE.

Examples:

avoid the traffic
raise money

In some descriptive models, the term is extended to cover:

(*a*) prepositional verbs:
I *refer to* your letter of 4th June
(*b*) phrasal prepositional verbs:
I won't *put up with* this sort of treatment
(*c*) verbs taking a clause:
I *admit* that I did not check

•• **monotransitive complementation**: the complementation of a monotransitive verb.

mood

1 One of the formal categories into which verb forms are classified, indicating whether the verb is expressing fact, command, hypothesis, etc.

Traditional grammar recognizes the INDICATIVE, IMPERATIVE, and SUBJUNCTIVE moods.

2 A distinction of meaning expressed by any one of the chief sentence types. According to this definition, INTERROGATIVE joins INDICATIVE (or *declarative*) and IMPERATIVE as a mood category.

Mood is an alteration, apparently in the 16th century, of the earlier *mode*, a borrowing of Latin *modus* 'manner', which was also used in this grammatical sense. The alteration may have been due to the influence of the unrelated word *mood* 'frame of mind', which has an evident semantic affinity with it.

mood adjunct See MODAL ADJUNCT.

morph

1 The actual (physical) realization of an (abstract) morpheme when that morpheme only has one realization.

For example, the present participle morpheme is always the morph *-ing*.

2 Any of the actual spoken forms of an abstract morpheme; the same as ALLOMORPH.

morpheme

1 The smallest meaningful unit of grammar.

According to a basically syntactic definition, the *morpheme* is an abstraction (comparable to the *phoneme*). Thus 'the plural morpheme' is realized in regular nouns by phonologically conditioned ALLOMORPHS (/–s/, /–z/, and /–ɪz/) but also has such realizations as the changes of vowel in *men* (from *man*) and *mice* (from *mouse*) and ZERO as in *sheep*. Similarly we can posit a past participle morpheme, which has various allomorphs (*-ed*, *-en*, etc.), and a negative morpheme prefix, again with different realizations (e.g. *in-* , *im-*, *il-*, *un-*, etc., as in *intolerable, impossible, illegible, unassailable*).

2 The smallest unit in word formation and morphology.

In a more phonological approach, the morpheme is the smallest meaningful part into which a word can be broken down. Thus, both STEMS (ROOTS) and AFFIXES are seen as morphemes, some of them BOUND and some FREE. According to this definition *looked* and *fallen* each consist of a free morpheme + a bound morpheme (*look* + *-ed*, *fall* + *-en*). But whereas in a syntactic definition *-ed* and *-en* are variants (i.e. allomorphs) of the same (abstract) past participle morpheme, in a phonological definition *-ed* and *-en* are different morphemes.

Compare BASE.

morphemic

Of or pertaining to a morpheme.

•• **morphemic alternant**: the same as ALLOMORPH; see ALTERNANT (3).
morphemic variant: the same as ALLOMORPH.

• **morphemically**: as regards morphemes.

morphemics

The study and analysis of language in terms of its morphemes.

morphology

The study of word formation.

Traditionally *morphology* (concerned with the internal rules of words) contrasts with SYNTAX (concerned with the rules governing the way words are put together in sentences). Morphology itself covers two main types of word formation: INFLECTION, concerned with changes to an individual lexeme (which remains 'the same word') for grammatical reasons (e.g. showing number or tense) and DERIVATION, which is concerned with the formation of one word from another (e.g. by the addition of an affix).

•• **morphological**. **morphologically**.

morphosyntactic

Linguistics. Combining morphological and syntactic properties.

Morphosyntactic properties (or *morphosyntactic features*) are properties of a linguistic unit that have effects on both morphology and syntax. Thus tense, person, number, etc. are morphosyntactic.

Plural number, for example, morphologically requires *-s* in a regular noun (e.g. *cats, readers*) to be followed, syntactically, by a plural verb (e.g. *are, have, want*, and not *is, has, wants*).

• **morphosyntactically**. **morphosyntax**.

multal

Implying a largish number or amount, in contrast to PAUCAL.

In some detailed classification of pronouns and determiners, the label *multal* is given to a subset of quantifiers: *many, much, more, most*.

multilingual

(*adj.*)

1 Speaking several languages.

2 Written in several languages.

(*n.*) A person who speaks several languages.

• **multilingualism**.

1953 U. WEINREICH Multilingualism, the practice of using alternately three or more languages.

multiple

(Used as a very general term.) Having several instances of a linguistic feature combined.

•• **multiple analysis**: an analysis of a linguistic unit that can be made in two or more satisfactory ways.

There are many instances where a clear-cut, right or wrong description is elusive. One classic problem concerns certain structures with the pattern verb

+ preposition + noun. Most people would argue that *I waited at the bus-stop* should be analysed as subject + verb + adverbial (*Where did you wait? At the bus-stop.*), but the superficially similar *I looked at the timetable* may alternatively be analysed as subject + verb (*look at*) + object (*What did you look at? The timetable.*).

multiple apposition: apposition involving three or more terms, rather than the usual two.

multiple coordination: coordination having more than two coordinated units.

multiple meaning: see HOMONYMY, POLYSEMY.

multiple negation: three or more negatives combined. Compare DOUBLE NEGATIVE.

multiple postmodification: postmodification involving more than one separate postmodification.

e.g. The lady *over there / in the red hat.*

multiple premodification: premodification involving more than one distinct premodifier, rather than a mere sequence of adjectives; for example, a premodifier modified by another.

In *poor quality leather*, *poor* premodifies *quality*, and the two together premodify *leather*. (Contrast *beautiful black leather*, where the premodifiers are simply in sequence.)

multiple sentence: a sentence that is a COMPLEX, COMPOUND, or COMPOUND-COMPLEX sentence; the same as CLAUSE COMPLEX in another analysis.

multiple subordination: the presence of more than one subordinate clause within a sentence.

multi-word lexical item

A LEXICAL ITEM composed of more than one word.

multi-word subordinator

multi-word subordinator See SUBORDINATOR.

multi-word verb

A verb together with a PARTICLE (or two particles) functioning as a single verb.

This is an umbrella term that covers PHRASAL VERBS, PREPOSITIONAL VERBS, and PHRASAL-PREPOSITIONAL VERBS.

multi-word verb construction

An idiomatic verb construction that, like a multi-word verb, forms a cohesive unit.

Examples:

break even (verb + complement)
let go, put paid to (verb + verb)

mutation

(Especially in historical linguistics.) A change in a phoneme in a particular word context under the influence of adjacent sounds.

In the history of English, the most important form of mutation was *i-mutation* (or *i/j-mutation* or *umlaut*). This phenomenon is common to all the surviving Germanic languages and continues to have an important functional role in modern German, in which, for example, some noun plurals are distinguished from their singulars by *umlaut* (e.g. *Bruder* brother, *Brüder* brothers). In English, the results of this mutation can be seen in

(*a*) the plurals of seven nouns (*foot, goose, louse, man, mouse, tooth, woman*) which are sometimes called *mutation plurals*

(*b*) the comparative and superlative *elder*, *eldest*

(*c*) derivative verbs such as *bleed* (beside *blood*), *fill* (beside *full*), *heal* (beside *whole*), etc.

(*d*) derivative nouns such as *breadth* (beside *broad*), *length* (beside *long*), *filth* (beside *foul*), etc.

This cannot be considered to have a live functional role in modern English, however.

Compare ABLAUT.

N

name See PROPER NOUN.

narrative present

The same as HISTORIC PRESENT.

narrow transcription

Phonetics. A method of representing the sounds of spoken language in fine detail; an example of this. Contrasted with BROAD TRANSCRIPTION.

A narrow transcription gives a much more accurate indication of actual speech sounds, but requires more symbols and diacritics. The word *tall* in a broad transcription could appear as /tɔːl/. A narrower transcription would show, for example, that the *t* is aspirated, and that the *l* is dark: /tʰɔːɫ/.

Compare PHONEMIC.

nasal

Phonetics. (*n.* & *adj.*) (A speech sound) made with an audible escape of air through the nose while the soft palate is lowered. Contrasted with ORAL.

English has three nasals, all of which are consonant phonemes:

bilabial /m/ as in *more, whim*
alveolar /n/ as in *no, win*
velar /ŋ/ represented by *ng* in *wing* and *n* in *wink* (and never world-initial in English)

Other nasal sounds occur in other languages: for example, a palatal nasal /ɲ/ in Spanish, as in the middle of *mañana*, and four nasal vowels in French, which contrast with homorganic oral vowels.

Nasal plosion (or *nasal release*) refers to the release of a normally oral plosive through the nose, usually under the influence of a following nasal at the same place of articulation. Thus nasal plosion may sometimes be heard in such words as

one-u*pm*anship, su*bm*erge, co*tton*, no*t n*ow, woo*den*

'Nasal twang', as a term for the accent of an individual speaker in which sounds are more nasal than in the average speaker's voice, is colloquial, but not favoured by phoneticians.

nasalize

Phonetics. Articulate with the air escaping through the nose rather than, as would be usual, through the mouth.

• **nasalization.**

English vowels can become *nasalized* under the influence of adjoining nasal consonants, e.g. in *manning* or *meaning*.

1888 H. SWEET Nothing is more common than the nasalizing influence of a nasal on a preceding vowel.

nationality word

A noun referring to a member of a nation or ethnic group, or a related adjective.

Grammatically these words are to some extent treated like names, being spelt with an initial capital, but at the same time, like common nouns, they are countable, and can have specific and generic reference. The commonest type has the singular noun identical to the adjective, and forms the plural (both specific and generic) with *-s*:

Italian (*adj.*): an Italian; Italians; the Italians; (similarly *Greek*, *Pakistani*)

Those with an adjective in *-sh* or *-ch* fall into two main types:

(*a*) Danish; a Dane; Danes; the Danes
(*b*) French; a Frenchman/Frenchwoman; Frenchmen/Frenchwomen; the French

Those formed with *-ese* (and *Swiss*) are invariable in all uses:

Chinese (*adj.*); a Chinese; Chinese; the Chinese

These patterns have numerous minor irregularities and variations (e.g. a *Briton*, *Britons*; *a Spaniard*, *Spaniards*, *the Spanish*).

native speaker See *first* LANGUAGE.

natural gender See GENDER.

natural language See LANGUAGE.

near negative

The same as SEMI-NEGATIVE.

near-RP

Any of a group of accents that are similar to Received Pronunciation, but differ from it in a number of relatively minor ways. (Also called MODIFIED RP.)

1982 J. C. WELLS The term *Near-RP* refers to any accent which, while not falling within the definition of RP, nevertheless includes very little in the way of regionalisms which would enable the provenance of the speaker to be localized within England . . . By the generality of the population it will be perceived as indeed 'educated', 'well-spoken', 'middle-class'.

Compare PARALECT.

necessity

One of the extreme parameters, the other being POSSIBILITY, used in describing the meanings of modal verbs.

Both deontic and epistemic meanings can be fitted on to such a scale, though moral (deontic) necessity may be designated as *obligation* or *duty* (e.g. *You must do your best*), leaving the term *necessity* to apply only to logical (epistemic) necessity (e.g. *It must be cold there in winter*).

negate

Make (usually, a clause or sentence) negative in meaning.

negation

The grammatical process by which the truth of an AFFIRMATIVE (or *positive*) clause or sentence is denied.

Typically an English clause or sentence is negated by adding *not* or *-n't* to the primary verb or to the first (or only) auxiliary:

This *is not* difficult
He *couldn't* have been thinking

In the absence of an auxiliary, the auxiliary *do* is introduced and takes over the finite function of the verb, while the latter takes on the bare infinitive form:

It *doesn't* matter

Generally the scope of the negation extends from the negative word to the end of the clause; hence the difference in meaning between such pairs as:

I didn't ask you to go; I asked you not to go
They aren't still here; They still aren't here

Sentences may also be negated through the use of other negative words:

There is nothing to do (*Compare* There isn't anything to do)
It's no trouble (*Compare* It isn't any trouble)
Nobody told me (*Compare* They didn't tell me/I wasn't told)

A negative affix makes a word negative, but not the whole sentence:

He looked worried and *un*certain
Perhaps they will sign a *non*-aggression pact

• • **transferred negation**: the positioning of a negative in the main clause, when logically it belongs in the subordinate clause.

Transferred negation is often found with verbs of opinion and perception, e.g.

I don't think you understand (= I think you don't understand)
It doesn't look as if they're coming now (= It looks as if they are not coming)

Compare DOUBLE NEGATIVE, MULTIPLE *negation*, NON-ASSERTIVE, SEMI-NEGATIVE.

negative

(*n. & adj.*) (An affix, word, clause, etc.) that expresses negation.

negative particle

A term sometimes used for the word *not*.

negator

A word expressing negation, particularly the word *not*.

neo-Firthian

(*adj.*) Of, pertaining to, or characteristic of a group of linguists who continued some of the principles developed by J. R. Firth.

(*n.*) A person belonging to the neo-Firthian group of linguists.

See FIRTHIAN.

neogrammarian

An adherent of (the views of) a loosely knit group of German scholars in the late nineteenth century who believed that SOUND LAWS must be entirely regular and without exceptions.

It soon became apparent that this theory was overstated, but a modified version of the principle is an important tenet of comparative and historical linguistics. The name is a translation of the German term *Junggrammatiker*.

neologism

(The coining or use of) a new word or expression.

Neologisms have various sources. They may be the result of

ABBREVIATION (e.g. *HIV*)
BACK-FORMATION (e.g. *ovate* from *ovation*)
BLENDING (e.g. *camcorder* from *camera* + *recorder*)
BORROWING (e.g. *karaoke* from Japanese)
CLIPPING (e.g. *cred* from *credibility*)
COMPOUNDING (e.g. *power dressing*)
CONVERSION (e.g. *to doorstep*, verb from noun)
DERIVATION (e.g. *fattism* from *fat* + *-ism*)

The term is also sometimes extended to include old words given new meanings (e.g. *wicked* 'marvellous').

Compare NONCE, NON-WORD.

nest

Place (a clause or phrase) within a larger structure of the same kind.

nesting

1 The inclusion of (usually) a clause or phrase within an endocentric construction.

The views of the man in the street is a noun phrase, consisting of a noun phrase head (*the views*) modified by an (exocentric) prepositional phrase (*of the man in the street*). But *the man in the street* is itself a noun phrase, consisting of head (*the man*) postmodified by a prepositional phrase (*in the street*). It is therefore exactly parallel to the one in which it is situated:

The views [of [the man [in the street]]]

Thus similarly constructed phrases are 'nested' inside each other.

Nesting can be almost infinitely RECURSIVE, as in this blurb for an article:

Why the pedestrian hates the cabby who hates the biker who hates the cyclist who hates the man who drives the coach that drives the lorry driver mad

2 (The fact of being) embedded in mid (medial) position.

1985 R. QUIRK et al. Initial clauses are said to be LEFT-BRANCHING, medial clauses NESTING, and final clauses RIGHT-BRANCHING . . . Nesting (medial branching) causes the most awkwardness, if the nested clause is long and itself complex.

In both definitions nesting is a type of embedding, but definition (1) excludes the embedding of many subordinate clauses (because they are structurally different from their superordinate clause), whereas definition (2) depends on the embedding of a subordinate clause, and the positioning of the clause, but implies no restriction on the type of clause.

neuter See GENDER.

neutral

Phonetics. Of the position of the lips: neither SPREAD nor ROUNDED.

The term is often used in describing the articulation of vowels. Although vowel quality is largely dependent on the height of the tongue, vowel sounds are affected by lip position.

Spread and *neutral* are sometimes lumped together as *unrounded*, but the two may be distinguished. Compare the typically spread lips required for English /iː/ in *meet, seed, eat* with the more neutral lip position that is typical of /æ/ as in *mat* or *sad*.

neutralization

Blurring of two distinct forms.

1 *Phonology*. The blurring of the usual distinction between two different phonemes in certain contexts, so that it is not possible to say which is being used.

English voiceless and voiced plosives normally contrast, and one of the distinguishing features, when they are in initial position, is that the voiceless series (/p/, /t/, and /k/) are aspirated. Contrast *pray* and *bray*, *tin* and *din*, *coal* and *goal*. But aspiration is lost after a preceding /s/, so that the distinction is neutralized. Hence there is no contrast in English of pairs such as *spare* and

**sbare*, *stem* and **sdem*, *sky* and **sgy*, and the /p/, /t/, /k/ words that actually exist could equally well be transcribed with a /b/, /d/, or /g/.

Similarly the vowels /iː/ and /ɪ/ must normally be distinguished, as in *seen* and *sin*, but in a weak syllable (e.g. at the end of a word like *city*) the distinction has no effect on meaning, and the sound actually used may be indeterminate.

2 The disappearance, in a particular context, of a grammatical distinction that is made in normal circumstances.

Contrasts of tense and aspect may undergo neutralization in certain circumstances. Those in the three different statements

She has gone
She had gone
She went

are neutralized after a past reporting verb:

They said she had gone

Similarly, the distinction between simple and progressive forms is neutralized in

Leaving the house, he tripped over the mat (= As he left *or* As he was leaving)

The lack of a distinction between the subject and object forms of the pronoun *you* is also a neutralization. More loosely, so is the double function of the noun phrase in a clause such as

We want Henry to go

where *Henry* is, arguably, both the object of *want* and the subject of *go*: but note that only an object pronoun is possible in such a clause:

We want him to go

neutralize

Subject (a contrast or distinction) to NEUTRALIZATION.

neutral vowel

Phonetics. A term sometimes applied to the central weak vowel /ə/, usually referred to as SCHWA.

new

(*n. & adj.*) (Designating) the not already known and therefore the important information in an utterance.

See GIVEN.

nexus See JUNCTION.

NICE properties

Four characteristics of English auxiliary verbs that distinguish them from full verbs.

NICE is an acronym standing for:

Negation.	Auxiliaries add *not* or *-n't* directly to the verb form (e.g. *don't, cannot*)
Inversion.	They invert with the subject in questions (e.g. *May* I go now?)
Code.	They can be used to avoid repetition of the whole verb phrase (e.g. I love going to concerts, and so *does* Jane)
Emphasis.	Auxiliaries can be used for emphasis (e.g. We *will* help you. I *do* remember)

node

Linguistics. Any point in a TREE DIAGRAM from which branches lead off. The term is particularly used in Generative Grammar.

1976 R. HUDDLESTON [A tree diagram] consists of a two-dimensional arrangement of labelled NODES connected by lines or branches. The topmost node . . is called the ROOT.

nominal

(*adj.*) Of (the function of) a word or phrase: nounlike because it is functioning as a noun or noun phrase.

The term is more comprehensive than the word *noun* strictly is, since not only nouns and pronouns but also other parts of speech can have nominal function. For example, in *The great and the good don't want to know*, *the great and the good* is nominal, functioning as the subject, but *great* and *good* are not nouns here: they are not pluralized (**The greats and the goods*) nor could we speak (except facetiously) of **a great and a good.*

(*n.*) A word or phrase functioning as a noun or noun phrase.

nominal clause

1 A clause functioning like a noun (or noun phrase). (Also called *noun clause.*)

Nominal clauses, other than nominal relative clauses, tend to be abstract in meaning. A nominal clause can be a subject, object, or complement in sentence structure:

[*What happened next*] remains a mystery (S)
He alleges [*(that) he doesn't remember a thing*] (O)
The question is [*how we should proceed*] (C)

and can function in various other ways:

I'm not sure [*if we should report this*] (complement of adjective)
It depends on [*what happens next*] (complement of preposition)
The question, [*whether this is a criminal matter*], is not easy to answer (apposition)

In grammatical analyses which allow non-finite clauses, nominal clauses include *-ing* clauses and infinitive clauses:

He's talking about *facing the music*
To err is human
All I did was *laugh*

2 (In some popular grammars.) Restricted in various ways, e.g. to mean NOMINAL RELATIVE CLAUSE.

In these descriptions, other nominal clauses (in sense 1) are simply called *that-clause*, *reported question*, and so on.

Compare NOMINAL RELATIVE CLAUSE.

nominal conjunction See CONJUNCTION.

nominal group

A group of words, usually with a noun as head, functioning in a clause like a single noun.

Nominal group is the term used in Systemic-Functional Grammar that corresponds to NOUN PHRASE in other models. The term is preferred because of the distinction, in that grammar, between the make-up of a GROUP and a PHRASE.

nominalization

A noun or noun phrase derived from, or corresponding to, another part of speech or a clause; the process by which such a phrase is derived.

The term is particularly useful in describing the way in which a clause can be turned into a noun phrase. For example,

talking heads
her determination to succeed
eating people (is wrong)

can be said to be nominalizations of

heads talk
she is determined to succeed
if you eat people

The derivation of single nouns from words belonging to other parts of speech is also called *nominalization*, and the nouns are examples of the process, e.g. *buyer* from *buy*, *actuality* from *actual*.

nominalize

1 Form a noun from (another part of speech).

Examples:

driver from *drive*
examinee from *examine*
shrinkage from *shrink*
accuracy from *accurate*
kindness from *kind*

The process often involves the use of a *nominalizing affix*.

2 Form a noun phrase from (a clause). See NOMINALIZATION.

• **nominalizable**.

nominal relative clause

A type of clause which has a nominal function, but which like many relative clauses begins with a *wh*-word, though unlike a relative clause it contains the antecedent within itself. (Also called *fused relative construction*, *independent relative clause*, or *free relative clause*. In popular grammar, nominal relative clauses are not distinguished from *nominal/noun clauses*.)

A nominal relative clause can refer to people and things, as well as to abstract ideas. Examples:

I don't know *what happened* (= I don't know [that which] happened)
Whoever told you that was wrong (= [that person who] told you that was wrong)

nominative

(*n. & adj.*) (In traditional grammar.) (The case) used for the subject of the verb.

The term was at one time applied to English nouns and pronouns. But its use implies a system of case inflection, lacking in English except in a few pronouns. As a label for the sentence function of a noun phrase, SUBJECT (or *subjective*) is preferred.

Compare ACCUSATIVE, CASE.

nominative absolute See ABSOLUTE.

non-agentive See AGENTIVE, PASSIVE.

non-anterior See ANTERIOR.

non-assertive

Designating a class of words and phrases that tend to be restricted to questions and negative contexts and to other tentative statements such as conditional clauses. Contrasted with ASSERTIVE.

In addition to the *any*-series of words (e.g. *any, anybody, anyone, anything, anywhere*) which contrast with the corresponding words in the *some*-series, predominantly non-assertive words include

either: Jane didn't know either (*compare* Jane knew too)
ever: Have you ever had a winter holiday? (*compare* I always have a winter holiday)
far: How far is it? Not far (*compare* It's a long way)
much: There isn't much food left (*compare* There's a lot of food left)
yet: Haven't you finished yet? (*compare* I have already finished)

•• **non-assertive territory**: see TERRITORY.

non-attributive

The same as PREDICATIVE.

nonce

(*adj.*)

Of a word, form(ation), etc.: deliberately coined for one occasion.

(*a*) (As originally used.)

> 1907 *New English Dictionary* *Nonce-word*, the term used in this Dictionary to describe a word which is apparently used only for the nonce.

The implication of the original use was that the word in question has only been used once (or a very few times) by a single author, or possibly once each by more than one author independently.

(*b*) The term is now frequently used of a word that, having been coined for a particular occasion or purpose, has become common. Quite a number of words can be traced back to their originators, although (as has been the case with Shakespeare) some authors have been credited with originating a word when they were merely the earliest known user, and have lost this distinction when subsequent research has unearthed an earlier example. But some coinages are reliably documented: T. H. Huxley invented *agnostic*; Jeremy Bentham gave us *international*; Horace Walpole coined *serendipity*; and more recently, Dr. M. Gell-Mann gave the name *quark* to the subatomic particle.

(*c*) The term is sometimes loosely used for jocular-sounding words that seem unlikely to last long (perhaps because a word that is generally adopted does not seem like a word for one occasion, or by association with 'non-sense').

> 1986 S. MORT On the matter of durability, this book can be judged only by the reader of the 1990s. Of course, some words are probably nonce.

Examples of this kind of word might be *jocumentary, nepotocracy, oldcomer, trendicrat*.

(*n.*)

A nonce-word.

> 1986 S. MORT Private Eye, master of the nonce, again demonstrates this kind of innovation.

Derived from the phrase *for the nonce* 'for the particular purpose; for the occasion, for the time being'. This is a Middle English METANALYSIS of the phrase *for than anes* 'for the one (thing, occasion, etc.)'

Compare NEOLOGISM, NON-WORD.

non-conclusive See CONCLUSIVE.

non-consonantal See CONSONANTAL.

non-count

Usually the same as UNCOUNT.

See also MASS.

non-defining

Of modification or a modifier: giving additional information about the headword. See DEFINING. (Also *non-identifying/non-restrictive.*)

•• **non-defining relative clause**: a relative clause that gives additional information about the noun phrase to which it belongs, but is not a *defining relative clause* because the noun phrase is already defined and identifiable.

A non-defining relative clause is usually separated from the rest of the sentence by a comma or commas, and if it is omitted the sentence will still make complete sense. e.g.

My mother, *who now lives alone*, does The Times crossword every day.

Contrast:

A woman I know does six crosswords a day.

which does not make the sense intended without the defining relative clause *I know*.

Compare *sentential* RELATIVE *clause*.

non-equivalence See EQUIVALENCE.

non-factive See FACTIVE.

non-factual See FACTUAL (I).

non-finite

(*n. & adj.*) (A verb form or a clause) without tense; contrasting with FINITE.

The term is wider than INFINITIVE, since it also covers the *-ing* form (e.g. *looking*, *knowing*) and the past participle (e.g. *looked*, *known*).

A *non-finite clause* is a clause whose verb is non-finite (an infinitive, an *-ing* participle, or an *-en* participle). E.g.

To expect a refund is unreasonable
All he ever does is *complain*
Having said that, I still hope he gets one
If consulted, I would have advised against

A non-finite clause may function as an integral sentence element (as in the first two examples here), or as a separate subordinate clause (as in the third and fourth). A non-finite clause may contain its own subject:

For him to expect a refund is unreasonable

Compare ABSOLUTE CLAUSE, HANGING PARTICIPLE, PARTICIPLE CLAUSE.

non-gradable See GRADABLE.

non-headed

The opposite of HEADED; the same as EXOCENTRIC.

non-identifying

The same as NON-DEFINING.

See IDENTIFY.

non-inherent See INHERENT.

non-linguistic

The same as EXTRALINGUISTIC.

non-past

(*n. & adj.*) Present (tense).

Morphologically, English has only two tenses, usually called the present simple (e.g. *look(s)*, *come(s)*, *can*, etc.) and the past simple (e.g. *looked*, *came*, *could*, etc.). The 'present' label fairly indicates the commonest use of the tense, but fails to cover its other regular meanings: for example, future meaning (e.g. *If you come tomorrow* . . .) or timeless reference (e.g. *Ice melts when you heat it*), and so on. For this reason, some linguists prefer the term *non-past* for this tense, claiming that it is not so much present as *unmarked* for time.

Compare MARKED.

non-perfective See PERFECT.

non-personal

Of noun or pronoun meaning: referring to something not regarded as having human personality, including inanimate things, abstract entities, and animals.

Non-personal constitutes a semantically based gender classification for dealing with certain distinctions, related to meaning, in nouns and pronouns.

Non-personal nouns include all nouns other than those referring to people, but the usage with regard to pronouns is not always straightforward.

Grammatically there is often a complication with nouns or actual referents and co-referential pronouns. Pronouns normally referring to people (e.g. *he/him*, *she/her*, *who/whom*) are sometimes used of animals and even things (e.g. *this ship and all who sail in her*), while *it* and *which* may refer to people (e.g. *A child needs its mother. Which of my cousins do you mean?*).

Non-personal is therefore a useful semantic label when there is some apparent mismatch. It should also be noted that, of the so-called PERSONAL pronouns, *it* is usually non-personal in meaning, while *they* and *them* can have personal or non-personal reference.

Compare IMPERSONAL.

non-plural

Not plural.

A useful term to describe the use of the demonstratives *this* and *that*, which can be used with both uncount nouns (*that food*) and singular nouns (*this apple*). Unfortunately most grammars inaccurately and confusingly use the term SINGULAR.

Compare NON-SINGULAR.

non-predicative

The same as ATTRIBUTIVE.

non-progressive See PROGRESSIVE.

non-proximal See PROXIMAL.

non-restrictive

The same as NON-DEFINING.

The term usually refers to relative clauses, but can be more widely applied:

> 1966 G. N. LEECH Proper nouns do occasionally combine with modifiers of non-restrictive force: 'fair Helen' . . ; 'beautiful Britain'.

See DEFINING.

non-rhotic See RHOTIC.

non-sentence

A group of words (in written or spoken language) functioning as a complete sentence, but lacking normal clause structure.

Similar terms, often not very precisely defined, include MINOR SENTENCE and UTTERANCE.

Non-sentences include formulae and interjections. Examples:

You and your headaches!
Nice one, Norman!
You fool
Of all the daft things to do!
Whatever next?
No way
No taxation without representation

non-sibilant See SIBILANT.

non-singular

Designating (the usage of) those determiners and pronouns that can have count or mass meaning.

The determiner and pronoun system of English is quite complicated. Some words relate only to *count singular* (e.g. *a*, *each*, *one*); some only to *uncount* (e.g. *much*); and some to *plural* only (e.g. *few*, *many*, *several*, *these*). There are also some words that overlap two categories: NON-PLURAL words can be used for both *count singular* and *non-count*; *non-singular* words are used for both *non-count* and *plural*. Non-singular words include *all*, *any*, *enough*, *some*, *most*.

non-specific See SPECIFIC.

non-standard See STANDARD.

non-vocalic See VOCALIC.

non-word

1 A word that is not recorded or not established.

This may be interchangeable with NONCE-WORD, but tends to be restricted to inventions that could be unintentional errors rather than deliberate coinages:

1963 *Punch* The aesthetically displeasing non-word 'annoyment'.

2 A string of letters (or sounds) that is not an English word.

1967 D. G. HAYS Almost every letter string is a nonword.

normative

Prescriptive.

Normative grammars or *normative rules* prescribe what is correct, rather than describing language as it is used. The term has largely been replaced by PRESCRIPTIVE as a pejorative applied to outdated or misconceived rules.

notation

A set of symbols used in a phonetic or phonemic TRANSCRIPTION. (Also called *script*.)

notional

Based on meaning.

(*a*) Traditional grammar, in which, for example, parts of speech are defined in terms of meaning (e.g. 'A verb is a doing word') rather than by syntax, is sometimes called *notional grammar*. *Notional* in this sense contrasts with FORMAL or GRAMMATICAL (1) and today has somewhat pejorative overtones, suggesting 'sloppy', 'unvigorous'.

(*b*) In the teaching of English as a foreign language, the term *notional* was applied in the 1970s to syllabuses aimed at developing COMMUNICATIVE competence. D. A. Wilkins's *Notional Syllabuses* (1976) advocated syllabuses based primarily on semantic criteria, in contrast to the older type of grammatical or 'situational' course, although this did not exclude 'adequate learning of the grammatical system'. Suggested *notional categories* covered three areas: semantico-grammatical categories (e.g. time and space), modal meaning, and functions (e.g. how to express disapproval, persuasion, or agreement).

Notional in this sense still contrasts with FORMAL, but is a term of praise: 'meaningful' and 'communicative' rather than formal and 'boring'.

In later developments in foreign language teaching, the term *notional* tended to be restricted to the first category (general concepts of time and space, etc.) which were explicitly contrasted with FUNCTIONS, such as agreement or suasion.

• **notionally**.

notional concord See CONCORD.

noun

1 A word other than a pronoun that belongs to the WORD-CLASS that inflects for plural, and that can function as subject or object in a sentence, can be preceded by articles and adjectives, and can be the object of a preposition.

In traditional grammar, nouns are defined notionally as 'the name of a person, place, or thing'. But this definition only partly works; abstract nouns like *criticism* or *tolerance* are hardly things, and it is syntax, not meaning, that decides that *think* is a verb in one sentence (*I must think*) and a noun in another (*I'll have a think*). Modern grammarians therefore prefer more formal, syntactical definitions.

Nouns are divided on syntactic and semantic grounds into PROPER NOUNS and COMMON *nouns*, and common nouns are further divided into COUNT and UNCOUNT. The division into ABSTRACT and CONCRETE is notional, and cuts across that between count and uncount.

See SUBSTANTIVE.

2 (Loosely.) A noun phrase or even a non-finite verb phrase in nominal function.

Compare NOUN-EQUIVALENT.

nounal

Of or pertaining to a noun; nominal. (Obsolescent.)

noun clause

The same as NOMINAL CLAUSE (1).

noun-equivalent

A word or words functioning like a noun.

This is a somewhat dated term, covering not only NOUN PHRASE but also NOMINAL CLAUSE.

noun modifier

1 A noun in attributive position. (Also called *noun premodifier*.)

e.g.

book review, *sun*-hat, *toffee* apple

2 A word modifying a noun.

1958 W. N. FRANCIS The most common noun modifier is the adjective.

noun phrase

A word or group of words functioning in a sentence exactly like a noun, with a noun or pronoun as HEAD.

A *noun phrase* (abbreviated *NP*) can be a noun or pronoun alone, but is frequently a noun or pronoun with pre- and/or post-modification:

the name
an odd name
the name of the game
the name he gave

Compare NOMINAL GROUP. See also INDEFINITE *noun phrase*.

NP

Abbreviation for Noun Phrase.

nuclear

1 *Phonetics & Phonology*. Being or constituting a NUCLEUS.

2 **Nuclear English**: a proposed simplified form of English, intended to be used as an international language.

1985 R. QUIRK et al. Following earlier attempts (such as 'Basic English') that were largely lexical, a proposal has also recently been made for constructing a simplified form of English (termed 'Nuclear English') that would contain a subset of the features of natural English; for example, modal auxiliaries such as *can* and *may* would be replaced by such paraphrases as *be able to* and *be allowed to*. The simplified form would be intelligible to speakers of any major national variety and could be expanded for specific purposes, for example for international maritime communication.

nucleus (Plural nuclei.)

Phonology. The one essential element in a TONE UNIT, characterized by a PITCH change.

By definition a *tone-unit* must contain a *nucleus*, which may occur on a single syllable, as in *Right!* or *Yes?* A single word said without marked pitch would be interpreted as an incomplete utterance. Nuclei are analysed into various types, such as FALL, RISE, FALL-RISE, RISE-*fall*, and these are sometimes further distinguished as *high fall*, *low fall*, etc.

In a clause or sentence said unemphatically, the *nuclear* (or *tonic*) *pitch* occurs on the last accented syllable, e.g.

What are you DOING?

but the nucleus may occur earlier for CONTRASTIVE stress, e.g.

WHAT are you doing? Did I HEAR correctly?

null anaphor See ANAPHOR.

number

1 A grammatical classification used in the analysis of word classes which have contrasts of SINGULAR and PLURAL.

Number contrasts in English are seen in nouns (e.g. *boy*, *boys*), pronouns (*she*, *they*; *myself*, *ourselves*; *this*, *these*), determiners (*this*, *these*; *each*, *all*), and verbs (*say*, *says*; *was*, *were*).

Compare AGREEMENT, COLLECTIVE, COUNT, DUAL (1), PLURAL.

2 A NUMERAL.

numeral

(*n. & adj.*) (A word) denoting a number.

As the word *number* is used as a grammatical category, *numeral* is often the preferred word for referring to the series *one, two, three,* etc. and *first, second, third,* etc. See CARDINAL, ORDINAL.

In traditional grammar, numerals (numbers) may be treated as a subclass of adjectives (and called *numeral adjectives*) or divided between adjectives and pronouns. Modern grammar prefers to treat numerals as determiners (strictly POSTDETERMINERS) and as pronouns.

numerative

(*n. & adj.*) (A word) denoting an amount or quantity.

This is used as a wider term than *numeral* (*cardinal* and *ordinal*) to include terms for indefinite quantity, e.g.

few, little, several

and also words related to ordinal numerals (ordinals) by reason of their 'ordering' function, e.g.

next, last, preceding, subsequent

Compare QUANTIFIER.

O

O

Object as an ELEMENT of clause structure.

object

(*n.*)

1 The DIRECT OBJECT.

The *direct object* is usually a noun phrase in form and, in an unmarked declarative sentence, follows the verb. It contrasts with INDIRECT object.

Traditionally, the object is said to be 'affected' by the verb, but essentially it is defined on syntactic, not semantic grounds; therefore 'affected' must be loosely interpreted. It must, for example cover *a flat*, *a house*, and *a mistake* in the following:

We've bought a flat. We really wanted a house. I hope we haven't made a mistake.

See also COGNATE *object*.

2 (In modern analyses.) One of the five elements in sentence structure (along with Subject, Verb, Complement, and Adverbial).

In this use, both *direct object* (DO) and *indirect object* (IO) may be represented simply by O. Thus

She + gave + the poor dog + nothing

would be represented as SVOO.

A direct object may be realized by a clause, finite or non-finite:

He said *(that) he was not coming*
They told us *not to go*

•• **object complement**: see COMPLEMENT.

3 The case taken by one of the pronouns that are morphologically marked when not in subject position.

In English there are only six distinct *object* (or *objective*) *pronouns*:

me, her, him, us, them, whom

4 **object of a preposition**: a word or words following (and 'governed' by) a preposition. See PREPOSITIONAL OBJECT.

e.g.

a bundle of *nerves*
look to *your laurels*

Compare ACCUSATIVE. See OBJECTIVE.

objective

1 (*n. & adj.*) (The case) expressing the object.

The differences between *subjective pronouns* (or *subject pronouns*) (e.g. *I*, *she*, *they*, etc.) and *objective* (or *object*) *pronouns* (e.g. *me*, *her*, *them*, etc.) can be described in terms of *subjective* and *objective case*.

2 Relating or referring to the object.

•• **objective complement**: see COMPLEMENT.

In *objective genitive* the reference is to a 'deep' object rather than to the object of the actual sentence or clause. Thus the genitive has objective meaning in

Caesar's assassination (cf. They assassinated *Caesar*)

Contrast SUBJECTIVE *genitive*.

3 (*n. & adj.*) (In Case Grammar.) (Designating) one of six original cases, defined by meaning, that underlie surface structure.

This case, sometimes also called AFFECTED or PATIENT, is narrower than the traditional object case.

object of result: see RESULT.

object-raising See RAISING.

object territory

Any noun phrase position after the verb. Contrasted with SUBJECT TERRITORY.

The term has been coined partly to explain the common tendency to use object pronouns where subject pronouns are considered grammatically correct, e.g.

That's him
You were quicker than me

Compare also the hypercorrection associated with PUSHDOWN:

They're looking for two men *whom* they think can help them with their inquiries

where the relative pronoun is felt to be in the object territory of *we think*.

obligation

One of the main meanings of DEONTIC modality, along with PERMISSION and VOLITION.

This covers the laying of a duty on someone (possibly oneself), e.g. *You must try harder*, *I must go now*.

obligatory

Of a word or structure: compulsory in a particular context. Contrasted with OPTIONAL.

Various words and structures can be analysed in terms of obligatory or optional elements. A finite verb phrase, for example, is obligatory in finite clause structure, whereas an object is not essential in the same way. Many individual verbs may be either transitive or intransitive (e.g. *He's cooking dinner*, *He's cooking*), but with some verbs an object is obligatory (e.g. *He's making dinner*, **He's making*) and some require an obligatory adverbial (e.g. *Put the food on the table*, **Put the food*).

In earlier Transformational Grammar, some transformational rules (needed to produce acceptable surface structures) are obligatory, while others are a matter of choice. For example, a *do*-support rule is obligatory if a simple tense of a lexical verb is to appear in a question or negative (**Like you it?*, **I like not it*) but the insertion of *do* for emphasis is optional (e.g. *I (do) like it*).

oblique

1 Designating any case other than subject(ive).

In an inflected language, all inflected cases of nouns, pronouns, and adjectives (other than subject(ive), i.e. NOMINATIVE) are covered by this umbrella term.

In English the term is occasionally applied to the object forms of those pronouns that have them.

2 (By extension.) Designating (the semantic case of) a noun phrase that does not directly follow a verb, but is the object of a prepositional verb, as in *I rely on my neighbours*, where *my neighbours* could be described as being an *oblique object* of *rely*.

obstruent

Phonetics. (*n. & adj.*) (A speech sound) made with a full or partial stoppage in the airflow. Contrasted with SONORANT.

This is a cover term for plosives, fricatives, and affricates.

occurrence See PRIVILEGE OF OCCURRENCE.

of-construction (Also called *of-phrase*.)

1 A phrase consisting of noun + *of* + noun.

Of-construction is a wider term than *of-genitive*. It is often equivalent in meaning to and interchangeable with a genitive construction (e.g. *the West End of London*, *London's West End*), but this is not always so, and *of*-constructions are sometimes preferred or essential: *the end of the road*, *a book of verse*, *an object of ridicule*, *a man of honour*.

2 A phrase consisting of *of* + noun.

The term is sometimes used to distinguish part of an *of*-construction from the whole, as when indefinite pronouns such as *all* or *many* are said to be able to 'take the partitive *of*-construction'.

The terms *of-construction* and *of-phrase* are both used somewhat loosely.

off-glide See GLIDE.

of-genitive

A phrase consisting of *of* + noun phrase, corresponding closely in meaning and function to a GENITIVE noun phrase.

For example

George V was the grandfather *of Queen Elizabeth II*

is roughly equivalent to

He was *Queen Elizabeth II's* grandfather

Similarly

the mother *of my friend* = *my friend's* mother
the West End *of London* = *London's* West End
the arrest *of the man* = *the man's* arrest
the message *of the book* = *the book's* message

***of*-phrase** See *OF*-CONSTRUCTION.

of-pronoun

A pronoun that can be followed by a PARTITIVE *of*-phrase.

For example

few (of those people)
much (of the time)
some (of our problems)

Old English

The form of English used in Britain from the earliest records until circa 1150; the earliest stage in the development of the English language. Abbreviated *OE*. (Also called *Anglo-Saxon*.)

Although Anglo-Saxon rule came to an end with the Norman Conquest of 1066, a written form of Old English continued in use until the twelfth century.

The Old English dialects were highly inflected. Nouns had grammatical gender and four cases, singular and plural; adjectives agreed with nouns; verbs inflected for person and number. Without study, Old English is largely incomprehensible to modern English speakers, but most of our core vocabulary is of OE origin, e.g. *man, woman, child, be, go, come, sit, stand, young, old,* as are our remaining inflections.

Compare MIDDLE ENGLISH, MODERN ENGLISH.

omission of words See ELLIPSIS.

on-glide See GLIDE.

onomatopoeia

The formation of a word with sounds imitative of the thing which they refer to; the use of such a word.

e.g.

cuckoo, cock-a-doodle-do, neigh, miaow.

The term is sometimes extended to cover words in which a sound is felt to be appropriate to some aspect of meaning, although the words do not necessarily denote sounds or sources of sound. The combination *sl-*, often occurring in words with unpleasant connotations, is sometimes cited as an example of such *secondary onomatopoeia* (e.g. *slag, slang, slattern, slaver, sleazy, slime, slop, sluggard, slurp, slut*). Other terms for onomatopoeia are PHONAESTHESIA and SOUND SYMBOLISM. Cf. ICON.

opaque

Not obvious in structure or meaning; not able to be extrapolated from surface structure; (of a phonological rule) not able to be extrapolated from every occurrence of the phenomenon. Contrasted with TRANSPARENT.

open

1 *Phonetics.* Of a vowel: made with the tongue low in the mouth, and the mouth somewhat open. (Also called LOW (1).)

Contrasted with CLOSE.

English RP /æ/ as in *hat* is the most open front vowel; /ɑː/ as in *father, car, heart, clerk, half, aunt* is the most open back vowel.

Compare HALF-OPEN.

2 Of a word class: capable of acquiring a theoretically infinite number of new words. (Also called MAJOR.) Contrasted with CLOSED (1).

The main open classes are nouns, lexical verbs, adverbs, and adjectives. Items belonging to these are sometimes called *open-class items*.

See INTERJECTION, NUMERAL.

3 Of a syllable: ending with a vowel sound (e.g. *see, new, hair, why*). Contrasted with CLOSED (2).

4 Of a condition or conditional clause or sentence: referring to an event, state, etc. that is capable of occurring. Contrasted with HYPOTHETICAL, CLOSED (3).

open juncture See JUNCTURE.

open-rounded See LIP-POSITION.

operator

The first or only auxiliary (or *be*, *have*, and *do* in certain uses) in a finite verb phrase.

The term emphasizes the fact that it is the *first* word in a verb phrase which 'operates' inversion (for questions) and the adding of *not* (for negation).

They *could* have been imagining things
Could they have been imagining things?
They *couldn't* have been thinking

When the verbs *be* or *have*, even as one-word main verbs, work in the same way, they too are operators, as *do* also is when introduced for *do*-SUPPORT:

Are you ready?
I *haven't* any money
Do you know something?

Modal auxiliaries therefore always function as operators; lexical verbs never do. Of the primary verbs, *be* is always potentially an operator by this definition, while *have* is sometimes an operator, even when it is the sole verb in a clause:

Are you serious?
I haven't a clue

have is not an operator when it uses *do*-support:

He doesn't have much fun

do is an operator when giving *do*-support, but as a main verb it needs *do*-support itself (**I didn't the housework*. **Did you the dishes?*)

Compare NICE PROPERTIES.

optative subjunctive

The same as *formulaic subjunctive*: see SUBJUNCTIVE.

optional

Not OBLIGATORY.

1 That can be omitted and still leave a grammatical structure. (Especially used of clause elements.)

Of the five elements of clause structure, Adverbial is the most often optional. Subject is apparently obligatory in most finite clause structure (**Is raining*); missing subjects in coordinated clauses are explained as ellipted. Subject is however optional in imperatives (e.g. *Go! You go!*).

2 *Phonology*. Of a sound: sometimes pronounced and sometimes omitted in certain words, either pronunciation being acceptable.

For example, *often* is traditionally /'ɒf(ə)n/ in RP, but /'ɒft(ə)n/ is frequently heard. Similarly whole syllables may be optional. *Medicine* is frequently pronounced as a two-syllable word, but three syllables are possible; *police* may be two syllables or one.

• **optionality**. **optionally**.

oral

1 *Phonetics*. Of a speech sound: articulated with the velum raised. Contrasted with NASAL.

All normal English sounds, except for the three nasal consonants, have oral 'escape' or 'release': that is, the air is expelled through the mouth, and there is no nasal resonance.

See VOCAL TRACT.

2 Using or pertaining to speech, as opposed to writing.

Oral competence, for example, may be contrasted with writing ability.

order of adjectives See ADJECTIVE ORDER.

order of words See WORD ORDER.

ordinal

(*n. & adj.*) (A number) defining position in a series. Contrasted with CARDINAL.

e.g. *first*, *second*, *third*, *fourth*, etc.

1892a H. SWEET Most of the ordinal numerals are derivatives of the cardinal ones.

ordinative

(*n. & adj.*) (A numeral or adjective) that indicates position in an order.

This is a wider term than *ordinal*, including words such as *next, last, preceding*.

See NUMERATIVE. Compare QUANTITATIVE.

organ of speech

A part of the mouth and adjoining organs involved in the production of speech sounds. (Also called *speech organ*.)

e.g. the lips, alveolar ridge, soft palate, larynx, etc.

***or*-relations**

PARADIGMATIC relationships. Contrasted with *AND*-RELATIONS.

orthographic

Of or pertaining to spelling.

•• **orthographic word**: a word as written or printed, with spaces on either side.

In general, a word, as commonly understood, is written with spaces on either side, although there is variability with COMPOUNDS. There are also problems sometimes with the use of the apostrophe. An advertisement some years ago said:

Four little words that can cost a tobacconist £400. THEY'RE FOR MY MUM.

They're for many people is two words, but it is a single *orthographic word*, so the advertisement is correct by the definition given here.

Compare LEXEME, WORD.

orthography

(The study or science of) how words are spelt. Contrasted with GRAPHOLOGY.

> 1873 J. EARLE When we use the word 'orthography', we do not mean a mode of spelling which is true to the pronunciation, but one which is conventionally correct.

Spelling being largely standarized, a word normally has only one recognized orthographic form, but in a few cases there are acceptable variants, e.g.

cipher/cypher
hallo/hello
mateyness/matiness
standardise/standardize

And there are also distinct British and American spellings

centre/center
colour/color
sceptical/skeptical
travelling/traveling

overcorrection

The same as HYPERCORRECTION.

overgeneralize

Use (a grammatical rule) in cases to which it does not apply.

The term is particularly used in connection with language acquisition by children. Thus a child who overgeneralizes the plural *-s* inflection might say *mans, mouses, sheeps.*

• **overgeneralization.**

overlapping distribution See DISTRIBUTION.

P

pair See MINIMAL.

palatal

Phonetics. (*n. & adj.*) (A speech sound) produced by the articulation of the FRONT of the tongue with the hard palate.

The term tends to be restricted to consonants. British (RP) English has one distinctly palatal phoneme, the sound /j/ which is heard at the beginning of *yes* /jes/ or *useful* /juːsf(ə)l/ and before the vowel in cure /kjʊə/. This is commonly classified as a SEMI-VOWEL, *approximant*, or *frictionless continuant* rather than as a full consonant.

Various palatal consonants exist in other languages, for example the palatal nasal symbolized as [ɲ], heard as the middle consonant in the Spanish word *mañana*.

Sounds combining a palatal articulation with an alveolar one are classified as *alveolo-palatal* (none in English) and PALATO-ALVEOLAR.

palatalize

Make (a sound) palatal by articulating it with the FRONT of the tongue raised towards the hard palate.

Use of this term is mainly confined to secondary articulations, that is, to speech sounds where this articulatory feature is secondary to the main position of the speech organs. This is in fact an essential part of four English phonemes which also have an alveolar articulation. See PALATO-ALVEOLAR.

• **palatalization**.

Compare LABIALIZATION, VELARIZATION.

palate

Phonetics. The roof of the mouth.

In the articulatory description of speech sounds the upper surface of the mouth, behind the alveolar ridge, is divided into the bony HARD PALATE and the soft palate or VELUM.

palato-alveolar

Phonetics. Designating a speech-sound in which the TIP (or TIP and BLADE) of the tongue articulates with the alveolar ridge, while at the same time the FRONT of the tongue (the part behind the tip and blade) is raised towards the hard palate.

English has two pairs of palato-alveolar consonants consisting of one voiced and one voiceless consonant each:

(*a*) the palato-alveolar affricates
/tʃ/ as in *ch*ur*ch*, na*t*ure
/dʒ/ as in *j*u*dg*e, *g*eneral
(*b*) the palato-alveolar fricatives
/ʃ/ as in *sh*op, ma*ch*ine, *s*ugar
/ʒ/ (never word-initial in English) as in u*s*ual, prestige

paradigm (Pronounced /ˈpærədaɪm/.)

1 A set of word forms produced by inflection from a single base form.

For example, *see, sees, seeing, saw, seen* constitute a paradigm.

The term comes ultimately from Greek *paradeigma* 'pattern, example'; the sense is based on the use in teaching of a set of forms from a particular word as a pattern for all the other words which inflect similarly.

2 A set of linguistic items such that any member of the set may (grammatically speaking) be substituted for another member; the relationship between these items.

The items in a paradigm are in an *or-relationship* (or CHOICE relationship), in contrast to members of a SYNTAGM, which are in an *and-relationship* or CHAIN. The English article system and the pronoun system are both paradigms. We can say *a book* or *the book*, but not **a the book*. Similarly, we can grammatically substitute one pronoun for another in *I told the truth* (e.g. *I/you/he/she/we/they/somebody*, etc.) but we cannot choose more than one pronoun unless they are coordinated (e.g. *You and I told the truth*).

• **paradigmatic**: forming, belonging to, or relating to a paradigm or paradigms. **paradigmatically**.

1977 J. LYONS From its very beginnings structural semantics . . has emphasized the importance of relations of paradigmatic opposition.

paragraph

(*n.*) A distinct section of a piece of writing, beginning on a new, and often indented, line.

(*v.*) Arrange in such sections.

Although the way a text is set out on the page may be an important factor in its intelligibility, the paragraph as such has no grammatical status comparable to that of a phrase, clause, or sentence.

Compare DISCOURSE.

paralanguage

The features forming the subject-matter of PARALINGUISTICS.

Somewhat rare.

paralect

An accent close to another (usually more prestigious) accent but retaining some marks of the speaker's original accent.

> 1989 J. HONEY To this category of 'almost-RP-but-not-quite' we can give the label PARALECT—the 'close-to' accent, whose closeness in this case is to the acrolect.

• **paralectal**.

Compare HYPERLECT, NEAR-RP.

paralinguistic

Of or pertaining to the non-verbal features of spoken language.

The term is used in a variety of ways to include or exclude certain non-verbal features of spoken communication. In analysis of non-verbal vocal phenomena, paralinguistic features are often contrasted with more measurable PROSODIC ones such as intonation and stress. Paralinguistic features can thus include tone of voice, and the distinctive characteristics of an individual's voice.

Non-vocal features accompanying spoken communication, such as eye movements, nodding, or other forms of body language are also frequently classified as *paralinguistic*.

There is some overlap between this term and EXTRALINGUISTIC.

• **paralinguistics**: the study of paralinguistic features.

paraphrase

(*n.*) A piece of text (especially a sentence) that expresses the 'same' meaning as another piece of text in a different way.

(*v.*) Make or constitute a paraphrase of (a piece of text).

In some linguistic theory, as in popular terminology, a paraphrase of a particular form of words 'means the same' as the original. Thus, in some Transformational-Generative Grammar, a passive sentence is regarded as a paraphrase of an active one. However, though meaning as denotation (objective, referential meaning) may be preserved, a paraphrase usually changes other types of meaning in some way. Sometimes it changes the emphasis and the way the information is structured (e.g. what the topic is, what the focus is), while a looser paraphrase may change the level of formality, the social implications, and so on.

parasynthesis

Morphology. Derivation from a compound or syntactic sequence.

red-faced from *red face* + *-ed* is an instance of parasynthesis.

• **parasynthetic**: formed (by derivation) from a compound or syntactic sequence of two or more elements.

paratactic See PARATAXIS.

parataxis

Equality of grammatical relationship between two linguistic units. Contrasted with HYPOTAXIS.

• **paratactic**: exhibiting parataxis.

Parataxis (literally 'arrangement side by side') is a very general term covering various kinds of juxtaposition of units of equal status, including the coordination of two (or more) equal clauses, phrases, or words, with or without coordinating conjunctions:

Go, (and) never darken my door again
by the grace of God (and) in the nick of time
poor but honest
mad, bad, and dangerous to know

However, some grammarians specifically use the term for coordination without overt coordinators, opposing it to COORDINATION, defined as having coordinators. (See ASYNDETIC.)

Others extend the term to include certain juxtapositions of two equal units which would not be regarded as coordination (since no coordinators could be inserted) such as

some kinds of apposition:
Oxford, city of dreaming spires
clausal linkage in tag questions:
It's a lovely day, isn't it?
or the relationship between a reporting verb and a direct quotation:
They keep shouting 'Go home'

(This last is certainly controversial, as many grammarians would consider '*Go home*' to be the object of a superordinate clause, in which case this particular relationship would be HYPOTACTIC.)

parenthesis (Plural **parentheses**.)

1 A word, clause, or sentence inserted as an explanation or afterthought into a passage which is grammatically complete without it, and in writing usually marked off by brackets, dashes, or commas.

2 (In plural.) A pair of brackets, usually round (), used for marking this.

Compare ANACOLUTHON.

parole See LANGUE.

paronym

A word derived from the same base as another, and used in a related meaning; a word formed from a foreign word with only a slight change of form (especially one used as a translation equivalent of the foreign word). Contrasted with HETERONYM (3).

Examples: *wise*: *wisdom*; *preface*: Latin *praefatio* (as contrasted with *foreword*).

• **paronymous**. **paronymy**: the use of morphologically related words in related senses.

parse

1 Describe (a word in context) grammatically, stating its inflection, relation to the rest of the sentence, etc.

2 Resolve (a sentence) into its component parts and describe them grammatically.

Parsing is unfashionable as a classroom exercise, but analysing clause and sentence structure is the basis of grammar and much linguistics.

partial apposition See APPOSITION.

partial conversion See CONVERSION.

participant

Linguistics.

1 One of the people involved in a text.

Whether written or spoken, every text involves at least two people: the speaker/writer and the addressee (the listener or reader). This is true whether the participants are mentioned (*I am asking you as a favour*) or not (*Keep off the grass!*). The relationship of the people involved in a text is a *participant relation* or *participant relationship*.

2 The same as CASE or role in Case Grammar.

The case or semantic function of a noun phrase can be called its *participant role*. For example, an agent is an *agentive participant*; an indirect object is typically a *recipient participant*; and so on.

participial

Of the nature of, of or pertaining to, a PARTICIPLE.

•• **participial clause**: the same as PARTICIPLE CLAUSE.

participial conjunction, participial preposition: a conjunction/preposition (according to function) that is a participle in form.

e.g.

> You can borrow it *providing/provided* (*that*) you return it in good condition (conjunction)
> *Following* the disclosures, the chairman resigned (preposition)

participial adjective

An adjective having the same form as the participle of a verb (i.e. formed with the same suffix). (Also called *verbal adjective*.)

A *participial adjective* is an adjective coincident with the *-ing* form or *-en* form of the verb to which it is related, e.g.

They are so *loving*	*frozen* assets
exciting times	*hurt* feelings
His symptoms were *alarming*	a *knitted* suit.

Some *-ing* forms in attributive position are not of course participial adjectives, but nounlike, e.g. *dining-room, planning permission*. See *-ING* FORM.

Participial adjectives also include words formed with the regular *-ed* ending, but which lack a corresponding verb, e.g.

booted and *spurred* — *unexpected* pleasure
honeyed words — *wooded* slopes
talented musicians

Parasynthetic adjectives of this sort are especially frequent:

able-bodied, half-hearted, one-legged, three-cornered, two-faced, white-haired

This kind (lacking a corresponding verb) is sometimes called a *pseudo-participle*.

Participial forms with complete adjectival status can, if they are gradable, take *very* (which verbs do not). Contrast

They are very loving
They are loving every minute of it (verb + object)

The distinction between adjectival and verbal use is not always, however, clear. An attributively used *-en* form of an intransitive verb may be analysed as verbal and active. For example, *an escaped prisoner* is 'a prisoner who has escaped' rather than 'a prisoner who has been escaped by someone'. And in some contexts the status of a participle-like form is ambiguous. *I was annoyed* can be interpreted verbally (e.g. *I was annoyed by their behaviour*) or as an adjective (*I was* (or *felt*) *very annoyed*) or even as both (*I was very annoyed by their behaviour*).

Some participial adjectives are gradable, e.g.

It was *more exciting* than usual
The *most talented* children were given scholarships

participle

A non-finite form of the verb which in regular verbs ends in either *-ing* or *-ed*.

Two participles are distinguished, traditionally labelled *present participle* (e.g. *being, doing, drinking, looking*) and *past participle* (e.g. *been, done, drunk, looked*). Neither name is accurate, since both participles are used in the formation of a variety of complex tenses and can be used for referring to past, present, or future time (e.g. *What had they been doing? This must be drunk soon*). Preferred terms are *-ING* FORM (which also includes *gerund*) and *-EN* FORM (or *-ED* FORM).

Compare FUSED PARTICIPLE.

participle clause

A non-finite clause with an *-ing* form or *-en* form as its principal verbal component. (Also called *participial clause*.)

Examples:

Looking to neither right nor left, he marched out

Treated like that, I would have collapsed
Having been warned before, he did not do it again

A participle clause can sometimes be introduced by a conjunction:

If treated like that, I would have collapsed

A participle clause that contains its own subject is a type of ABSOLUTE clause. See HANGING PARTICIPLE.

particle

1 An adverb or preposition used in the formation of a MULTI-WORD VERB. Also called *adverb(ial) particle*.

This is a neutral term covering any adverb or preposition used in such a verb. Most particles are high-frequency words which can function both as adverbs and prepositions. Compare

She looked up the word She looked it up in her dictionary	(*up* = adverb)
She looked up the road *She looked the road up	(*up* = preposition)

2 The word *to* when used before an infinitive. Sometimes called *infinitival particle*.

To is generally a preposition (as in *ten to six*, *go to Oxford*) and less frequently an adverb (e.g. *Brandy might bring him to*). Traditionally the *to* before an infinitive is classified as a preposition, but its grammar is different from that of the preposition *to*—most notably in that the latter, if followed by a verb, requires this verb to be an *-ing* form, and does not permit the infinitive. Contrast

We *look forward to* your visit We *look forward to* seeing you	(*to* = preposition)
We hope *to see* you soon	(*to* = particle))
*We look forward to see you	

3 **pragmatic particle**: see FILLER (2).

4 (Formerly.) A member of a set of words including adverbs, prepositions, and conjunctions.

Membership of this set varied according to different writers. It was often reckoned to include articles, sometimes other determiners, and by some even affixes or interjections.

See also NEGATIVE PARTICLE.

partition See PARTITIVE.

partitive

(*n. & adj.*) (Denoting) (a word that is the first element in) a phrase or construction that expresses the relationship of part to whole (*partition*), and having the essential form 'X of Y'.

(*a*) The partitive construction often refers to quantity or amount (*quantity partitive*):

a piece of paper — a bit of a problem
two pieces of paper — an item of clothing

but also indicates quality (*quality partitive*):

a sort of clown
different kinds of cheese
that type of person

Count nouns that can act as the first element in such a structure (e.g. *piece, bit, sort*, etc.) are *partitive nouns* or *partitives*. Some words that form the second part of the construction take specific partitives (also called UNIT NOUNS)

a *blade* of grass — a *loaf* of bread
a *flock* of sheep — a *speck* of dirt

Partitives are useful because they provide a means of counting uncount nouns:

three *slices* of bacon (*three bacons)
an interesting *piece* of information (*an interesting information)

(*b*) When the second noun in a similar structure is plural, the first one may denote not a part but a larger group

a *bunch* of flowers — a *crowd* of people
a *clump* of trees — a *flock* of sheep

Some of these are in fact COLLECTIVE nouns, and are so classified in some grammatical models. Partitives can also denote containers:

a *packet* of cigarettes
a *sack* of potatoes
a *teaspoon(ful)* of sugar

(*c*) Pronouns with indefinite meaning are sometimes labelled *partitives* or *indefinite partitives*. This set includes not only those that can be followed by an *of*-phrase (e.g. *many of, few of*, etc.) but even such pronouns as *someone, anything*.

•• **partitive genitive**: (*a*) the *of* + noun part of the partitive construction; (*b*) a genitive that indicates that of which something is part (see below).

The term is used in various ways, sometimes referring to various meanings of the 'X of Y' construction, and sometimes to a particular use of the genitive case to indicate that of which something is a part rather than strictly a possession (e.g. *the baby's eyes, the earth's surface*).

• **partitively**.

part of speech

(A member of) a class of words with shared grammatical characteristics that distinguish them from others.

This is the traditional term for different types of word, which are distinguished largely by syntactic (and sometimes morphological) criteria, but which are often defined in NOTIONAL terms.

There is no single correct way of analysing words into parts of speech (more usually today called WORD CLASSES or *form classes*). The boundaries between different groups are not always clear. Generally recognized are nouns, verbs, pronouns, adverbs, adjectives, conjunctions, prepositions, and interjections. Other categories are articles and determiners.

> 1711 J. GREENWOOD I have not made the Article (as some have done) a distinct Part of Speech.

Compare MAJOR *part of speech*.

passive

(*adj.*) Designating the VOICE of the verb whereby the grammatical subject 'suffers', 'experiences', or 'receives' the action of the verb; also, (of a verb) in the passive voice; (of a construction) involving a passive verb. Contrasted with ACTIVE.

(*n.*) The passive voice; a passive verb.

In formal terms, a passive tense is formed with (minimally) a form of the verb *be* and a past participle, as in

> Dinner is served
> Trespassers will be prosecuted
> Mistakes cannot afterwards be rectified

Contrast (with the active voice)

> Someone is/We are serving dinner
> We will prosecute trespassers
> We cannot rectify mistakes afterwards

Only transitive verbs can have passive forms, since the grammatical subject of a passive verb corresponds to the object of an active verb. Even so, not all active structures have a passive counterpart:

> Blue suits you. *You are suited by blue
> Those people lack confidence. *Confidence is lacked by those people

See MIDDLE VERB.

There are also passive constructions which do not have active counterparts:

> They are said to be very intelligent. (*People say them to be very intelligent)

If an agent is mentioned with a passive verb, this is usually by means of a *by*-phrase (e.g. *Dinner was served by uniformed staff*). In some cases, mention of an agent is unlikely or impossible (*Churchill was born in 1874*). A passive construction with no agent mentioned is labelled a *non-agentive passive* or an *agentless passive*.

• **passively**. **passivity**.

Compare STATAL. See also DOUBLE PASSIVE, *GET* PASSIVE, PSEUDO-PASSIVE, SEMI-PASSIVE.

passive articulator: see ARTICULATOR.

passive infinitive: see INFINITIVE.

passivize

Convert into the passive; be subject to conversion into the passive.

1984 F. R. PALMER We can passivize the main clause with PERSUADE, but not with WANT:

The doctor was persuaded to examine John
**The doctor was wanted to examine John*

• **passivizable. passivization.**

This group of terms was coined by Chomsky.

past

(*n. & adj.*) (A tense or form) relating to time gone by. Contrasted with PRESENT.

The grammatical sense relates essentially to the usual meaning of the term, but this association can be misleading. It is more accurate to say that past tenses are MARKED as *non-present*. While they refer primarily to past time, they can also be used for *hypothesis*, i.e. as tenses marked for unreality:

I wish I *knew*
If I *had* my way, I would . . .

and for *social distancing* (the ATTITUDINAL *past*):

Could you lend me some money?
I *wanted* to ask you something

With no further label, the past of a verb means that morphologically marked form or that tense (the *past tense* or *past simple*) which in regular verbs always ends in *-ed*, and whose form is normally listed second when verb forms are given, e.g.

see, *saw*, seen; drive, *drove*, driven

Compare BACKSHIFT, FUTURE IN THE PAST.

past definite See DEFINITE

past-in-the-past See PAST PERFECT.

past participle

That part of the verb which is used in perfect and passive tenses and sometimes adjectivally.

Examples:

Have you *looked*?
Were you *seen*?
lost property

It is usually the third form listed when verb forms are given in dictionaries, e.g.

see, saw, *seen*

In regular verbs the past participle ends in the same *-ed* inflection as the past tense, and is called the *-ed form* (or *-ed participle*) by some grammarians; others prefer the label *-en form* (based on the distinctive ending of certain irregular verbs such as *spoken, driven*) so as to distinguish it more clearly.

The past participle signifies 'perfectiveness' or completion, but is not restricted to past time (e.g. *You'll have forgotten by this time next year*). It can also have a passive meaning; contrast *bored* (passive) and *boring* (active).

Compare PSEUDO-PARTICIPLE, PARTICIPLE.

past perfect

(A tense) formed with *had* + a past participle. (Also called *pluperfect, before-past*, and *past-in-the-past*.)

With no further label, *past perfect* refers to a simple active tense:

I had forgotten (until you reminded me).

Past perfect progressive tenses, past perfect passive tenses, and combinations of the two also occur:

We *had been wondering* about that, when the telegram arrived
The matter *had been overlooked*
It *had been being compiled* by hand.

In general, past perfect tenses refer to a time earlier than some other past time. But like other so-called past tenses, the past perfect in a subordinate clause may signify *hypothesis* (something contrary to fact):

If you had told me before now, (I could have helped)
If you had been coming tomorrow, you would have met my mother.

The past perfect may also stress perfectiveness or completion

They waited until I had finished.

Compare ASPECT.

past progressive

(The tense) formed with a past form of the verb *be* + an *-ing* form.

Examples:

We were waiting. It was raining.

past simple

The morphologically marked tense of the verb.

See PAST.

past subjunctive See SUBJUNCTIVE.

past tense See PAST.

patient

A semantic role taken by a noun phrase which is acted upon or affected in some way by the verb.

In some Case Grammar, *patient* is distinguished from *goal* (see GOAL). In other analyses, *patient* is equated with AFFECTED.

pattern

The (possible or necessary) way in which elements of language combine to form larger units.

(*a*) At the syntactic level, clause structure can be analysed in terms of a comparatively small set of overall patterns that are essentially determined by the type of verb involved, namely SV (with an intransitive verb), SVO (transitive verb), SVC (linking verb), SVOO (ditransitive verb), SVOC (complex transitive verb).

> 1984 S. CHALKER The primacy of the verb is shown by the fact that it is the verb that dictates the basic patterns of the simple sentence or clause. Modern grammar has reduced the patterns underlying all simple sentences to as few as five . . . Notice that adverbial is largely optional.

Looked at another way, individual verbs are said to have their own verb patterns. For example, both *want* and *wish* can be followed by (object) + *to*-infinitive: *I want (you) to go*, *I wish (you) to go*. But only *wish* can be used in the pattern of verb + *that*-clause; so *I wish that you would go* but not **I want that you (would) go*. Thus *want* and *wish* exhibit different verb patterns. Similarly we can talk of noun patterns (some can be followed by a *to*-infinitive, e.g. *determination to succeed*; others by prepositions, e.g. *love of money* etc.) and adjective patterns (e.g. *keen to help*, *thoughtful of you*, *sorry (that) I spoke*).

(*b*) The concept of *pattern* can also be applied to phonology.

> 1921 E. SAPIR Every language, then, is characterized as much by its ideal system of sounds and by the underlying phonetic pattern (system, one might term it, of symbolic atoms) as by a definite grammatical structure.

paucal

Implying a smallish number or amount. Contrasted with MULTAL.

In some detailed classifications of pronouns and determiners, the label *paucal* is given to a subset of quantifiers: *a few*, *fewer*, *fewest*, *a little*, *less*, *least*. The term is taken from general linguistics, where it is specifically used when describing languages that have a number system with more categories than *singular* and *plural*.

pause

Phonetics. A break in speaking.

Connected speech is much more of a continuum than written language suggests by its spaces between words. Pauses do however occur in speech; obviously for breathing and also for communicative reasons at grammatical boundaries. Various efforts have been made to incorporate an analysis of pauses into a theory of speech.

The term FILLED PAUSE has been coined for the hesitation noises (*er*, *um*, etc.) that speakers use, probably to give themselves time to think out what they are going to say next.

pedagogical grammar See GRAMMAR.

perception verb

One of a set of verbs denoting the use of one of the physical senses, e.g. see, hear, feel, taste, smell. (Also called *perceptual verb* or *verb of perception*.)

Grammatically, a subset of *perception verbs* are important because they form a group that shares two verb patterns, as shown in:

I heard him sing
I heard him singing

Other verbs that fit both these patterns are *feel, notice, observe, see,* and *watch*. Though not actually a perception verb *have* also fits the pattern. Some other verbs take one or other of these patterns, but not both.

perfect

(*n. & adj.*) (A verb form or tense) expressing an action or state accomplished before some stated or implied time.

Linguists prefer to speak of *perfect* (or *perfective*) ASPECT, but in common parlance we talk of the *perfect tense*, the *perfect infinitive*, etc. *Perfect* verb forms consist of a part of the verb *have* plus a past participle, e.g.

he has/had won
I will have finished by next week
having said that, . . .

Perfect or *perfective aspect* contrasts with PROGRESSIVE *aspect*; *perfect tenses*, with both *progressive tenses* and *simple tenses* (which could be described as *non-perfective*).

See FUTURE PERFECT, PAST PERFECT, PRESENT PERFECT.

perfect infinitive

An infinitive formed with *have* + a past participle.

Examples:

(to) have wanted, (to) have forgotten.

Like other infinitive forms, the perfect infinitive may be used in a non-finite clause (e.g. *To have forgotten was unforgivable*) and as part of a finite verb phrase (e.g. *You can't have forgotten*).

perfective

(*n. & adj.*) (The aspect) expressing an action or state accomplished before some stated or implied time. The same as PERFECT.

A rather different contrast is sometimes made between *perfective* and IMPERFECTIVE (*progressive*) in the non-finite verb. Here the *to*-infinitive which often refers to unreal or potential events (as in *To open, cut along the dotted line*) is labelled *perfective*; while the *-ing* form, referring to real or actual events (e.g. *Opening the carton, I hurt my hand*) is labelled *imperfective*.

See IMPERFECTIVE, IRREALIS.

performance See COMPETENCE.

performative

(*n. & adj.*) (Designating or belonging to) an utterance that constitutes an action in itself.

(*a*) The term is used in Speech-Act Theory (first developed by the philosopher J. L. Austin) to describe statements, whether spoken or written, that themselves 'do' something and so constitute a kind of action, e.g.

I advise you to reconsider
I promise to pay the bearer on demand the sum of five pounds
You are hereby notified that . . .
Patrons are advised not to leave valuables in their cars

Performative utterances can be divided into *explicit performatives*, where the action performed is stated as here (e.g. *I advise*, *I promise*, etc.) and *primary performatives*, where the action (e.g. promising) is merely implicit (e.g. *The bank will pay the bearer* . . .).

Performative utterances were originally contrasted with *constative* statements, which state that something is or is not so, and which therefore, unlike performatives, can be tested for truth-value. Later it was realized that in a sense every utterance does something and has an underlying performative statement, such as *I tell you*, *I am asking*, which rather weakens the theory. However *performative verb* remains an important concept.

Compare ILLOCUTION.

(*b*) In ordinary grammatical usage the term is restricted to verbs that explicitly perform an action, e.g. *advise, apologize, beg, confess, promise, swear, warn*. The term includes verbs that in a more detailed analysis of speech acts are covered by the term declaration (as in *I name this ship . . . , I declare the meeting closed*).

It is usually said that such verbs are performatives only when used in the first person of the present tense (e.g. *I advise you to reconsider*, *I name this ship* . . .), and that in other contexts such a verb is merely reporting a performative act (e.g. *She advised him to reconsider*). But as the examples above show, this is an overnarrow definition of what constitutes a performative utterance.

Austin first used the term *performatory* (1949), but later substituted *performative* (1955).

period

The same as FULL STOP.

periphrasis

The use of separate words to express a grammatical relationship that is also expressed by inflection.

Periphrasis is a common feature of adjective and adverb comparison, where periphrastic phrases with *more* and *most* are an obligatory alternative to forms with *-er* and *-est* for longer adjectives and most adverbs (*more beautiful, most oddly*). A choice between inflection and periphrasis is possible with some two-syllable adjectives (e.g. *It gets lovelier/more lovely every day*).

The term is sometimes applied to the formation of English tenses, which mostly consist of several words (a verb phrase) rather than a single inflected form.

• **periphrastic**.

perlocution

(In Speech-Act Theory.) An act effected by means of saying something. Contrasted with LOCUTION, ILLOCUTION.

Perlocution is concerned with an act, a change of state, etc. in someone other than the speaker, that results from the speaker's *illocution* (a speech act viewed in terms of the speaker).

• **perlocutionary**: designating an act of this kind.

> 1977 J. LYONS By the illocutionary force of an utterance is to be understood its status as a promise, a threat, a request, a statement, an exhortation, etc. By its perlocutionary effect is meant its effect upon the beliefs, attitudes or behaviour of the addressee and, in certain cases, its consequential effect upon some state-of-affairs within the control of the addressee . . . It is the intended perlocutionary effect that has generally been confused with illocutionary force.

permanent See TEMPORARY.

permission

One of the main meanings of DEONTIC modality, along with *obligation* and *volition*.

e.g.

Can/may I go now?

person

A category used, together with NUMBER, in the classification of pronouns, related determiners, and verb forms, according to whether they indicate the speaker, the addressee, or a third party; one of the three distinctions within this category.

A threefold contrast, between *first*, *second*, and *third person* is made, and is particularly distinct in the PERSONAL PRONOUNS. *Be* is unique among English verbs in having three distinctive forms in the present tense (*am, is, are*) and two in the past (*was, were*). Other verbs have a distinctive form only for the third person singular of the present tense (e.g. *has, does, wants,* etc., as opposed to *have, do, want,* etc.), and for 'tenses' formed with *be* and *have*.

• **personal**: denoting one of the three persons.

Compare IMPERSONAL, NON-PERSONAL. See FIRST PERSON, SECOND PERSON, THIRD PERSON.

personal pronoun

A pronoun belonging to a set that shows contrasts of person, gender, number, and case (though not every pronoun shows all these distinctions).

The personal pronouns are

1st person: I, me, we, us
2nd person: you
3rd person: he, him, she, her, they, them, it

Reflexive pronouns (*myself*, *ourselves*, etc.), possessive pronouns (*mine*, *ours*, etc.) and possessive determiners (*my*, *our*, etc.) are sometimes included.

pharyngeal (Pronounced /færın'dʒi:əl, fə'rındʒəl/.) (Also **pharyngal**, pronounced /fə'rıŋgəl/.)

Phonetics. Of a speech sound: articulated with the root of the tongue pulled back in the pharynx, the cavity behind the nose and mouth connecting them to the oesophagus.

There are no pharyngeal consonant phonemes in standard English. (Arabic has pharyngeal fricatives.)

The English vowel /ɑ:/ can be described as pharyngeal; but place of articulation is not usually part of the description of vowels, and so this vowel is normally described simply as an *open back vowel*.

• **pharyngealize**: articulate (a speech sound) with the root of the tongue retracted so as to obstruct the air-stream at the pharynx.

The term is particularly applied to secondary articulations, i.e., where this feature is not an intrinsic part of the phoneme. Pharyngealized sounds may be heard in English, but they are not significant.

Compare LABIALIZE, VELARIZE.

phase

1 The meaning relationship between a CATENATIVE verb and the following verb.

Such a relationship can be analysed as:

(*a*) a *time phase*, a time relationship, as in *keeps on complaining*; or

(*b*) a *reality phase*, a relationship of reality, as in *seems to misunderstand* or *proved to be wrong*.

2 The structure of a catenative verb + a following verb. Also called **phase structure** or **phase verb**.

3 In some modern terminology *phase* is preferred to *aspect* in describing the meaning of perfect tenses.

phatic

Sociolinguistics. Of speech: used to convey general sociability, rather than to communicate any real meaning.

Observations about the weather (*Nice day, isn't it?*) are often phatic.

The term is loosely derived from the anthropologist Bronisław Malinowski's coinage *phatic communion* 'speech communication used to establish social relationships' (1923), in which *phatic* has its etymological sense 'of or pertaining to speech'.

phenomenon See SENSER.

philology

The science of language.

This is a traditional term, used particularly for the study of historical linguistic change and comparison between languages. Terms favoured today include *comparative* and *contrastive linguistics*. In the wider sense it has been superseded by LINGUISTICS.

• **philological. philologically. philologist.**

> 1935 J. R. FIRTH The evolutionary and comparative method had been used by philologists in the eighteenth century. Comparative Philology was, in fact, the first science to employ this method.

phonaestheme

Linguistics. A phoneme or group of phonemes with recognizable semantic associations due to recurrent appearance in words of similar meaning.

A rather grand term to convey the fact that some sounds seem to have meaningful associations. Commonly cited are *sn-* which often seems to have nose-related unpleasantness (e.g. *snarl, sneer, sneeze, snide, sniff, snigger, snipe, snivel, snook, snooty, snore, snort, snotty, snout*) and *sl-* with various unpleasant connotations, mainly of moisture and its effects (*slime, slink, slippery, slob, slop, slurp*, etc.).

• **phonaesthesia, phonaesthesis**: the use of phonaesthemes, secondary onomatopoeia. **phonaesthetic. phonaesthetically.**

The term was coined by J. R. Firth.

> 1930 J. R. FIRTH A much bigger group of habits we may call the *sl* phonæstheme. *Ibid.* Play on phonæsthetic habits gives much of the pleasure of alliteration, assonance, and rhyme.

Compare ONOMATOPOEIA.

phonation

Phonetics.

1 Production or utterance of a vocal sound.

A very general term for referring to vocal activity, not specifically to the articulation of individual speech sounds. *Phonation types* include the voiced versus voiceless opposition, glottal constriction, the fortis versus lenis opposition, and so on.

2 (In the usage of some phoneticians.) The production of voiced sounds.

> 1991 P. ROACH If the vocal cords vibrate we will hear the sound that we call voicing or phonation.

phone

1 An elementary sound of spoken language, the smallest recognized by a listener as a separate sound.

2 The same as ALLOPHONE.

Phones are actual realizations of phonemes as produced by individual speakers. The concept of *phone* can therefore be distinguished from *allophone*, which, though more accurately described in physical terms than the phoneme is still an idealized abstraction. Loosely, however, *phone* is synonymous with *allophone*.

phoneme

Phonetics & Phonology. A speech sound which contrasts meaningfully with other sounds in the language.

Conventionally the phoneme is either (*a*) the smallest meaningful unit in the sound system of a language (though this unit may also be viewed as a bundle of DISTINCTIVE FEATURES) or (*b*) a 'family' of similar sounds. But however it is viewed, the phoneme is an idealized abstraction, embracing any variant realizations (ALLOPHONES) that do not change meaning. Thus /p/ or /b/ are separate phonemes in English, differentiating such words as *pan* and *ban*, *dapple* and *dabble*, *hop* and *hob*. But the actual articulations of these phonemes are conditioned by their syllable positions: pre-vocalic /p/ is aspirated in *pan*, but completely unaspirated in *span*.

Phonemes and the ways in which their allophones are conditioned vary from language to language, from dialect to dialect, and even from one accent to another. English RP is currently analysed as having 22 consonants, 2 semi-vowels, 12 pure vowels, and 8 diphthongs, making a total of 44 phonemes.

• **phonemic** (also **phonematic**): of or pertaining to a phoneme or phonemes. **phonemically**. **phonemics**: the study of phonemes.

The term *phoneme* was introduced by the Polish linguist Kruszewski (1879); the same general concept was independently developed by Henry Sweet, Otto Jespersen, and others. *Phonemic* seems to have been used first by Bloomfield. [See table p. 446]

phonemic transcription See TRANSCRIPTION.

phonetic

1 Relating to actual speech sounds.

2 Representing a speech sound or speech sounds.

•• **phonetic alphabet**: a set of symbols based on place and manner of articulation and on acoustic and auditory qualities, such that each symbol has one characteristic sound. Thus the phonetic symbol [f] represents the same sound however it is spelt: /faɪn/, /fəʊn/, /kɒf/ (*fine*, *phone*, *cough*). In practice, of course, the context of a sound may affect its exact articulation, and if required these differences can be shown by DIACRITICS.

Compare INTERNATIONAL PHONETIC ALPHABET. See TRANSCRIPTION.

phonetics

The science or study of speech sounds.

> 1902 H. SWEET My own subject, Phonetics, is one which is useless by itself, while at the same time it is the foundation of all study of language, whether theoretical or practical.

Three areas of study are common—ACOUSTIC, ARTICULATORY, and AUDITORY *phonetics*.

• **phonetician**: a specialist working in (one of) these fields.

phonic

Of sound, etc.: produced by the human speech organs.

The *phonic medium* means spoken language, in contrast to the *graphic* (or *written*) *medium*.

phonological

Of, pertaining to, or relating to phonology.

In models of Transformational-Generative Grammar, the *phonological component* contrasts with the *syntactic* and *semantic components* of the grammar.

phonology

(The study of) the way in which speech sounds are used in a particular language. (Sometimes called *linguistic phonetics* or *functional phonetics*.)

phonotatics

Phonology is concerned not only with the meaningful contrasts (the PHONEMIC system) and the regular ways in which the phonemes are realized (the predictable allophonic variations), but also with the possible combinations of phonemes, the PHONOTACTICS. These aspects of phonology together are sometimes labelled *segmental phonology*, and contrasted with *suprasegmental phonology*, which is concerned with features of speech stretching over more than one sound, such as intonation.

See also GENERATIVE PHONOLOGY.

phonotactics

That part of phonology which comprises or deals with the rules governing the possible phoneme sequences of a particular language.

• **phonotactic**: of or pertaining to phonotactics (*phonotactic rule*, any of the rules governing the possible phoneme sequences in the language).

Various quite complicated consonant clusters are possible in English: see CONSONANT CLUSTER. Many single phonemes too are restricted in their positions. The consonant /ŋ/, usually spelt *ng*, never starts an English syllable, while several vowels (/e/, /æ/, /ʌ/, /ɒ/) are never syllable-final.

phrasal

Consisting of a phrase.

• **phrasally**.

phrasal auxiliary verb

A two- or three-part verb based on an auxiliary and having some of the same grammatical characteristics.

e.g.

have to, had better, be about to, be able to

The term is an alternative to the more usual SEMI-AUXILIARY, although possibly not exactly synonymous.

phrasal-prepositional verb

A multi-word verb containing a lexical verb and two particles. Also called (*three-part verb* or *three-part word*.)

In this type of verb, the first particle is adverbial and the second a preposition, e.g.

We're *looking forward to* the holidays
You shouldn't *put up with* that sort of treatment

Some phrasal-prepositional verbs need two objects, e.g.

I *put* it *down to* his ill-health

phrasal verb

1 A multi-word verb consisting of a verb plus one or more particles and operating syntactically as a single unit. (Also called *compound verb*, *verb-particle construction*, and sometimes *verb phrase* (but see VERB PHRASE).)

Thus defined, *phrasal verb* is an umbrella term for different kinds of multi-word verbs. Some analysts make metaphorical (or idiomatic) meaning a criterion for phrasal verbs, excluding combinations which have a transparent literal meaning: but it is not always easy to draw the line. For example, *get in* seems literal when the implied object is a vehicle (e.g. *I got in and drove off*); but is it literal or metaphorical when the meaning is 'into one's own home' (e.g. *I usually get in by 7 pm*) or 'into Parliament' (e.g. *He got in by a tiny majority*)?

2 (More narrowly.) A multi-word verb consisting of a verb plus an adverb particle, in contrast to a PREPOSITIONAL VERB or a PHRASAL-PREPOSITIONAL VERB.

In this narrower definition, phrasal verbs are of two main types

(*a*) intransitive:

The plane *took off*
I don't know—I *give up*!

and (*b*) transitive, where the particle can follow the object:

Take off your coat. *Take* your coat/it *off*
I've *given up* chocolate. I've *given* chocolate/it *up*

phrase

1 A linguistic unit at a level between a word and a clause.

(*a*) In traditional grammar, phrases include what are now often called non-finite clauses (e.g. *to come* in *things to come*).

Various kinds of phrase are recognized in modern grammar: ADJECTIVE, ADVERB, NOUN, PREPOSITIONAL, and VERB PHRASE.

(*b*) In Systemic-Functional Grammar, *phrase* is distinguished from GROUP. A *group* is a HEAD word expanded with modification, while a *phrase* is a reduced clause. Thus *A small town in Germany* is a *nominal group* (not a *noun phrase* as in other models), being an expansion of *town*, but *in Germany* is a *prepositional phrase*, because it could be expanded into a clause *which is in Germany*.

2 (In Generative Grammar.) One of the elements at the level immediately below that of the sentence, to which most of the lower-level elements are assigned.

The major elements at the highest level are the noun phrase, which is generally used in the same way as in other kinds of grammar, and the VERB PHRASE (2), which is used rather differently. A *phrase-marker* is part of a method of showing the syntactic linear structure of a sentence, and is sometimes equated with a *tree diagram*. A *phrase-structure grammar* (with *phrase-structure rules*) is a type of simple generative grammar that lacks 'transformational' rules, but is more 'powerful' than a *finite-state grammar*.

See also FIXED *phrase*.

phrase word

Morphology. A word formed from a phrase.

This is a term occasionally used to label a variety of lexical items derived from phrases, especially phrases used attributively, e.g.

> his *down-to-earth* manner, a *couldn't-care-less* attitude, a *once-in-a-lifetime* offer, the *carrot-and-stick* method

It may also include words derived from phrases by affixation (e.g. *no-Goddism*, *up-to-dateness*) and PHRASAL VERBS.

pidgin

A grammatically simplified form of a language (e.g. English), with a restricted vocabulary taken from several languages, that is used as a means of communication between people not sharing a common language and which is no one's mother tongue.

Pidgins based on various European languages developed in the heyday of colonial expansion. English-based pidgins evolved particularly on Pacific and Atlantic trade routes. A pidgin that gains a wider currency may develop into a sort of lingua franca for a region. One of the best known is Tok Pisin, the pidgin of Papua New Guinea, said to be the most widely used language in that country. A pidgin that becomes a mother tongue is called a CREOLE.

The term derives from *pidgin English*, the name of a trade jargon used from the 17th century onward between the British and Chinese. It is believed to be an alteration of *business*.

pied piping

The placing of a relative or interrogative pronoun immediately after its preposition, instead of deferring the preposition until later in the clause.

e.g.

> To whom are you talking? (Cf. Who are you talking to?)
> Here's the book about which I was telling you. (Cf. Here's the book which I was telling you about)

The term is used in some Generative Grammar, playfully based on the story of the Pied Piper, who was closely followed by the rats.

pitch

Phonetics & Phonology. The perceived 'height' of the human voice, depending on the rapidity of the vibrations of the vocal cords.

Speech analysts use the same metaphors of *high* and *low* as are used in music. The slower the frequency of vibration, the lower the pitch; the higher the frequency, the higher the pitch.

In tone languages (e.g. Mandarin), identical syllables with different pitch patterns or tones form words with totally different meanings. In non-tone languages (e.g. English and most other European languages), basic word meaning is not affected by pitch variations (though emotional attitudes may be distinguished) and intonation patterns are studied over sequences of words.

Various typical pitch changes or tones have been identified, e.g. FALL, RISE, and LEVEL; these are often further distinguished in terms of the pitch levels they span. The term *pitch contour* is sometimes used instead of *tone*, particularly in the description of a tone as it affects several syllables.

See LOW (2), HIGH (2), LEVEL, TONE.

place

1 A semantic category used in the classification of adverbs (adverbials) and prepositions, answering the question 'where?'.

This is one of the traditional categories, along with *time* and *manner*, that is still used today. An alternative label is *space*.

2 **place of articulation**: (*Phonetics*) (a part of) one of the vocal organs primarily involved in the production of a particular speech sound.

Place of articulation, along with MANNER *of articulation*, is a major part of the framework for describing the production of speech sounds, especially consonants. For this purpose, the vocal organs are diagrammatically divided up and the places labelled, as BILABIAL, LABIO-DENTAL, ALVEOLAR, PALATAL, VELAR, UVULAR, PHARYNGEAL, and GLOTTAL. *Place of articulation* is less satisfactory as a parameter for vowels, which are more dependent on tongue-height, lip-rounding, etc. Stylized diagrams of the mouth are, however, used with scales from the back (of the mouth) to the front, and close to open.

pleonasm

The use of more words than are needed to give the sense.

e.g.

see with one's eyes, at this moment in time

• **pleonastic**.

1898 H. SWEET The pleonastic genitive, as in *he is a friend of my brother's*

(The construction described in this example would now be called DOUBLE GENITIVE.)

plosion

Phonetics. Sudden expulsion of air as the final stage of a PLOSIVE; the *release* stage.

Compare LATERAL PLOSION, NASAL *plosion*.

plosive

(*n. & adj.*) (A consonant sound) that has total closure at some place in the vocal organs, followed by a 'hold' or compression stage, and a third and final release stage. (Also called *stop* or *stop consonant*.)

The English plosives consist of three pairs of sounds (each pair a corresponding voiceless and voiced sound):

/p/ and /b/ as in *poor, bore*; *tap, tab* (bilabial plosives)
/t/ and /d/ as in *true, drew*; *cat, cad* (alveolar plosives)
/k/ and /g/ as in *cold, gold*; *whack, wag* (velar plosives)

The glottal stop /ʔ/, which (though heard in RP) is not a phoneme, is a voiceless plosive.

A plosive sound made with a slow release stage involving friction is an AFFRICATE.

pluperfect

The same as PAST PERFECT.

pluperfect

(*n. & adj.*) (Designating or consisting of) a verb phrase that contains an additional, superfluous auxiliary, and is used as an alternative to the past perfect.

This is heard colloquially and is occasionally written, but is regarded as non-standard. The tense in full is *had have* + past participle, but is commonly used in shortened form:

If we'd have found an unsafe microwave oven, we would have named it (advertisement in *Daily Telegraph* 9 Dec. 1989).
We all take TV for granted, but if it hadn't have been for the pioneers at Alexandra Palace it might never have happened (speaker quoted in *Daily Telegraph* 13 Dec. 1992).

The term was coined by I. H. Watson (1985).

1985 I. H. WATSON Please comment—however briefly—on the proliferation of the use of the plupluperfect tense (I cannot think what else to call it). An example is: 'If he had have gone . . .'

plural

(*n. & adj.*) 1 (A word or form) denoting more than one. Contrasted with SINGULAR.

The term is one of several covered by the more general term NUMBER. In English, *plural* applies to certain nouns, pronouns, and determiners, and to verbs. In general, COUNT nouns have distinct plural forms, which in regular nouns end in *-s* or *-es*. Nouns with irregular plurals include some of Old English origin (*feet, children*, etc., and ZERO plurals such as *sheep, deer*) and some FOREIGN PLURALS (*crises, errata*, etc.).

A few nouns are *plural only* (see PLURALE TANTUM). Many end in *-s* (e.g. *premises*). But some plural-only words are unmarked (e.g. *cattle, people*).

With nouns and pronouns, the plural versus singular contrast leads to various grammatical constraints. Some DETERMINERS are restricted to use with one or the other. In particular, there must be AGREEMENT between subject and verb.

2 (A noun) ending in *-s*. (Also called *plural in form*.)

Some grammarians, and many dictionaries, label nouns ending in *-s* 'plural' or 'plural in form', even those which are never plural in syntax (e.g. *news*, *measles*, *physics*).

Compare COLLECTIVE NOUN.

plurale tantum (Pronounced /plʊə'reɪlɪ 'tæntəm/: plural **pluralia tantum**.)

A noun used only as a plural.

This is sometimes defined as a noun which at least in a particular sense is used only in the plural, which would theoretically include words such as *people* or *police*. Usually, however, it is exemplified by words with the plural ending *-s*. They are of three types:

(*a*) words for tools, articles of clothing, etc. that consist of two parts: *secateurs, trousers, binoculars*

(*b*) words that are never singular: *clothes, riches, thanks*

(*c*) words that with a particular meaning are always plural, though there may be a singular form with a different meaning: *arms* = weapons, *regards* = best wishes.

Some grammarians label type (*a*) *binary nouns* or SUMMATION PLURALS, reserving *pluralia tantum* (or AGGREGATE NOUN or, more simply, a term such as *plural-only*) for (*b*) and (*c*).

pluralize

Make (a word or clause) plural; (of a word etc.) become plural, take a plural form.

• **pluralization**.

polarity

Linguistics. A condition in which two opposites exist.

The term is particularly applied to the difference between POSITIVE and NEGATIVE in clauses, but is sometimes extended to cover more general oppositions (e.g. *good* versus *bad*, *up* versus *down*).

polyseme

Semantics. A word having several or multiple meanings.

Many English words have multiple meanings, which are all uses of the same word that have grown apart over time (e.g. *draw* 'cause to move in a certain direction', 'produce a picture', 'finish a game with an equal score'; *flat* 'apartment', 'note lowered by a semitone', 'piece of stage scenery'; *plain* 'unmistakable', 'unsophisticated', 'not good-looking'). Theoretically, a *polyseme*, with meanings which are all ultimately related, is distinguished from a set of *homonyms*, which are different words (with different meanings) which have all come to have the same form (e.g. *pile* (i) 'heap', (ii) 'beam driven into ground', (iii) 'soft surface of fabric'). In practice it is very difficult for a person who is not a historical linguist to tell whether a word with several meanings is a case of polysemy or homonymy or a mixture of both; and in some cases evidence is lacking by which even the scholar could decide.

• **polysemantic, polysemic, polysemous**: having several or multiple meanings.

polysemy (Also **polysemia**.)

The fact of having several meanings; the possession of multiple meaning.

> 1972 M. L. SAMUELS The effect of polysemy is in principle the same as that of homonymy—the representation of two or more meanings by a single form.

polysyllabic

Having more than one syllable.

Compare DISYLLABIC, MONOSYLLABIC.

popular etymology

The same as FOLK ETYMOLOGY.

> 1926 H. W. FOWLER It is true .. that *-yard* [in *halyard*] is no better than a popular etymology corruption.

portmanteau word

The same as a morphological BLEND.

The term originates with Lewis Carroll's explanation (in *Through the Looking-Glass* (1872)) of the invented word *slithy* as a combination of 'lithe' and 'slimy': 'It's like a portmanteau—there are two meanings packed up into one word.'

position

Any of the places within a larger unit in which a particular linguistic element can appear.

See FINAL, INITIAL, MEDIAL. Compare WORD ORDER.

• **positional**: of, pertaining to, or determined by position. **positionally**.

> 1970 B. M. H. STRANG Positional conditioning [of allophones of *l*] .. can lead to a considerable articulatory gap, if, say, a high, front vowel is followed by a dark *l*.

positive

(*n. & adj.*) **1** (A verb, clause, or sentence) having no negative marker. (Sometimes called AFFIRMATIVE.)

2 (Designating) the unmarked DEGREE in the three-way system of comparing adjectives and adverbs. (Sometimes called ABSOLUTE.)

e.g.

good, beautiful, soon (as contrasted with *better/best*, *more beautiful/most beautiful*, *sooner/soonest*)

possessive

(*n. & adj.*) (A word or case) indicating possession or ownership.

The *possessive case* of nouns is also called the GENITIVE *case*, e.g. *boy's, boys', Mary's, the Smiths'*.

Pronouns in the possessive case are the series *mine, yours*, etc.; the corresponding determiners are *my, your*, etc. Some grammars include these determiners under the label *possessive pronouns*; more traditional ones classify them as *possessive adjectives.*

The basic meaning of the verb *have* is sometimes described as possessive (e.g. *We have a house*) in contrast to its other meanings, especially the dynamic ones such as *have a bath*, *have dinner*, *have an operation*, *have a holiday*, *have fun*.

Compare APOSTROPHE.

possibility

One of the main semantic categories used in the classification of modal verbs (particularly EPISTEMIC modality), together with NECESSITY.

Post-Bloomfieldian

(*adj.*) Of, pertaining to, or characteristic of a group of linguists who continued principles developed by Bloomfield.

(*n.*) A person belonging to the Post-Bloomfieldian group of linguists.

1992 P. H. MATTHEWS With the triumph of Chomsky's theories in the early 1960s, many of the preoccupations that had been dear to the Post-Bloomfieldians were abandoned.

Compare BLOOMFIELDIAN.

post-creole

Designating a community whose speech has developed beyond the CREOLE stage.

Coined by D. Decamp (1968).

postdeterminer

1 A determiner that must follow any predeterminer or central determiner in its noun premodification. See DETERMINER (1).

2 One of a set of adjectives that are used rather like determiners, and must come before any other adjectives.

e.g.

certain important new ideas
the *other* big red cardboard box

post-genitive

The same as DOUBLE GENITIVE.

postmodification

The act or an instance of a dependent word or words affecting the meaning of a preceding HEAD word. Contrasted with PREMODIFICATION.

Postmodification can occur in different kinds of phrases (groups), e.g. in an adjective or adverb phrase, and may take the form of a single word or (sometimes) a clause:

Is that warm *enough*?
He speaks too quietly *for me to hear*

Postmodification is, however, particularly discussed in relation to noun phrase structure. Postmodification types include

a prepositional phrase: A Question *of Upbringing*
a relative clause: The Way *we Live Now*
a non-finite clause: Virgin Soil *Upturned*, a book *to read*
an adjective (especially after an indefinite pronoun): Nothing *Sacred*

postmodify

Follow and modify the meaning of (a head).

• **postmodifier**.

postpone

Put (a word or words) later in the clause. See POSTPONEMENT.

• • **postponed identification**: see DISLOCATION.

Compare POSTPOSE.

postponement

The placing of a word or words later in a clause than is necessary.

A word or words may be postponed for the stylistic effect of giving end-focus to a particular part of the message. Grammatical devices for postponement include:

discontinuity, which often postpones part of a noun phrase
Everyone was delighted *except the chairman*
extraposition, often in order to postpone a clausal subject
It is hardly surprising *that he did not like the architect's original plans*

passive structures, which postpone the agentive subject to the end

The building is to be opened *by the Prince*

Compare CLEFT, PSEUDO-CLEFT, POSTPOSITION.

postpose

Place (a word or words) after the word modified. See POSTPOSITION.

Although *postposition* is distinguished from *postponement*, there is occasionally confusion between *postpone* and *postpose*.

postposition

1 The positioning of a word (or words) after the word that it modifies.

2 (Rarely in English.) A word that follows (instead of preceding) the word it modifies. Contrasted with PREPOSITION.

In some languages (e.g. Japanese) the sorts of meaning and function that prepositions have in English are exhibited by words that follow their objects or complements and which are appropriately called *postpositions*. Such a class of words does not exist in English, though some words, such as *ago* (as in *a month ago*) and ENCLITICS (*-n't*, *-'s*) are sometimes so described.

A preposition does not become a postposition just because in some non-basic structure it apparently follows its object or complement (e.g. *What are you looking at?*).

More generally, any word or words following the words they modify can be said to be *postposed* or *in postposition*:

I am a man *more sinned against than sinning*

the astronomer *royal*

• **postpositional**: (*n. & adj.*) (an element) that is positioned after the word modified by it.

postpositive

Characterized by postposition; (of position) immediately following, postposed.

Postpositive and *postposed* are virtually synonymous, but the former is the preferred term, to contrast with ATTRIBUTIVE and PREDICATIVE in describing the position of adjectives. Postpositive position is obligatory for adjectives modifying indefinite pronouns and adverbs (e.g. *nobody special*, *somewhere quiet*); in certain set expressions (e.g. *heir apparent*, *the body politic*); and with some adjectives in particular meanings (e.g. *the members present*, *the parents involved*).

• **postpositively**.

1961 R. B. LONG Superlatives in *most* [e.g. *innermost*, *uppermost*, etc.] are now felt as compounds in which a modifying auxiliary pronoun has been united, postpositively, with a basic-form adjective head.

postvocalic

Phonetics. Of a consonant: occurring after a vowel.

The articulation of a phoneme is affected by its phonetic context, which may condition the use of different allophones. Thus in RP a postvocalic /l/ followed by silence or another consonant is always dark.

• **postvocalically**.

1964 R. H. ROBINS In Scots English, /r/ occurs both prevocalically and postvocalically.

Compare INTERVOCALIC.

pragmatic

Of or pertaining to pragmatics.

•• **pragmatic particle**: the same as FILLER (2).

• **pragmatically**.

pragmatics

The branch of linguistics dealing with language in use. Contrasted with SYNTACTICS (or *syntax*) and SEMANTICS.

The term is defined in a variety of ways, but it is often used in connection with the communicative functions of language, in contrast to the more precise grammatical or syntactical meaning of the text. Thus *'I've borrowed this book from the library'* in different contexts might have the pragmatic implication *'so I don't want to watch TV'*, or *'so I don't want to go out'* or even *'I wasn't going to buy it'*.

1937 C. MORRIS Analysis reveals that linguistic signs sustain three types of relations (to other signs of the language, to objects that are signified, to persons by whom they are used and understood) which define three dimensions of meaning. These dimensions in turn are objects of investigation by syntactics, semantics, and pragmatics.

Prague School

The name of a group of linguists belonging to the Linguistic Circle of Prague, founded in 1926, and others whom it has influenced, used with reference to linguistic theories and methods initiated by them.

The effects of this school on the analysis of English have been considerable. It has been influential in the development of DISTINCTIVE FEATURE analysis in phonology, and in the analysis of the COMMUNICATIVE DYNAMISM of sentences, which is concerned with the differing amount that each element contributes to the overall information content of a sentence.

• **Praguian**: of or pertaining to the Prague School.

predeterminer

A determiner that must precede any central determiner or postdeterminer in a phrase. See DETERMINER (1).

predicate

(*n.*) (Pronounced /ˈpredɪkət/.)

1 All that part of a sentence which is not the SUBJECT.

Traditionally sentences are divided into two parts, the *subject* and *predicate*:

subject	*predicate*
All good things	must come to an end
Attack	is the best form of defence
Familiarity	breeds contempt

The division is grammatically valid, but the terms *subject* and *predicate*, being notionally based, can be misleading. The subject is not necessarily what the sentence is about, and the rest is not always 'what is predicated about' the subject. For example, the grammatical subject in *It is raining* is empty of meaning.

Modern grammarians analyse predicate structure in various ways; for example, into components such as Verb, Object, Complement, and Adverbial, and into elements of information structure such as *new* (versus *given*) or *comment* (versus *topic*). Grammatical distinctions are also made between the operator and the rest of the *predication*.

Compare PREDICATOR.

In some older grammar, *predicate* rather than *predicative* is used to describe an adjective, noun, or pronoun when such a word is 'predicated of the subject', i.e. is used in predicative position. For example:

He became mad (*mad* = predicate adjective)
Croesus was king (*king* = predicate noun)
I am he (*he* = predicate pronoun)

In modern terminology such a word functioning after a linking verb is said to be a *subject complement* or possibly a *predicative complement*.

2 The term in a proposition which provides information about the individual or entity.

In logical semantics the term has a semantic rather than a syntactic meaning. Grammarians who pursue this line of analysis may make this clear by using the term *syntactic predicate* for the element of syntax (contrasted with *subject*); they can then talk of propositions in terms of a semantic predicate with ARGUMENTS. A semantic predicate taking one argument is a *one-place predicate*, and so on.

(*v.*) (Pronounced /ˈpredɪkeɪt/.) Assert (something) about the subject of a sentence; make (something) the predicate.

predication

1 What is predicated.

The term may be used as a rough synonym of PREDICATE, but *predication* is often used more theoretically to suggest the proposition expressed by the actual words of the predicate.

2 (Specifically, in some models.) A predicate minus its operator.

1985 R. QUIRK et al. Simple sentences are traditionally divided into two major parts, a SUBJECT and a PREDICATE ... A more important division, in accounting for the relation between different sentence types, is that between OPERATOR and PREDICATION as two subdivisions of the predicate.

This analysis serves to emphasize that the operator and the rest of the predicate have distinct characteristics. For example, the operator is separated from the rest of the predicate in questions (e.g. *Are your parents flying home tomorrow?*); an operator can allow coordination of two predications (e.g. *I must phone the airport and ^ check flight times*); an operator can stand for a completely ellipted predication (e.g. *Yes, you should ^*).

•• **predication adjunct**: an adjunct applying to the predication. See SENTENCE ADJUNCT.

verb of incomplete predication: an older term for a LINKING VERB.

predicative

1 Occurring after a linking verb. Contrasted with ATTRIBUTIVE.

The term is particularly used in the classification of adjectives. Most adjectives can be used in both attributive and predicative positions (e.g. *a fine day*, *the day was fine*; *an expensive restaurant*, *the restaurant looks expensive*). But some adjectives are used in only one of these positions.

Among predicative-only adjectives are a group beginning with *a-* (e.g. *afraid, alone*, etc.: see *A*-WORD) and those that usually or always require complementation by a prepositional phrase (e.g. *answerable, conducive, devoid, loath*, etc.).

Predicative adjectives can also take POSTPOSITION (e.g. *anybody aware of these facts*, *those people impatient with the slow progress*).

2 Occurring in the predicate or predication.

In some models subject(ive) and object(ive) complements, whether adjectives or noun phrases, are labelled in more detail as:

subjective predicative complement (e.g. *I felt a fool*)
objective predicative complement (e.g. *They called me a fool*)

• **predicatively**: in the predicate or predication.

predicator

1 A verb phrase in its functional capacity as part of sentence structure.

Most of the functional constituents of clause structure (Subject, Object, Complement, Adverbial) are clearly distinguished by the terminology from the forms (whether single words or longer phrases) that these functional constituents may take. For example, an adverbial may be realized by an adverb (e.g. *quickly*), a prepositional phrase (e.g. *under the table*), or a noun phrase (e.g. *last night*). But the terms *verb* and *verb phrase* can be ambiguous, meaning either the functional constituent or the formal category. Therefore some linguists use the term *predicator*, instead of *verb* or *verb phrase*, as the functional label. Thus the predicator in *Books Do Furnish a Room* is *do furnish*, giving a structural pattern symbolized by SPO rather than SVO.

2 (In some models.) The verb phrase (verbal group) minus the actual operator.

For example, in *I wouldn't have thought so*, the predicator is *have thought*. This analysis emphasizes the fact that the first word is temporal and finite, while the predicator is concerned with the other meanings of the phrase, such as 'secondary' time in relation to that of the operator, voice, and whatever 'process' (e.g. action, mental activity) is predicated of the subject.

prediction

The assertion that an event etc. will happen in the future.

One of the meanings used in the classification of EPISTEMIC modality, and particularly uses of the verb *will*.

predictive

Relating to prediction.

•• **predictive conditional**: the commonest type of conditional sentence, in which there is a causal link of the type 'If X, then Y follows'.

e.g.

If you drive like that, you'll have an accident

prefix

Morphology.

(*n.*) An affix added before a word or base to form a new word. Contrasted with SUFFIX.

Prefixes are primarily semantic in their effect, changing the meaning of the base. Common prefixes include:

counter-productive (M20)
*de*frost (L19)
*fore*warn (ME)
*dis*connect (L18)
*hyper*active (M19)
*inter*national (L18)
*mal*function (E20)
*mini*skirt (M20)
non-event (M20)
*re*build (L15)
*sub*zero (M20)
*un*natural (LME)
*under*nourished (E20)

(*v.*) Place before a word or base, especially so as to form a new word.

• **prefixation**.

1974 P. H. MATTHEWS Processes of affixation may then be divided into *prefixation*, *suffixation* or *infixation* . . . In English the commonest processes are those of suffixation . . . Examples of prefixation are found, however, in the negative formations of *happy* → *un* + *happy* . . or of *order* → *dis* + *order*.

preglottalization

Phonetics. The same as GLOTTAL REINFORCEMENT.

• **preglottalize**.

prehead

Phonetics. Any syllable in a TONE UNIT occurring before the head.

In a common description of a tone unit the (possible) elements occurring before the (essential) nucleus can be analysed as HEAD and *prehead*. The *prehead* consists of unaccented syllables before the head, e.g. the word *I* in:

I 'THOUGHT it was AWful.

With different intonation, if the *I* were accented, there would of course be no prehead.

premodification See MODIFICATION.

preparatory *it* See ANTICIPATORY.

prepose

Put (a linguistic element) before something else.

Not a very common use.

1946 O. JESPERSEN *Well-to-do* = 'well off, living in easy circumstances' is often preposed, generally written with hyphens: *a well-to-do farmer*.

(This type of compound element can be called a PHRASE-WORD in attributive position.)

preposition

A traditional word class, comprising words that relate two linguistic elements to each other and that generally precede the word which they 'govern'.

Simple prepositions are predominantly short words (e.g. *at, by, down, for, from, in(to), to, up*), some of which also function as adverbs (see PARTICLE). There are some longer prepositions (e.g. *alongside* the quay, *throughout* the period), and also COMPLEX *prepositions* consisting of combinations of two or three words that function in the same way (e.g. *according to*, *regardless of*, *in front of*, *by means of*, *in addition to*).

There was at one time considerable prejudice against putting a preposition later than the word it belongs to(!). Rewording is possible in some contexts (e.g. *the word to which it belongs*) but a *deferred* (or *stranded*) *preposition* is sometimes unavoidable without major rewriting:

What did you do that *for*?
The problem is difficult to talk *about*
It's not to be sneezed *at*

This prejudice goes back to Latin grammar, in which the characteristic placing of the particle is indicated by its name *praepositio*, from *praeponere* put before.

Prepositions overlap not only with adverbs but with other word classes: e.g. *near* is like an adjective in having comparative and superlative forms (*nearer* the window); *since* can be preposition (*since the war*), adverb (*I haven't seen them since*), and conjunction (*since the war ended*).

See MARGINAL *preposition*, PIED PIPING.

prepositional

Of, pertaining to, or expressed by a preposition; formed with a preposition; serving as, or having the function of, a preposition.

prepositional adjective

An adjective that obligatorily or optionally takes a prepositional phrase as complementation.

Examples:

I am not *averse to* subsidizing them
He is *fond of* opera

prepositional adverb

A PARTICLE, a word identical or closely related in form to a preposition, but functioning as an adverb.

Words of this kind can perfectly well be classified either as adverbs or as particles. But this term serves to distinguish the function of a particular particle in a particular context.

I fell *down* the stairs (preposition)
I fell *down* (prepositional adverb)

prepositional complement

The word or phrase 'governed' by a preposition. (Also called *object of preposition.*)

A *prepositional complement* is usually a noun phrase

in *the end*
before *the war*

but can be other parts of speech or a non-finite clause

in *short*
afraid of *being killed*

prepositional object

The object of a prepositional verb; any complement following a preposition.

The terms *prepositional complement*, *object of preposition*, etc. are often used interchangeably, but some grammarians reserve the term *prepositional object* for those complements that follow a prepositional verb, e.g.

Listen to *this*

James has come into *a fortune*
We were hoping for *a breakthrough*

prepositional phrase

A preposition plus its object (or complement).

Prepositional phrase is a formal class rather than a functional one. Prepositional phrases function mainly as

(*a*) adverbials:

Come *into the garden*, Maud

and (*b*) postmodifiers in noun phrases:

the Lady *of Shalott*

Compare PREPOSITION GROUP.

prepositional verb

A verb consisting of a lexical verb plus a preposition. Also called *two-part verb*.

As the particle is (by definition) a preposition, it generally comes before its object:

I am looking after the children/looking after them

not

*looking the children after/looking them after

But within certain structures the particle may be deferred:

They need looking after

See PREPOSITION. Compare PHRASAL VERB, PHRASAL-PREPOSITIONAL VERB.

preposition group

(In Systemic-Functional Grammar.) A structure consisting of a preposition with modification.

Examples:

immediately after
right in front of

Such a group functions exactly like a preposition. It is not to be confused with a *prepositional phrase* and can of course form part of one:

immediately after lunch

prescriptive

Linguistics. Concerned with or laying down rules of usage. Contrasted with DESCRIPTIVE.

The term is generally used pejoratively, and the inappropriateness of misconceived Latin-based rules is held up to ridicule (for example 'Don't end a sentence with a preposition'; 'Say *It's I*'; 'Never split an infinitive'; etc.).

In reality most descriptive statements about language are based on some value-judgement of what is acceptable and normal, however objectively descriptive they try to be.

• **prescriptivism**: the practice or advocacy of prescriptive grammar. **prescriptivist**.

present

(*n. & adj.*)

1 (Occurring in) the time now existing. Contrasted with PAST.

2 (A tense or form) referring to the time now existing, or not referring particularly to the past or future. Also, specifically, the same as PRESENT SIMPLE.

As applied to tenses the term can be misleading. Although present tenses predominantly refer to some sort of present time, they would more accurately be described as being UNMARKED for time. Thus the *present simple tense* is used for general or 'eternal' or 'timeless' truths (e.g. *The earth goes round the sun*); and present tenses can also be used where the pastness of the event time is unimportant or irrelevant (e.g. *Shakespeare says . . .*) and with reference to a future event that is certain or taken for granted (e.g. *We are leaving tomorrow*, *If you come next week, we can . . .*).

See also HISTORIC *present*.

present participle

The *-ing* form of the verb when used in a verblike way (in contrast to its nounlike use).

This participle is used in the formation of progressive tenses and, alone, in non-finite clauses:

We are/were listening
Sitting here, I haven't a care in the world

See *-ING* FORM.

present perfect

(A tense) formed with *have* or *has* + past participle.

Present perfect tenses generally refer to some state or event or series of events already achieved in a period up to the moment of speaking and often relate them in some way to the present. With no further label, *present perfect (tense)* means the present perfect simple tense:

I *have known* her since she was twenty
James *has come* (i.e. he is here now)
She's *written* literally hundreds of novels

Contrast the simple past:

My father *knew* Lloyd George
James *came* (but perhaps is no longer here)

Ivy Compton-Burnett *wrote* 19 novels in a uniquely bizarre uncompromising style

American English however often uses a simple past (e.g. *Did James come yet?*) where British English still prefers the present perfect (e.g. *Has James come yet?*).

Present perfect progressive and *present perfect passive* tenses are also possible:

It's been raining for two days now
He has been working hard today/recently
She has been told

present progressive

(A tense) formed with the present tense of *be* + the *-ing* participle. (Also called *present continuous.*)

The meaning is often an action in progress at the time of speaking, or of limited duration at the present time. e.g.

I *am asking* you a question
We *are listening*
I'*m living* with friends until my flat is ready

It can also refer to arrangements for the near future, e.g. *I'm lunching with Margaret tomorrow.*

present simple

(The tense) that is identical to the base of the verb (except in the case of *be*) and adds *-s* for the third person singular. (Also called *present (tense)* simply.)

e.g.

I know and he knows too

present subjunctive See SUBJUNCTIVE.

presupposition

Semantics. An assumption underlying a statement or other utterance.

The term is taken from philosophy and logic and is used in various ways, sometimes related to truth-conditions. Thus a sentence such as *Did you pass your driving test?* presupposes that a driving-test exists and that you took it: without this the sentence can have no truth value.

Presupposition is sometimes contrasted with ENTAILMENT, which is more closely based on the actual words. *I passed my test* at its simplest entails *I passed something. I didn't pass the test* may presuppose or imply that I tried and failed but it does not entail this: *I didn't pass it for the simple reason that I didn't take it.* The differences between *presupposition*, *entailment*, and IMPLICATURE are the subject of much discussion.

preterite (American English **preterit.**)

(*n. & adj.*) (In traditional grammar.) The past simple tense.

prevocalic

Phonetics. Of a consonant: occurring before a vowel.

• **prevocalically.**

> 1970 B. M. H. STRANG RP selects two qualities . . and distributes them positionally—dark *l* post-vocalically, and clear *l* pre- or inter-vocalically.

Compare INTERVOCALIC, POSTVOCALIC.

primary

1 *Phonetics & Phonology.* Designating the principal stress in a word.

Primary stress (or *primary accent*) (marked with a superior vertical bar preceding the relevant syllable [ˈ]) contrasts with *secondary stress* (marked with an inferior vertical bar [ˌ]), and even *tertiary stress.* The difference can be heard in longer words, which always have their own basic patterns, even though the pattern may be modified by the overall intonation of the utterance in which it occurs.

> /ˌpɒlɪˈteknɪk/ polytechnic, /ˈæpɪˌtaɪzɪŋ/ appetizing

Primary stress is always on a syllable where pitch change can potentially occur.

2 (*n. & adj.*) (In some older grammar.) (Designating the RANK or level of) the most important word in a group. Contrasted with SECONDARY (3) and TERTIARY (2). See RANK.

Chiefly used by Jespersen.

> 1924 O. JESPERSEN We may, of course, have two or more coordinate adjuncts to the same primary: thus in *a nice young lady*, the words *a*, *nice*, and *young* equally define *lady*.

3 (In some modern grammar.) Designating a tense expressed by one word; the same as SIMPLE.

In this system, COMPOUND tenses (containing more than one word) are secondary tenses.

primary performative See PERFORMATIVE.

primary verb

One of the three verbs which can function either as main verbs or auxiliaries, namely *be*, *do*, and *have*.

Verbs are traditionally divided, according to function, into main verbs and auxiliaries. But as *be*, *do*, and *have* can function as both, it can be useful to make further distinctions. *Primary* is used as a label for these verbs, however they are functioning.

Compare AUXILIARY, FULL, MAIN.

primitive

Linguistics. (*n. & adj.*) (A term) that is taken as basic and 'given', but not strictly defined.

1968 J. Lyons When the linguist sets out to describe the grammar of a language on the basis of a recorded corpus of material, he starts with a more primitive notion than that of either the word or the sentence (by 'primitive' is meant 'undefined within the theory', 'pre-theoretical'). This more primitive notion is that of the *utterance.*

principal clause

The same as MATRIX CLAUSE.

principal part

One of the three forms of a verb from which all the other forms can be derived.

These are the three forms of a verb given in a dictionary, namely, the base, the past form, and the past participle. e.g.

blow, blew, blown
come, came, come
hurt, hurt, hurt
like, liked, liked
swim, swam, swum

The only verbs outside this pattern are *have* (where *has* is exceptional) and *be* (where the base, i.e. *be* itself, does not indicate the present tense forms *am/are/is*).

private verb

A verb expressing an intellectual state and taking a *that*-clause. Contrasted with a PUBLIC VERB.

The majority of verbs introducing *that*-clauses are sometimes loosely described as *reporting verbs*. More technically, they are classified as FACTUAL verbs and divided into two groups: those which involve a public statement, and those which refer to private thinking only, e.g. *believe, know, realize, understand.*

privilege of occurrence

Linguistics. The functional slot(s) in which a word or phrase can be used.

Words of the same class, in a general way, share the same *privileges of occurrence*, and this is in fact one of the tests used in assigning a word to a particular class. Thus *along*, *back*, *down*, *out* could all replace the adverb *round* in the sentence *I hurried round.* On the other hand *at* or *into* could not (e.g. **We hurried at*, **we hurried into*) and are not adverbs.

1933 L. Bloomfield The lexical form in any actual utterance, as a concrete linguistic form, is always accompanied by some grammatical form: it appears in some function, and these privileges of occurrence make up, collectively, the grammatical function of the *lexical* form.

Compare CO-OCCURRENCE.

process

A continuous action or series of actions, events, changes, stages, etc.

(*a*) The word is used in its everyday sense in the classification of various parts of speech on semantic grounds.

Process adverbs or *adverbials* are a category of adverbs that includes adjuncts of manner, means, and instrument. E.g.

We went *quickly* (manner)
We travelled *by car* (means)
He walks *with a stick* (instrument)

Process verbs are dynamic verbs that indicate changing states; e.g. *change, deteriorate, grow.*

The evenings are *drawing out*

(*b*) In Systemic-Functional Grammar, the concept of process occupies a much more important place in the interpretation of the clause as a 'representation of experience'.

1985 M. A. K. HALLIDAY A process consists potentially of three components:
(i) the process itself;
(ii) participants in the process;
(iii) circumstances associated with the process.

The process itself is normally realized through a verb, the participants through nominal groups (noun phrases), and the circumstances through an adverbial. Processes are analysed into *material processes* (often with an Actor and Goal), *mental processes* (in which the participants are labelled SENSER and PHENOMENON), and various RELATIONAL *processes.*

proclitic

(*n. & adj.*) (A word) pronounced with very little emphasis, and attached, usually in shortened form, to the following word. Contrasted with ENCLITIC.

This phenomenon is much rarer in English than the *enclitic*. Arguably, the articles (*a/an*, *the*) are proclitics; likewise *do* and *it* when reduced to a single consonant sound (e.g. *D'you know?*, *'Twas brillig and the slithy toves . . .*).

productive

Describing a linguistic process that is still in use, in contrast to **unproductive**.

Many kinds of affixation are highly productive in forming new words. For example, dozens of new words are formed with *anti-, Euro-, dis-, un-, -ee, -ness* and so on, but few with *-dom* or *-hood.*

pro-form

A word or other linguistic unit that can co-refer to, or substitute for, another.

Pronouns, as the name implies, are commonly used as pro-forms for nouns. But pro-forms also include:

adverbs (e.g. *here, there, then*):

All her life she dreamt of Paris, but she never got *there*

phrases with the 'pro-verb' *do* (e.g. *do it, do so, do that*) replacing a predicate or predication, e.g.

They said they would *do it/do so*

so and *not* as pro-forms replacing object *that*-clauses, e.g.

Are there any survivors? I hope *so*/I fear *not*

A pro-form may relate to its antecedent in two grammatically different ways. Contrast

I went to the library for a book yesterday, but they hadn't got *it* (a relationship of CO-REFERENCE)

I went to the library for a book yesterday, but didn't borrow *one* (SUBSTITUTION for a noun phrase)

• **proformation**: substitution by pro-forms for other words or linguistic units.

progressive

1 (*n. & adj.*) (A verb form) expressing ongoing, durative ASPECT; contrasted with a SIMPLE tense or the PERFECTIVE (PERFECT) aspect. (Also called CONTINUOUS.)

Linguists classify *progressive* as an aspect of the verb, rather than as part of tense. It expresses activity in progress, and therefore often of limited duration, rather than merely temporal meaning.

In common parlance we refer to *progressive tenses*. They are formed with a part of the verb *be* plus an *-ing* form of a lexical verb, e.g.

I am staying with friends until the 30th (*present progressive*)

I was wondering what to do (*past progressive*) when this job cropped up

We will have been waiting two whole hours (*'future' perfect progressive*) by the time their train arrives

Passive progressive tenses are also possible (e.g. *It's being repaired*).

2 **progressive assimilation**: see ASSIMILATION.

projection

(In Systemic-Functional Grammar.) A type of relationship between clauses. Contrasted with EXPANSION.

Projection is largely concerned with the reporting of speaking and thinking 'events' (dealt with as direct and indirect speech in traditional grammar). But it also embraces such 'rankshifted' remarks and thoughts as

The question *whether the old man would survive*

(traditionally treated as appositional clauses); and 'facts', e.g.

That he was seriously ill was not disputed

(traditionally dealt with as noun clauses).

prominence

Phonetics. The perceived importance or conspicuousness of speech sounds.

What the listener perceives as 'loudness' may be due to other factors, such as stress, pitch, phoneme quality, and duration, rather than simply greater volume of sound.

> 1962 A. C. GIMSON It is better to use a term such as *prominence* to cover these general listener-impressions of variations in the perceptibility of sounds.

pronominal

Of, pertaining to, or of the nature of, a pronoun.

In some traditional grammar, adverbs of vague general meaning (e.g. *here, where, why, somewhere, anyhow*) are called *pronominal adverbs*. This category overlaps with the more general modern category of PRO-FORM.

• **pronominalization**: the fact or process of replacing a noun phrase by a pronoun. **pronominalize**: render pronominal.

pronoun

A member of one of the closed word classes (parts of speech) having nominal function.

Traditional grammar defines *pronoun* as a word that can replace or stand for a noun. But this is not strictly accurate. For example in *I asked my neighbours if they would cut their hedge*, *they* certainly replaces *my neighbours* (note that this is a noun phrase rather than a noun), but *I* can hardly be said to replace a noun. Secondly, a pronoun (e.g. *that*) often replaces a clause, as in *Why did you ask that*?

Pronouns constitute a closed class, which can be subdivided in various ways. Commonly accepted subdivisions are PERSONAL PRONOUN, RELATIVE *pronoun*, INTERROGATIVE *pronoun*, DEMONSTRATIVE *pronoun*, and INDEFINITE PRONOUN. The class overlaps with DETERMINERS, many words being used as both; and indeed the possessive determiners (*my*, *your*, etc.) are sometimes classified as pronouns.

pronunciation See ACCENT, GENERAL AMERICAN, INTERNATIONAL PHONETIC ALPHABET, PHONETICS, RECEIVED PRONUNCIATION.

proper noun

A noun referring to a particular unique person, place, animal, etc. Contrasted with COMMON *noun*. (Also called *proper name*.)

The traditional distinction between common and proper nouns is both grammatical and semantic. Proper nouns do not freely allow determiners or number contrasts (e.g. **my Himalaya*, **that New York*, **some Asias*), and article usage tends to be invariant (e.g. *the Chilterns*, *the (River) Thames*, *Oxford*, not **Chilterns*, **Thames*, **the Oxford*).

However, the categories are not watertight; since more than one referent may share the same name, names may in some circumstances be treated like common nouns (e.g. *the Smiths*, *a Mr. X. Y. Z. Smith*). Nationality words and names of days of the week are also distinctly borderline (e.g. *three Scots*, *an Australian*, *three Mondays in succession*).

The terms *proper noun* and *(proper) name* are often used interchangeably. But a distinction is sometimes made between *names*, which can include ordinary dictionary words (e.g. *the United States*, *New York*, *the Daily Telegraph*, *the South Downs*, *A Midsummer Night's Dream*) and *proper nouns*, which are then single words (e.g. *Dorchester*, *Elizabeth*, *England*).

prop *it*. See DUMMY.

proportional clause

One of two joined parallel clauses involving a kind of comparison.

This is a semantic label for a particular type of clause. As well as covering such fairly standard sentences as

As he grew older, (so) he worried less

it also covers the more unusual pattern exemplified by

The more he thought, the less he spoke
The more, the merrier

proposition

An utterance in which something is affirmed that can be proved or disproved; (the fact of) making such an utterance.

• **propositional**: having the nature of a statement or assertion about something; literally expressed by an utterance.

The term is often used to explain how two different sentences logically 'mean' the same. For example *He is hard to persuade* and *It is hard to persuade him* can be said to have the same *propositional meaning*. This kind of meaning can then be contrasted with the *communicative* or *illocutionary* function of an utterance.

The terms, taken from logic, are used in a number of ways.

1977 J. LYONS The term 'proposition', like 'fact', has been the subject of considerable philosophical controversy . . . Further difficulties are caused by the use of 'proposition' in relation to 'sentence' and 'statement': some writers identify propositions with (declarative) sentences, others identify them with statements, and others with the meaning of (declarative) sentences; and there is little consistency in the way in which 'statement' is defined.

pro-predicate, pro-predication.

The use of a pro-form to stand for a predicate or predication.

prop word

(In some traditional grammar.) The pronoun *one* used as a substitution pro-form.

The pronoun *one* (plural *ones*) can be used to substitute for part or all of a noun phrase:

Have you any navy blue leather handbags? Yes, we have *some nice ones*. I could also show you a smart red *one*.

It is confusing, however, to single out this particular proform. Other words (e.g. *another*, *the same*, etc.) can also substitute in similar ways.

The term was coined by Henry Sweet (1892b) and much used by Jespersen.

prosodic

Phonetics. Of phonetic features: extending beyond individual phonemes. (Also called *suprasegmental*.)

1962 A. C. GIMSON A sound has not only quality, whose phonetic nature can be described and function in the language determined, but also length, pitch, and a degree of stress. Such features may extend in time beyond the limits of the phoneme and embrace much higher units of the utterance. Indeed, in the case of the pitch variation characteristic of an intonation pattern, analysis may involve the whole of a lengthy utterance. In this sense, such features are *prosodic*, or supra-segmental.

• **prosodically**.

prosody

A phonological feature having as its domain more than one segment.

Prosodies, in some models, seem to be synonymous with the class of *suprasegmental* features such as intonation, stress, and juncture. But some authors extend the term to include some features which are regarded as *segmental* in phonemic theory, e.g. palatalization, lip-rounding, and nasalization.

The term originated with J. R. Firth and has been used by his followers. Before the nineteenth century, *Prosody* was counted as one of the four branches of grammar (the others being Orthography, Etymology, and Syntax), and was itself in two parts; the first part dealt with pronunciation, especially accent, quantity, emphasis, pause, and tone, and the second with the laws of versification.

1948a J. R. FIRTH We can tentatively adapt this part of the theory of music for the purpose of framing a theory of the prosodies. Let us regard the syllable as a pulse or beat, and a word or piece as a sort of bar length or grouping of pulses which bear to each other definite interrelations of length, stress, tone, quality—including voice quality and nasality.

prospective

Referring to the future.

A term used in some traditional grammar in the discussion of time clauses with future meaning.

protasis (Pronounced /'prɒtəsɪs/; plural **protases.**)

The clause expressing the condition in a conditional sentence. Contrasted with the APODOSIS.

- **protatic** (pronounced /prə'tætɪk/) (not common).

pro-verb

The verb *do* used as a pro-form, i.e. as an operator, or as a SUBSTITUTION item.

proximal

Indicating things that are near. Contrasted with *non-proximal* (or *distal*). This is a label sometimes applied to the deictics *this* and *these*; *non-proximal* or *distal* are applied to *that* and *those*.

proximity agreement, proximity concord See AGREEMENT.

pseudo-cleft

Designating a type of sentence with, as subject, a nominal clause introduced by a *wh*-relative, usually *what*. See CLEFT.

- **pseudo-clefting.**

> 1980 E. K. BROWN & J. E. MILLER Another structure of this general type is pseudo-clefting, illustrated by sentences like: . . What John bought was a screwdriver.

pseudo-coordination

Apparent coordination of two words where, however, the relationship between them is not one of equality.

> 1985 R. QUIRK et al. When they precede *and*, members of a small class of verbs or predications have an idiomatic function which is similar to the function of catenative constructions . . and which will be termed *pseudo-coordinations*.

Examples of this (which tend to be colloquial) include *try and come*, *went and complained*. Other pseudo-coordinations are found with adjectives, e.g. *nice and warm*, and the very colloquial adverbial *good and proper*.

pseudo-participle See PARTICIPIAL ADJECTIVE.

pseudo-passive

A construction, consisting of part of the verb *be* + a past participle, that resembles a passive, but which neither has an active counterpart nor permits an agent.

The active sentence corresponding in meaning to

My homework is finished now

is

I have finished my homework now (*not* *I finish my homework now)

Is finished refers to a state. Some passive sentences are ambiguous, especially in the past tense, e.g.

The job was finished at two o'clock

This is a pseudo-passive, with a statal verb, if the meaning is 'By the time I arrived at two o'clock it was already finished'. This contrasts with a true dynamic central passive (with *actional*) verb where an agent can be given, and the verb can be progressive.

The job was finished at two o'clock (by Bill, who had just arrived)
The job was (being) finished at two o'clock (just as the clock was striking)

Compare PARTICIPIAL ADJECTIVE.

pseudo-subjunctive

A hypercorrect use of the *were*-subjunctive, where an indicative is acceptable. e.g.

He tried to drop in sometimes on his way to his constituency if he were alone.

psycholinguistics

The study of the psychological aspects of language and language learning.

• **psycholinguist. psycholinguistic**.

psychological subject

An older term for the TOPIC OR THEME of a clause, i.e. the 'subject-matter' or what the clause is about; contrasted with the *grammatical subject* and the *logical subject*.

public verb

A verb whose meaning includes or implies the idea of 'speaking', taking a *that*-clause expressing a FACTUAL proposition; contrasted with a PRIVATE VERB.

Examples:

affirm, announce, boast, confirm, declare

pulmonic

Phonetics. Of (the articulation of) a speech sound: made with lung air.

All normal English speech sounds are pulmonic and egressive. See AIR-STREAM MECHANISM.

1988 J. C. CATFORD Glottal stop is, of course, *pulmonic*, since the pressure buildup below the glottis is initiated by the lungs.

punctual

Of (the action signified by) a verb: having no duration; contrasted with DURATIVE.

Semantic classifications of the verb vary in their amount of detail, the main division usually being between STATIVE and DYNAMIC. But these broad categories may be further analysed. *Punctual verbs* refer to an action or event that is momentary (e.g. *The bomb exploded*), or at any rate is felt to be 'transitional' (e.g. *The concert's beginning*), or to a series of such actions or events (e.g. *I was sneezing*).

punctuation

The practice or system of inserting various marks in written text in order to aid interpretation; the division of written or printed matter into sentences, clauses, etc. by means of such marks.

Earlier punctuation reflected spoken delivery, marking especially the pauses where breath would be taken. Since the eighteenth century it has been based on grammatical structure, marking sentences, clauses, and some types of phrase. Broadly speaking, it has the function either of linking items (e.g. *three potatoes, two carrots, and an onion*) or of separating them (e.g. *I don't know. Ask someone else*).

In a series of adjectives, a comma is used when the meanings are not linked (e.g. *a small, neat room*) but is not used when they are (e.g. *a silly little boy*). Various marks (brackets, dashes, and commas) are used to separate off passages that interrupt the main structure of the sentence (e.g. *This tile (despite its fresh colours) is more than two hundred years old*).

Grammatically complete units can be separated off by lighter punctuation than the normal full stop, either to link parallel statements (semicolon) (e.g. *I wasn't going to leave; I'd only just arrived*), or to lead from one thought to the next (colon) (e.g. *I wasn't going to leave: I stood my ground*).

Punctuation (the hyphen, the comma) is also used to avoid grammatical or semantic ambiguity (e.g. *a natural-gas producer* versus *a natural gas-producer*; *The quarrel over, the friendship was resumed* versus *The quarrel over the friendship was resumed*). The apostrophe could be said to have a quasi-morphological role in distinguishing the 'possessive' (singular and plural) from the plural in most nouns (e.g. *girl's*, *girls'*, versus *girls*).

pure vowel

A vowel made without a glide; contrasted with DIPHTHONG.

It is not in fact possible for a vowel to be held without any movement of the speech organs involved, but some vowels change relatively little during articulation.

English (RP) has twelve pure vowels:

/iː/ see, me, wheat, piece, mach*i*ne
/ɪ/ fit, pr*e*tty, priv*a*te, b*u*ild

/e/ bed, head, m*an*y
/æ/ pan, plait
/ɑː/ far, bath, heart, clerk, calm, aunt
/ɒ/ dog, what, cough, s*au*sage
/ɔː/ force, saw, bought, d*augh*ter
/uː/ food, who, soup, rude, blue, chew
/ʊ/ put, w*o*man, good, could
/ʌ/ hut, son, en*ou*gh, blood, does
/ɜː/ bird, earn, turn, word, j*ou*rnal
/ə/ [always unstressed] *a*go, moth*er*

purpose

The motive, the intention behind an action.

The term is used in its usual sense particularly in the semantic description of adverbials and conjunctions. e.g.

They only do it *to annoy* (infinitive of purpose)

Similarly *in order (not) to, so as (not) to.*

A finite clause of purpose, introduced by *so (that)* or *in order that* normally requires a modal verb, e.g.

They shredded the evidence so that no one would discover the truth.

Contrast:

They shredded the evidence so that no one discovered the truth (result)

Negative purpose is suggested by *lest* and *in case.*

pushdown

(*n. & adj.*) (Designating) a type of embedding in which a linguistic element that is part of one clause operates indirectly as part of another.

Example:

What do you think happened? [*What* do you think? / *What* happened?]
He earns a lot more money than he admits. [*He earns* a lot more money than . . . / He admits *he earns*]

Confusion over pushdown with a relative clause is a common cause of hypercorrection:

He was searching for his parents whom he hoped were still alive [He was searching for his parents *who* . . were still alive / He hoped *they* were still alive]

putative *should*

The use of the word *should*, particularly in a subordinate clause, to refer tentatively to a possible situation, rather than to assert the situation as fact.

The term is particularly applied to the use of *should* in subordinate noun clauses where the *should* does not express obligation, but emphasizes an emotional reaction to a possible or presumed fact:

It is a pity (*or* It is sad) that you should think that

Here it is not even certain that you do think that. Contrast

It is a pity that you think that

Putative should also occurs in subordinate clauses as an alternative to the subjunctive after expressions of suggesting, advising, etc.:

They insisted that I (should) stay the whole week

Putative should also occurs as a main verb in a few expressions of the following type:

How should I know?
Who should walk in but James?

Q

qualification See QUALIFIER.

qualifier

A word (or group of words) that attributes a quality to another word, or that modifies another word or phrase in some way.

(*a*) *Qualifier* (together with *qualify*, *qualification*) is sometimes used with much the same meaning as MODIFIER (and the related *modify, modification*). But in much modern grammar MODIFIER etc. are the preferred terms.

(*b*) In some traditional grammar, distinctions were made between the two terms. *Qualify* etc. were largely reserved for words, especially adjectives, that assign *qualities* to a noun. MODIFY etc. were used for the way adverbs affect verbs. In this kind of usage, general adjectives (in today's terms, the entire adjective class) could be labelled *qualifiers*, while all or most of today's determiners were contrasted as QUANTIFIERS.

> 1933b O. JESPERSEN *Little* is sometimes a qualifier (*a little girl*), sometimes a quantifier (*a little bread*).

(*c*) In Systemic-Functional Grammar, *qualifier* is contrasted with *modifier* as in traditional grammar (see (*b*)), but in a completely different way. Here *qualifier* and *qualification* describe the function of whatever follows the head in a nominal group, thus being virtually synonymous with the *postmodifier* and *postmodification* of other models. The terms *modifier* etc. can then be used to describe modification before the head—the *premodification* of other models.

qualify

Attribute a quality to another word.

> 1892b H. SWEET Thus *very* in *a very strong man* qualifies the attribute-word *strong*.

The term was at one time particularly used of the way adjectives affect nouns, although usage could be wider, as in the quotation.

In many present-day grammatical models, the terms *qualify* etc. are often replaced by *modify* etc.: see QUALIFIER.

In Systemic-Functional Grammar, *qualify* is used to describe the effect that words following the headword in a noun group (noun phrase) have on that head. Such words do not, of course, have to be adjectives, but can include relative clauses and prepositional phrases, so in this area *qualify* has an extended meaning (i.e. it is equivalent to *postmodify* in other models).

> 1971 M. L. SAMUELS *Son* is usually either modified by *my/his/her*, etc. or qualified by an *of*-group, whereas *sun* is normally preceded by the definite article.

qualitative

1 *Phonetics & Phonology*. Relating to sound quality: see QUALITY (1).

2 **qualitative adjective**: an adjective that is gradable and describes a quality, in contrast to a CLASSIFYING ADJECTIVE.

The division of adjectives into *qualitative* and *classifying* is just one of many ways of categorizing adjectives. *Qualitative adjective* is roughly the same as EPITHET.

> 1990 *Collins Cobuild English Grammar* Adjectives that identify a quality that someone or something has, such as 'sad', 'pretty', 'small', 'happy', 'healthy', 'wealthy' and 'wise', are called qualitative adjectives.

Compare ATTRIBUTIVE, GRADABLE.

quality

1 *Phonetics & Phonology*. The distinguishing characteristic(s) of a sound.

The distinctive features of a sound, which make it recognizable as a particular phoneme, constitute its *sound quality*, which is distinct from such features as length, pitch, or loudness. Hence the difference between two phonemes (e.g. between the vowels of *pat* and *part*) can be said to be a QUALITATIVE difference.

2 **quality partitive**: a partitive indicating categorization by quality or kind.

In the classification of partitive nouns and phrases a distinction is sometimes made between *quality partitives*, e.g.

> a *sort* of menu, two *kinds* of pudding

and the more usual *quantity partitives*, e.g.

> a *piece* of cake, a *lot* of food

quantifier

1 A determiner or pronoun expressing number or amount.

The term normally refers to indefinite quantity and covers such *quantifying* or QUANTITATIVE words as *much, many, (a) few, (a) little, several, enough, lots (of),* etc. It is sometimes extended to include indefinite pronouns such as *everyone, somebody, nothing,* etc., and sometimes also open class words such as *heaps (of)*, *lashings (of)*. It may also be extended to include nouns of definite quantity, i.e. CARDINAL NUMBERS.

2 (In one grammatical model.) A partitive phrase including the word *of*.

> 1990 *Collins Cobuild English Grammar* **Quantifier**, a phrase ending in 'of' which allows you to refer to a quantity of something without being precise about the exact amount; e.g. *some of, a lot of, a little bit of*.

Compare IDENTIFIER.

quantify

Indicate quantity. See QUANTIFIER.

quantitative

1 Relating to quantity or amount as part of a word's meaning. See QUANTIFIER and QUANTITY (1).

2 (*n. & adj.*) (A word) belonging to a class consisting of (indefinite) QUANTIFIERS (e.g. *few*, *several*, etc.), definite CARDINAL NUMBERS (e.g. *one*, *two*, *three*), etc., and measurement terms such as *a couple of*, *half*, etc.

In this classification, *quantitative* contrasts with ORDINATIVE, and both are subclasses of NUMERATIVE.

3 *Phonetics*. Relating to phonetic length.

1962 A. C. GIMSON It may be objected that a quantitative as well as qualitative difference distinguishes /iː/ from /ɪ/; but in the examples given—*seat* and *sit*—the phonetic context imposes a quantity on /iː/ which is practically the same as that of /ɪ/.

quantity

1 Number or amount.

For number and amount as semantic categories, see PARTITIVE and QUANTIFIER.

•• **quantity partitive**: see QUALITY (2).

2 *Phonetics & Phonology*. The relative time taken in the articulation of speech sounds.

This is length as perceived by the listener; the same as LENGTH (1).

question

A sentence seeking information; a sentence that is interrogative in form.

Some grammarians make a distinction between sentences that are questions in form and those that are questions in meaning, using INTERROGATIVE for the syntactical classification and *question* as the functional label (but in some cases, the applications are exactly reversed!). In practice, many grammarians use the word *question* with both meanings on different occasions.

In general, sentences that are interrogative in form are also genuine questions in meaning. Formally (and usually functionally) there are three main types:

(*a*) yes-no questions, beginning with an operator (e.g. *Is it raining?*)
(*b*) *wh*-questions, beginning with a *wh*-word (e.g. *What has happened?*)
(*c*) alternative questions (e.g. *Should I telephone now or later?*)

EXCLAMATORY and RHETORICAL *questions* are in fact interrogative in form only, and are not used to ask genuine questions:

Isn't it a lovely day! (= It is a lovely day)
What's that got to do with me? (= That has nothing to do with me)

By contrast, sentences that are declarative in form may sometimes be used to ask questions (*declarative questions*):

You've already spent all that money?

Compare ECHO UTTERANCE, QUESTION-TAG. See INDIRECT QUESTION.

question mark

A punctuation mark ⟨?⟩ chiefly used to show that the preceding word, phrase, or sentence is a question.

question-tag

A phrase that is interrogative in form, consisting of an operator + a pronoun or *there*, tagged on to the end of a declarative sentence. (Also called simply *tag*.)

In the most usual type of sentence, a negative tag is added to a positive statement, and a positive tag is added to a negative one:

It's been cold this week, hasn't it?
You're not really going to walk all the way, are you?
You usually drive, don't you?

Said with a falling tone on the tag, the whole sentence is more like an exclamation, assuming the listener's agreement; with a rising tone the tag becomes a question inviting a response.

Less usual are positive statements followed by positive tags (often with a note of criticism):

So that's what you think, is it?

Question tags can also be added to imperatives and exclamations:

Keep in touch, won't you?
What a wonderful thing to do, wasn't it!
Oh, stop complaining, will you!

A complete sentence plus tag is sometimes called a TAG QUESTION.

question word See *WH*-WORD.

quotation mark

A punctuation mark ⟨‘ ’⟩ or ⟨“ ”⟩ used as one of a pair to mark the beginning and end of a form, word, phrase, or longer stretch of text that is being quoted by the writer from another context.

R

radical

Morphology. (*n. & adj.*) (Of, belonging to, connected with, or based on) the root of a word.

The term has a long history, but words such as BASE, ROOT, and STEM (along with *morpheme* and *minimum free form*) are more usual today.

raise

1 *Phonetics*. Move (part of the tongue) to a position higher in the mouth; pronounce (a vowel) doing this; (of a vowel) move in this way.

1991 P. ROACH We need to know in what ways vowels differ from each other. The first matter to consider is the shape and position of the tongue. It is usual to simplify the very complex possibilities by describing just two things: firstly, the vertical distance between the upper surface of the tongue and the palate, and secondly, the part of the tongue, between front and back, which is raised highest.

In historical linguistics, vowels that took on a new and closer (higher) quality are said to have been *raised* (or to have raised). See GREAT VOWEL SHIFT.

2 Move (a noun phrase) from a position in an embedded clause to a position in the main clause. See RAISING.

1984 F. R. PALMER Once again the subject .. is raised and the remainder of the embedded clause .. is moved to the end of the sentence.

raising

1 *Phonetics*. The action of RAISE (1).

1909 O. JESPERSEN The great vowel-shift consists in a general raising of all long vowels with the exception of the two high vowels.

2 (In some forms of Transformational Grammar.) The moving of a noun phrase from a position within an embedded clause to a position in the main clause.

Transformational Grammar is concerned, among other things, with the different syntactical ways of 'saying the same thing'. A rule of *raising* is a rule that seeks to explain how the subject or object of a clause is raised to form a constituent of a higher clause, while preserving approximately the same meaning.

In the sentence

The board considers / that Mr Smith is unsuitable for the post

Mr Smith is the subject of the subordinate clause. If we reword it as

The board considers Mr Smith (to be) unsuitable for the post

we have (by *object-raising*) raised *Mr Smith* to object position in the main clause.

In a different sort of transformation, *subject-raising* is achieved:

It appears / that Mr Smith does not mind

can become

Mr Smith appears not to mind

with *Mr Smith* now subject of the main clause.

rank

A level of structure related to others on a scale.

(*a*) Grammar can be seen as a system of levels or ranks, going from the 'highest' rank of sentence (not clearly defined), through *clause*, *phrase* (or *group*), and *word* down to *morpheme*, each smaller unit being included in the larger one.

1964 M. A. K. HALLIDAY et al. The term used to name the hierarchical relation among the units is *rank*; they can be arranged on a scale, and this is known as the 'rank scale'.

1971 R. A. HUDSON What a grammar will contain . . is not a number of different system-networks, each for a different 'rank' (clause, phrase, etc.) or a different environment (subject, main verb, etc.) but a single network which includes all the grammatical systems needed for the language.

(*b*) In earlier grammar, the term *rank* is used as a way of classifying words in a phrase or clause on a scale of importance.

1933b O. JESPERSEN Take the three words *terribly cold weather*. They are evidently not on the same footing, *weather* being, grammatically, most important, to which the two others are subordinate, and of these again *cold* is more important than *terribly*. We have thus three ranks: "weather" is Primary, "cold" Secondary, and "terribly" Tertiary in this combination.

rankshift

(*n.*) A downward shift of a linguistic unit into a lower rank.

(*v.*) Assign an inferior rank to (a unit in a grammatical structure).

1964 M. A. K. HALLIDAY et al. If I say 'the house where I live is very damp', the sentence consists, *in its structure*, of only one clause; the clause 'where I live' is rankshifted and operates in the structure of the *group* 'the house where I live'.

r-colouring

The same as RETROFLEXION.

real condition

The same as *open* CONDITION.

realis See IRREALIS.

reality phase See PHASE.

realization

The (phonetic, grammatical, etc.) manifestation of a word class, grammatical category, or linguistic unit in an actual context.

Realization is applicable at all levels of analysis. For example, in *cats and dogs* /kæts ən dɒgz/ the phonemes /s/ and /z/ are both realizations of the abstract plural morpheme. On another level, a *lexeme* (a word in the dictionary sense) can have several realizations (or be *realized* in several ways) e.g. as *break, breaks, breaking, broke, broken*. Again, in

Break, break, break, on thy cold grey stones, oh sea!

we could say that *break* here is a realization of verb, or verb phrase, or imperative, or predicate.

realize

Cause or be the REALIZATION of.

1980 E. K. BROWN & J. E. MILLER There is no very satisfactory way to identify part of the word *wrote* as realizing the lexeme WRITE and some other part of the word as realizing the syntactic description 'past tense': rather, the whole form *wrote*, as a unity, realizes the description 'WRITE + past'. *Ibid.* 'Morphological realization rules' . . . These are those rules that realize the various morphemes, lexical and grammatical, introduced by the grammar.

Received Pronunciation

The pronunciation of that variety of British English widely considered to be least regional, being originally that used by 'educated' speakers in southern England. (Also called *Received Standard (English)*. Abbreviated *RP*.)

The use of *received* in the context of pronunciation variety was initiated by the phonetician A. J. Ellis (1869); the term *Received Standard* as the label for the most prestigious British variety was introduced by H. C. Wyld; and *Received Pronunciation* was given pedagogical and quasi-academic status in the studies and dictionaries of the phonetician Daniel Jones (1881–1967):

1962 D. JONES I do not consider it possible at the present time to regard any special type [of pronunciation] as 'Standard' or as intrinsically 'better' than other types. Nevertheless, the type described in this book is certainly a useful one. It is based on my own (Southern) speech, and is, as far as I can ascertain, that generally used by those who have been educated at 'preparatory' boarding schools and the 'Public Schools'. This pronunciation is fairly uniform in these schools and is independent of their locality. It has the advantage that it is easily understood in all parts of the English-speaking countries; it is perhaps more widely understood than any other type.

Alternative terms are BBC ENGLISH and even 'Oxford English', never a very accurate term.

Within RP further distinctions are sometimes made:

> 1962 A. C. GIMSON The *conservative* RP forms used by the older generation and, traditionally, by certain professions or social groups; the *general* RP forms most commonly in use and typified by the pronunciation adopted by the BBC; and the *advanced* RP forms mainly used by young people of . . the upper classes, but also, for prestige value, in certain professional circles.

With the greater egalitarianism of recent years, Received Pronunciation has lost some of its prestige. RP as described by Jones was never the accent of more than a tiny percentage of the British population (albeit a socially and economically powerful group). Today, as ever, many educated speakers in fact use some sort of *modified RP* or *near-RP*.

> 1990 J. C. WELLS RP itself inevitably changes as the years pass. There is also a measure of diversity within it. Furthermore, the democratization undergone by English society during the second half of the twentieth century means that it is nowadays necessary to define RP in a rather broader way than was once customary . . . Within England, RP is not a local accent associated with any particular city or region. Rather, it is a social accent associated with the upper end of the social-class continuum.

Compare STANDARD and see NEAR-RP.

recipient

Semantics. The role of the animate being that 'receives' some object or event. The role of the recipient is often that of the indirect object of traditional grammar:

> I bought *my cousin* a present
> I gave *him* your letter

A distinction is sometimes made between an intended (or *benefactive*) recipient (e.g. *I bought a present for him*) and an *actual* recipient (e.g. *I gave the letter to him*).

Semantically it is possible for the grammatical subject of some verbs to be assigned a recipient role, as in *I heard a noise*, though this role may alternatively be described as EXPERIENCER.

Compare BENEFACTIVE, DATIVE, EXPERIENCER, GOAL.

reciprocal

1 Expressing mutual action or relation.

The term is mainly applied to the pronouns *each other* and *one another*. There is no real difference in usage, despite the existence of a prescriptive rule that *one another* should be used when the reference is to more than two.

Occasionally the term is extended to cover other words, such as the verbs *meet* (e.g. *John and Mary met*, which can be regarded as combining *John met Mary* and *Mary met John*) or *exchange* (e.g. *They exchanged addresses*, where A gave his address to B and B gave his address to A).

2 *Phonetics*. The same as COALESCENT.

reclassification

The same as CONVERSION.

recover

Deduce (information) not made explicit so as to make sense of an utterance. See RECOVERABILITY.

recoverability

The possibility that a piece of information can be deduced or retrieved by the hearer or reader.

Recoverability, enabling us to make sense of otherwise incomplete utterances, is essential to the use both of ELLIPSIS and PRO-FORMS.

In *precise recoverability* the exact words ellipted are completely recoverable from the surrounding text, just as the exact referent of a pro-form is. Thus in

The man said he would telephone this morning but didn't

we can recover *the man* from the pro-form *he*, and we can recover the ellipted words

but *he* didn't *telephone this morning*

A distinction can be made between *structural recoverability* that relies (as in this example) on our knowledge of grammar, and *situational recoverability*, where we recover what is missing from some extralinguistic context, e.g. in a sentence like *Look, she's found it*, where the referents of *she* and *it* may be physically obvious to the hearer.

Compare UNDERSTOOD.

recurrence

Repeated or frequent occurrence, repetition.

Recurrence is used in its everyday sense in discussing the meaning of some verbs and adverbials.

1985 R. QUIRK et al. Here, the type of recurrence in which we are interested depends both on the semantics of the verb and also on its aspect. Compare:
She *usually* smiles. [recurrent activity; eg 'When she sees me . . .']
She is *usually* smiling. [continuous activity; eg 'Whenever one sees her . . .']

recursion

The repeated use of a particular type of linguistic unit; the use of an element recursively.

The term is applied to the generative feature of language that makes it possible for some structures to be repeated endlessly. For example, there is theoretically no limit to the length of an adjective string:

What fascinates them is his maddening, engaging, dotty, intelligent, charming, . . and humourless personality or to the number of prepositional phrases that can follow each other:

The ring was in a bag in a box in the corner of a drawer in a chest in the corner of the room in (etc.)

or to the number of relative clauses:

1975 F. R. PALMER The structure of language involves 'recursion' of the kind illustrated by 'This is the house that Jack built', 'This is the mouse that lived in the house that Jack built' and so on—if necessary *ad infinitum*.

Compare EMBEDDING, NESTING.

recursive

Of a linguistic feature or a grammatical rule: that can be used repeatedly in sequence.

1972 R. A. PALMATIER A recursive rule is a rule which reapplies indefinitely to its own output . . . The recursive power of a grammar . . is its ability to generate an infinity of sentences.

The concept was introduced from mathematics into linguistics by Chomsky (1955a).

• **recursively. recursiveness.**

reduce

1 Abbreviate (a clause or other linguistic form) by omitting some elements.

The concept of grammatical *reduction* embraces both ELLIPSIS and SUBSTITUTION. Thus in reply to

You should write to your bank manager

a reduced response could be

I have (ellipted)

or

I've done so (with a substitute form)

•• **reduced clause**: a shortened clause, particularly a non-finite or verbless postmodifying structure which can be interpreted as a relative clause with its pronoun and finite verb omitted, e.g.

Anyone *scared of heights* is advised not to attempt this climb (= anyone *who is scared of heights* (etc.))

2 *Phonology*. Articulate (a vowel sound) with little effort—as /ə/ or /ɪ/; WEAKEN.

A noticeable feature of English pronunciation is the way that vowels in unaccented positions often undergo a *reduction* to /ə/ or /ɪ/. Contrast

photo /ˈfəʊtəʊ/ : photograph /ˈfəʊtəgrɑːf/
present (noun, adjective) /ˈprez(ə)nt/ : (verb) /prɪˈzent/
telephone /ˈteləfəʊn/ : telephonist /təˈlefənɪst/

Words and syllables containing sounds that have been entirely omitted are sometimes described as *reduced*, but more precise terms for such phenomena are CONTRACT, CONTRACTION and ELIDE, ELISION.

3 *Phonology*. Shorten (a vowel).

The actual time taken in the articulation of a vowel depends to some extent on context. In stressed syllables, English vowels (particularly the so-called long vowels and also diphthongs) are shortened or *reduced* when preceding a voiceless consonant. Compare the fuller length of the vowels in *sea* or *seed* as against the reduced length of the vowel (representing the same phoneme) in *seat*. Similarly *car*, *card* versus *cart*; *saw*, *sword* versus *sort*; *fur* versus *first*; *five* versus *fife*; *plays* versus *place*; *lab* versus *lap*.

reduction

1 The omission of syntactic elements, especially by ellipsis and substitution.

2 *Phonology*. The weakening of a vowel in an unaccented syllable to /ə/ or /ɪ/.

See REDUCE.

redundancy

The superfluity of a linguistic feature due to its predictability within the overall structure.

Redundancy is to some extent a normal and necessary feature of linguistic communication, enabling the 'message' to be understood even if there is some 'interference', for example, in spoken delivery owing to noise, or in the written language owing to the occasional misspelling or the erroneous omission of a word.

A degree of redundancy is also built into the syntax. For example, in *The sun rises* there are two markers of singular, where one might be sufficient (as indeed it is in the simple past, *The sun rose*). On the other hand, there is a counter-tendency to avoid tedious redundancy by the REDUCTION of easily RECOVERABLE features. There are, however, grammatical limits on the avoidance of redundancy. See ELLIPSIS.

The concept originated in theories of information transfer and telecommunication.

• **redundant**: exhibiting or characterized by redundancy.

> 1954 G. A. MILLER et al. If a language is highly redundant, the relative information per symbol is much lower than it would be if successive symbols in a message could be chosen independently.

reduplicate

Form (a word, phrase, etc.) by repetition of an element; repeat (an element) so as to form a word, phrase, etc.

> 1985 R. QUIRK et al. It is curious that analogous reduplicated phrases are virtually restricted to informal use: *for months and months*, *for years and years*.

• **reduplication**.

See REDUPLICATIVE.

reduplicative

(*n. & adj.*) (A compound word) having two identical or very similar parts, often rhyming. (Sometimes called *iterative*.)

Most reduplicatives are fairly informal, e.g.

goody-goody (M19) hugger-mugger (E16)
happy-clappy (L20) wishy-washy (E18)
harum-scarum (L17)

reference

Relationship between one expression and another or between an expression and what is spoken of.

The term is used in its normal way. Thus we speak of *anaphoric* and *cataphoric reference* within a text (one expression, usually a pro-form, referring back or forwards to another):

Henry says *he* is coming (anaphoric)
Before *she* leaves, I must write to Mary (cataphoric)

A distinction is however to be made between *reference* and SUBSTITUTION (although there is some overlap). Co-reference items may appear to refer to each other, but they share the same REFERENT in the real world, and this still applies when a pro-form refers to a 'fact', which is given the status of being real, as in

Listen to *this*! Car prices are coming down

Substitution, by contrast, like ellipsis, is a relationship between words. Contrast

Is that your paper? May I borrow *it*? (reference)
Is that your paper? I didn't get *one* today (substitution)

or

Car prices are coming down
Who told you *that*? (reference to a fact)
It says *so* in the paper (substitution for a clause)

In logic and linguistics *reference* is further used in more technical ways. Sometimes it contrasts with other kinds of meaning (see REFERENTIAL), and sometimes with SENSE.

1977 J. LYONS Expressions may differ in sense, but have the same reference; and 'synonymous' means "having the same sense", not "having the same reference". A rather better example than Frege's [i.e. of the Morning Star and the Evening Star both being the planet Venus] is Husserl's, 'the victor at Jena' and 'the loser at Waterloo' . . , both of which expressions may be used to refer to Napoleon.

See also UNIQUE *reference*.

reference grammar See GRAMMAR.

referent

The person, thing, etc. referred to by an expression.

The victor at Jena and *the loser at Waterloo* are *referring expressions*; Napoleon is the *referent*.

referential meaning

Semantics. That aspect of meaning that can be expressed in terms of a referent; objective, cognitive, denotational meaning.

Referential meaning may be contrasted both with subjective, emotive, connotational meaning, and with situational meaning.

referring expression

An expression used to refer to some individual person, object, place, event, etc. See REFERENT.

reflexive

(*n. & adj.*)

1 (A pronoun) that refers back to the subject of the same clause.

The *reflexive pronouns* end in *-self* or *-selves* (e.g. *myself, themselves*).

A reflexive pronoun is not usually considered acceptable as a subject in standard English (e.g. **James and myself intend to help*) but can be used as emphatic reinforcement (e.g. *I myself believe that he's telling the truth, despite what the others say*).

2 (A verb or structure containing it) taking (or having) a reflexive pronoun as object.

English has very few verbs that require such an object. Some examples are:

absent oneself, demean oneself (usually), perjure oneself, pride oneself

Other verbs may be understood *reflexively*, but a reflexive pronoun object is optional:

He washed, shaved, and dressed

• **reflexivity**: the property of being reflexive. **reflexivization**: the action of making reflexive; the process or fact of being made reflexive. **reflexivize**.

reformulation

Rewording; expressing again in clearer words.

The relationship of the second term in an apposition may be that of *reformulation*, a rewording perhaps in order to explain the first term more accurately.

• **reformulatory conjunct**: a conjunct that introduces a reformulation.

e.g.

The city of dreaming spires, *in other words* Oxford
The Press, *that is to say* Oxford University Press

register

A variety of language used in particular circumstances.

Register is used differently by different linguists.

(*a*) A variety of language related to a level of formality, anywhere on a scale from the extremely formal or ceremonial to the colloquial or slangy, and manifested in syntax, vocabulary, and, possibly, pronunciation.

(*b*) A variety of language related to a particular FIELD OF DISCOURSE, a particular subject or occupation; e.g. advertising language or the language of the law.

Compare DOMAIN.

regressive

The same as ANTICIPATORY (2).

See ASSIMILATION.

regular

Of a linguistic form: following the general rules for its class. Contrasted with IRREGULAR.

A regular verb has a past tense and past participle formed by adding *-ed* to the base (or *-d* if the base ends in *-e*). For example

look, looked, looked; race, raced, raced.

Similarly, regular noun plurals are formed by adding *-s* (or *-es* after sibilants), as

books, boxes

reinforcement

The strengthening of the meaning (of a word or words).

Reinforcement is used in its everyday sense to describe the way additional words are used to emphasize meaning. This may be by the use of conjuncts:

It was expensive; *furthermore/what's more/in addition* I thought it was ugly

or by repetition:

It's much *much* too early to decide

rejected condition

The same as HYPOTHETICAL condition.

1947 E. PARTRIDGE Rejected Condition, as in 'If wishes were horses, beggars would ride'.

relation

Connection between linguistic units. (Also **relations, relationship**.)

These are very general and loosely used terms describing connection (syntagmatic, paradigmatic, semantic, or phonetic) between units at any level. Thus we can speak of a constructional relation (or of structural relations) between clauses in the same sentence; of a dependency relation between a subordinate clause and another clause; of relations of apposition, correlation, etc.; of the relationship between different functional elements in a clause, e.g. a subject-verb relationship; and of transformational relations between different structures in Transformational-Generative Grammar; and so on.

1968 J. LYONS The most characteristic feature of modern linguistics—one which it shares with a number of other sciences—is 'structuralism' (to use the label which is commonly applied, often pejoratively). Briefly, this means that each language is regarded as a *system of relations* (more precisely, a system of interrelated systems), the elements of which—sounds, words, etc.—have no validity independently of the relations of equivalence and contrast which hold between them.

•• **relation word**: the same as RELATIONAL *word*.

relational

(*adj.*)

Indicating relation(s) or relationship.

Relational (like *relation* and *relationship*) is used of both syntactic and semantic connections.

In the discussion of texts, *relational* describes ways in which sentences and other elements are connected in grammatical, semantic, and pragmatic ways.

1985 R. QUIRK et al. But irrespective of the various purposes and general intentions of a text, there are a few relationships within texts that constantly recur, which involve particular connective devices . . . They can be seen as basic relational structures.

Specifically in syntax, prepositions and conjunctions are sometimes classified as *relational words* (or *relation words*), their function being to indicate the relation between other words.

More semantically, a *relational verb* is a verb indicating a *relational process*, such as 'being' (expressed by a linking verb) or possession (expressed by *have*, *belong*, *lack*, etc.). In Systemic-Functional Grammar, such verbs are overtly contrasted with verbs of material and mental processes.

•• **relational grammar**: a theory of grammar developed as an offshoot of Transformational-Generative Grammar from the mid-1970s on, in which clauses are analysed as networks of relationships rather than in terms of phrase structure.

(*n.*)

(The less usual, and probably outdated term for) a relational word.

1969 E. A. NIDA Relationals are any units which function primarily as markers of relationships between other terms e.g. *at, by, because, and, or*.

relationship See RELATION.

relative

(*n. & adj.*) (A word, especially a pronoun) referring to an antecedent and attaching a clause to it.

The *relative pronouns* are *who, whom, whose, which* (which, as a group, express differences of gender and case) and the (invariable) *that*.

When and *where* can be used as *relative adverbs*:

I remember, I remember, the house *where I was born.*

relative clause: a clause attached to another clause usually by a relative pronoun (also called *adjectival clause*).

Relative clauses are of two main kinds (sometimes subsumed under the label *adnominal*):

(*a*) DEFINING, in which any relative pronoun can be used, or sometimes none at all (see CONTACT CLAUSE):

the woman who/whom/that I love

the woman I love

(*b*) NON-DEFINING, where a *wh*-pronoun must normally be used.

Algernon, whom I greatly admire, has really put himself out

A third type of relative clause (though treated as non-defining in some models) is the *sentential relative clause*, which refers back to a part or the whole of the previous clause. It is usually introduced by *which*:

The hotel is very expensive. Which is a pity.

Compare CONTINUATIVE RELATIVE CLAUSE, NOMINAL RELATIVE CLAUSE, REDUCED *clause*. Compare also APPOSITIVE CLAUSE.

relativity See SAPIR-WHORF HYPOTHESIS.

relator

1 Syntactically, a word or phrase that serves to relate one part of a sentence to another, e.g. a conjunction or preposition. The same as RELATIONAL *word*.

2 Semantically, a word or phrase that contextualizes an utterance with regard to time and place.

Place and time relators can thus include adverbs (e.g. *here, downstairs, now*); noun phrases (e.g. *last night*); complete prepositional phrases (e.g. *to the lighthouse, before the flood*); and adjectives (e.g. *previous, later*).

release

The same as PLOSION.

relevance See CURRENT RELEVANCE.

replacive

(*n. & adj.*)

1 *Morphology*. (A linguistic element) that replaces or substitutes for something else.

The term is particularly used in the label *replacive morph* or *replacive morpheme* to enable irregular forms such as *men* from *man* and *sang* or *sung* from *sing* to be described in morphemic terms, despite falling outside the straightforward rules for forming noun plurals or past verb forms by the addition of inflections.

> 1974 P. H. MATTHEWS *Men*, for example, would be said to consist of the regular allomorph *man* of the morpheme MAN plus a 'replacive morph' ('replace *a* with *e*' or '*a* → *e*') which was assigned as yet another allomorph of PLURAL. The PLURAL morpheme would thus be regarded as a class of morphs with [z] (in *seas*), [s] (in *masts*), zero (in *sheep*) . . and '*a* → *e*' among its members . . . This was nonsense, of course. A process of replacement is no more a 'morph' than zero is a 'morph'.

2 *Semantics*. (A conjunct) that introduces a statement replacing a previous statement.

Replacive conjunct is one among many semantic categories in which the meaning of conjuncts is analysed. Replacive conjuncts include *alternatively*, *rather*, *on the other hand.*

reported speech

1 The same as INDIRECT SPEECH.

The reporting of speech uses an introductory reporting verb (e.g. He *says* . ., She *told* us). This is the usual meaning of the term; it contrasts with DIRECT SPEECH.

2 (More generally.) Any of the ways in which a speaker or writer reports what someone else has said..

In this sense, reported speech includes both DIRECT and INDIRECT SPEECH.

reporting verb See PRIVATE VERB.

resonance

Phonetics. Transmission of air vibrations in the vocal tract.

The significance of this term is that resonances at different frequencies in the vocal tract help to give speech sounds, and particularly vowels, their distinct and characteristic patterns.

resonant

(*n. & adj.*) (Designating) a consonant which is either a liquid or a nasal.

> 1943 K. L. PIKE The *sonorants* are nonvocoid resonants and comprise the lateral resonant orals and resonant nasals (e.g. [m], [n], and [l]).

restricted code See CODE.

restriction See COLLOCATION, CO-OCCURRENCE, SELECTIONAL RESTRICTION, SEMANTIC RESTRICTION.

restrictive

1 The same as DEFINING. Contrasted with NON-RESTRICTIVE, NON-DEFINING.

The term is particularly applied to relative clauses, but has wider applications.

> 1990 S. GREENBAUM & R. QUIRK Modification can be restrictive or non-restrictive. That is, the head can be viewed as a member of a class which can be linguistically identified only through the modification that has been supplied (*restrictive*).

2 **restrictive adjective**: a member of a subcategory of ATTRIBUTIVE adjective, semantically defined because it restricts or limits the meaning of the following noun (also called LIMITING ADJECTIVE, LIMITER ADJECTIVE).

e.g. *a certain person*, *the main chance*.

The term is potentially ambiguous, because most adjectives can be used in both restrictive and non-restrictive modification.

See DEFINING.

result

(Expressing) the outcome of some action.

One of the categories used in the semantic description of subordinators and adverb clauses.

result clause: a clause expressing a result, introduced by *so . . . that*, *such . . . that*, and *so*:

> It was so hot (that) I nearly fainted
> It was such a hot day (that) I nearly fainted
> The weather was too hot, so I cried off.

Result clauses are often contrasted with those of PURPOSE.

• **resultant, resultative (resultive), resulting**: expressing, indicating, or relating to result.

The concept of result is so general that many language elements can be described as relating to it, and consequently adjective usage is somewhat fluid.

A grammatical object that comes into existence only as a result of the action of the verb is sometimes called a *result object* (or *object of result*; also *effected object*) (e.g. *They built their own house*).

A subordinator introducing a *result clause* (see above) is variously described as a *subordinator of result* or a *resultive/resultative subordinator/conjunction*.

A conjunct with the same meaning is called a *resultive conjunct* (e.g. *consequently*, *hence*).

An infinitive with this meaning can be called an *infinitive of result* (e.g. *He arrived to find the place on fire*).

An adverbial may sometimes be said to have *resultative meaning*:

I want everyone *back here* by ten (= Everyone will be back by ten)

Some LINKING VERBS are described as *resulting verbs* or *verbs of resulting meaning* (e.g. *He became president/powerful*), and their complements (*president*, *powerful*) are variously described as *resulting attributes*, or as showing *resultant* or *resulting states*.

Compare FACTITIVE.

retroflex

Phonetics. Articulated with the tip of the tongue turned back behind the alveolar ridge.

• **retroflexed**: the same as *retroflex*. **retroflexion**: retroflex articulation (also called *r-colouring*).

A retroflex articulation is characteristic of the pronunciation of the phoneme /r/ in many accents of English (e.g. in the West Country and in Ireland), though not generally in RP. In some rhotic accents (that is, where a post-vocalic /r/ is pronounced in such words as *bird*, *heard*, *term*) anticipatory retroflexion may affect the vowel, making it an *r-coloured* vowel. Alternatively such words may be articulated with a single retroflex vowel sound.

Retroflexed /t/ and /d/ are characteristic of the pronunciation of these consonant phonemes by some Indian speakers.

retrospective

Expressive of looking back in time.

Verbs such as *forget, regret, remember* are sometimes singled out as *retrospective verbs*, or verbs with *retrospective meaning*, when they are followed by an *-ing* form.

I'll never forget hearing Sutherland
I remember wondering how she did it
I regret saying that

But in fact 'looking back' is part of the meaning of many verbs when followed by an *-ing* form, in contrast to a 'forward-looking' meaning in other verbs when a *to*-infinitive follows. Contrast *I enjoy meeting people* with *I want to meet them*.

Revised Extended Standard Theory See STANDARD THEORY.

rewrite

(In Generative Grammar.) Write (an abstract symbol or symbols) in a different, usually more explicit, form.

rewrite rule

(In Generative Grammar.) A 'rule' that takes the form 'Rewrite X as Y'. (Also called *rewriting rule*.)

r-ful(l)

The same as RHOTIC.

rhematic See RHEME.

rheme

Linguistics. The second part of clause structure, in which information is given about the THEME.

The linguistic use of *theme* and *rheme* comes from the Prague School of linguists. *Rheme*, however, was previously used in Logic.

> 1959 J. FIRBAS Those sentence elements which convey something that is known, or may be inferred, from the verbal or from the situational context . . are to be regarded as the communicative basis, as the theme of the sentence. On the other hand, those sentence elements which convey the new piece of information are to be regarded as the communicative nucleus, as the rheme of the sentence.
>
> 1975 M. A. K. HALLIDAY As a message structure . . a clause consists of a Theme accompanied by a Rheme; and the structure is expressed by the order—whatever is chosen as the Theme is put first.

It should be noted that in the first definition of *theme* ('something that is known or may be inferred') the theme can communicatively be rather unimportant, corresponding to what is 'given' (in a GIVEN and NEW analysis). The second definition is somewhat different: the theme is important, and is put first to attract the reader or listener's attention, although even here the rheme is more important. This is more like the TOPIC and COMMENT analysis, with the topic being what the sentence is about.

Some grammarians use the term FOCUS rather than rheme, and so contrast *theme* and *focus*. Theme and rheme (and some of these other terms) may coincide with the traditional syntactical binary division into subject and predicate, but they are concerned with information structure rather than syntax.

• **rhematic**: of or pertaining to a rheme.

rhetorical

Spoken or written for effect.

When applied to a conditional clause, *rhetorical* means 'not to be taken literally'. A *rhetorical conditional clause* may look like an open condition, but is actually strongly assertive, e.g.

> If he wins, I'll eat my hat (= he will not win)
> She's sixty, if she's a day (= she is at least sixty)

A *rhetorical question* is similarly assertive:

> What's that got to do with you? (= it has nothing to do with you)
> Who am I to complain? (= it is not for me to complain)

rhotic

Phonology. Designating a pronunciation in which the consonant sound /r/ has not been lost before another consonant or a pause. (Also called *r-pronouncing*, *r-ful(l)*.)

In Scottish, Irish, General American, and a number of regional English accents, /r/ is pronounced before a consonant (as in *bird*, *are fine*) and in final position before a pause (e.g. *That's not fair*). Such accents are rhotic.

By contrast RP, Australian, and New Zealand accents and most varieties of 'English English' have lost the /r/ sound in such positions. Such accents are *non-rhotic* or *r-less*.

Compare INTRUSIVE *R*, LINKING *R*.

rhythm

Phonetics. The characteristic movement or 'timing' of connected speech.

The fundamental speech rhythms of languages are usually classified as either STRESS-TIMED or SYLLABLE-TIMED.

right-branching See BRANCHING.

right dislocation See DISLOCATION.

rim

Phonetics. The edges of the tongue, in particular the sides (excluding the tip).

The term is used in describing the pronunciation of the lateral /l/.

rise

Phonetics.

(*n.*) In the intonation of a syllable or longer utterance, a nuclear pitch change from relatively low to relatively high; contrasted with a FALL. (Sometimes called a *rising tone*.)

Various kinds of rise are distinguished, such as the *low rise* [ˏ], starting near the bottom of an individual speaker's pitch range, and the *high rise* [ˊ], starting higher, and of course going higher still.

•• **rise-fall**: *Phonetics*. A tone in which the pitch rises and then falls [ˆ].

This tone often conveys feelings of surprise, approval, or disapproval.

(*v.*)

1 Of pitch: to change from relatively low to relatively high. Usually in the term RISING *tone*.

2 Of a diphthong: see RISING (2).

rising

Phonetics.

1 **rising tone**: the same as RISE (*n.*).

2 Of a diphthong: having most of the length and stress, the greater prominence, on the second element.

This type of diphthong is unusual in English.

***r*-less** See RHOTIC.

r liaison

An umbrella term covering INTRUSIVE *R* and LINKING *R*.

roll

Phonetics.

(*n.*) An articulation characterized by a series of rapid closures or taps of the tongue (or the uvula). (Also called *trill.*)

(*v.*) Articulate (the sound /r/) with a roll.

The /r/ phoneme, normally a frictionless continuant in RP, is sometimes pronounced with a *lingual roll* (rapid taps of the tongue against the alveolar ridge) or with a *uvular roll* (with the uvula tapping against the back of the tongue.)

root

1 *Morphology*. A core element left when all affixes have been removed from a complex word. (Sometimes called *radical.*)

In discussions on word formation, terms such as *base*, *morpheme*, *root*, and *stem* are used in varying ways. The *root* of a word is commonly a morpheme which cannot be further analysed, and which underlies related derivatives of the word. Thus *go* is the root of *goes*, *going*, *goer*, etc., and also of *undergo*. (In this definition, *root* may be contrasted with BASE: the base of *undergoes*, *undergoing*, etc. is *undergo*.)

Root in this sense may be less than a complete word. For example, *-duce* (as in *conduce*, *deduce*, *reduce*) or *jeal-* (as in *jealous*) are roots or *root morphemes*. (But such a form may also be described as a *lexical morpheme*, a *stem morpheme*, a *stem*, or even a *base*.)

Compare BASE, STEM.

2 (In historical linguistics.) An element (either a word (a *root-word*) or a root in sense 1) that is the ancestor of a more recent word.

For example, Latin *magister* is the root of both *master* and *magistrate*; Latin *moneta* is the root of both *money* and *mint*.

3 *Linguistics*. The topmost NODE in a tree diagram.

4 **root modality**: the same as DEONTIC *modality*.

5 *Phonetics.* The extreme back part of the tongue.

The root of the tongue has no particular linguistic function in the English sound system.

rounded, rounding See LIP POSITION.

royal we

The use of *we* by a king or queen to mean 'I'.

An example is Queen Victoria's 'we are not amused'. The style is now restricted to formal documents.

Compare EDITORIAL WE.

RP

Abbreviation for RECEIVED PRONUNCIATION.

***r*-pronouncing**

The same as RHOTIC.

rule

A principle regulating or determining the form or position of words in a sentence, morphemes or phonemes in a word, etc.

(*a*) It is perhaps worth pointing out that terms such as *grammatical rule* or *rule of grammar* are used with two somewhat different meanings, which are often conflated in people's minds. In an ideal world, a *descriptive rule*, describing objectively how some feature of syntax or morphology works would be the same as a *prescriptive rule* for a user or foreign language learner. In reality, some prescriptive rules are based on misunderstanding, while other prescriptive rules, though useful guidelines, are oversimplifications and are inaccurate if taken to be a complete description.

(*b*) In Generative Grammar, rules are viewed as predictive, often in a rather abstract or mathematical form, and are meant to be capable of generating an infinite number of grammatical sentences.

See also CATEGORIAL RULE, CO-OCCURRENCE *rule*, PHONOTACTIC *rule*, REWRITE RULE, TRANSFORMATIONAL *rule*.

S

S

1 Subject as an element in clause structure. See ELEMENT.
2 Sentence (in a tree diagram or phrase structure rule).

sandhi

The process whereby the form of a word (especially the beginning or ending) changes as a result of its position in an utterance; inter-word assimilation.

The term (meaning 'joining') was originally applied by Sanskrit grammarians to changes in the final and initial sounds of words in an utterance (*external sandhi*) and in the final sounds of stems (roots) in word formation (*internal sandhi*), caused by a type of assimilation. By extension the term is sometimes used of similar phenomena in other languages. In English the phenomenon of *R*-LIAISON is sometimes described in terms of *sandhi*.

> 1982 J. C. WELLS We may expect to find sandhi /r/ used frequently in mainstream (native) RP, but sparsely in speech-conscious adoptive RP.

Compare ASSIMILATION.

Sapir-Whorf hypothesis

A theory developed by the American anthropologists and linguists Edward Sapir (1884–1939) and Benjamin Lee Whorf (1897–1941), which holds that a people's language conditions the way they view the world (*linguistic determinism*), and that different perceptual distinctions are made by different languages (*linguistic relativity*).

> *ante* 1941 B. L. WHORF Concepts of 'time' and 'matter' are not given in substantially the same form by experience to all men but depend upon the nature of the language or languages through the use of which they have been developed. They do not depend so much upon ANY ONE SYSTEM (e.g. tense, or nouns) within the grammar as upon the ways of analyzing and reporting experience which have become fixed in the language as integrated 'fashions of speaking' and which cut across the typical grammatical classifications, so that such a 'fashion' may include lexical, morphological, syntactic, and otherwise systemically diverse means coordinated in a certain frame of consistency.

Whorf illustrated his theory particularly through Hopi, an Amerindian language, which, for example, has no tense system. But this does not mean that Hopi speakers cannot talk about time, and the fact that he could explain Hopi concepts in English militates against the theory in its extremer forms. Similarly it is often said that various words in one language cannot be translated into another; but even though there may be no 'one for one' equivalent, a word can always be explained by a phrase. It has also been pointed out that language users are not totally naive: English speakers who talk of the sunrise and sunset are not flat-earthers.

Saussurean

(*adj.*) Of, pertaining to, or characteristic of the Swiss linguist Ferdinand de Saussure (1857–1913) or his theories.

(*n.*) An adherent of de Saussure's theories.

De Saussure was influential particularly in his concept of a language as a system of interrelated parts that mutually affect each other. His notions of *langue* contrasted with *parole*, and of *paradigmatic* versus *syntagmatic*, underlie much modern linguistics. His *Cours de linguistique générale* (1916) is a collection of papers made by his students after his death.

Saxon genitive See GENITIVE.

scale-and-category grammar

An early version of Systemic-Functional Grammar, formulated by M. A. K. Halliday.

Scale-and-Category Grammar is so called because language is analysed as an interrelationship between three (or four) scales and four categories.

The original scales included RANK, DELICACY, EXPONENCE, and possibly 'depth'.

> 1985 G. D. MORLEY Scale-and-category grammar seeks to account for any stretch of language as it actually occurs, in either written or spoken form . . . By contrast with transformational grammar, however, scale-and-category is designed to analyse structures as they appear rather than to generate them.

Compare SYSTEMIC. See CATEGORY.

schwa (Also **shwa**.)

The weak central vowel /ə/, heard at the beginning of *ago* and the end of *mother*.

This phoneme is always unstressed. It occurs frequently in English and is sometimes called the *neutral vowel*, presumably because other vowel phonemes are often reduced to this sound when in unstressed position. Compare *photo* /ˈfəʊtəʊ/, *photograph* /ˈfəʊtəgrɑːf/, and *photography* /fəˈtɒgrəfɪ/.

The term comes from Hebrew grammar, in which it is the name of a vowel symbol traditionally given this phonetic value.

scope

The range over which a particular linguistic item meaningfully extends its influence.

Some words affect the meaning of their clause or sentence. An operator or a *wh*-word, at the beginning of a sentence, usually marks the whole sentence as a question. Similarly, a negative word often negates everything that follows it and may therefore necessitate non-assertive forms (e.g. *Nobody seemed to know anything about anyone*).

Adverbials vary greatly in their scope. Some refer to the entire sentence:

Frankly, I don't want to talk to him

others only to part of the predication:

I dread having to talk *frankly* to him

others to just one word:

Telling him the truth is going to be *incredibly* difficult

•• **scope of negation**: see NEGATION.

secondary

1 *Phonetics & Phonology*. Designating the next most important stress after the primary stress. See PRIMARY (1), STRESS.

2 *Phonetics*. Designating (the place of) an articulation which adds some quality to the main articulation.

Thus in the so-called *dark l* allophone [ɫ], while the main articulation is lateral, as for any *l* sound, there is an additional feature, the raising of the back of the tongue towards the velum (VELARIZATION). Other *secondary articulations* are LABIALIZATION, PALATALIZATION, PHARYNGEALIZATION, and NASALIZATION.

3 (*n. & adj.*) (In some older grammar, mainly Jespersen's.) (Designating the RANK or level of) a less important word in a unit. Contrasted with PRIMARY (2) and TERTIARY (2).

For example, in the phrase *terribly cold weather*, *cold* is (according to Jespersen) a *secondary*. The analysis of verbs as secondaries, in this model, is controversial, since verbs are analysed in modern grammar as the most essential element in clause structure:

1933b O. JESPERSEN If we compare the two expressions *this furiously barking dog* and *this dog barks furiously* . . the verb *bark* is found in two different forms, *barking* and *barks*; but in both forms it must be said to be subordinated to *dog* and superior in rank to *furiously*; thus both *barking* and *barks* are here secondaries.

second person

(Denoting, or used in conjunction with a word indicating) the person addressed, in contrast to the speaker or writer and any other person.

The second person pronoun in modern standard English, for singular and plural, subject and object, is *you*; the related reflexives are *yourself*, *yourselves* and the possessives are *your* and *yours*.

The lack of singular/plural distinction (except in the reflexive form) is a noteworthy feature of the standard dialect; in many non-standard forms of English it has been remedied by the creation of forms such as *yous(e)* (L19), *yez* or *yiz* (E19), *you-all* (E19), *y'all* (E20).

segment

(*n.*) (Pronounced /ˈsegm(ə)nt/.)

A unit forming part of a continuum of speech or (less commonly) written text; an isolatable unit in a phonological or syntactic system.

The term is particularly used in descriptions of speech and the analysis of a language into its phonemes.

> 1973 J. D. O'CONNOR We can say that there are as many segments in a stretch of speech as there are changes in position on the part of the vocal organs, but this requires qualification. Firstly, the changes must lead to perceptible differences . . . Secondly, since the change of position of various organs is continuous and gradual rather than instantaneous it is the extremes of such movements that are important.

(*v.*) (Pronounced /segˈment/.)

Isolate (a word, sequence of sounds, etc.) into such units.

• **segmentable**: capable of being divided into segments. **segmentability**. **segmentation**.

segmental

Phonetics & Phonology. Of or pertaining to a segment or segments. Contrasted with SUPRASEGMENTAL.

The term is frequently used in the terms *segmental phonology* and *segmental phoneme*.

> 1966 S. BELASCO et al. Pitch levels are not always suprasegmental features. When they are short enough, they can be considered one of the distinctive features of a segmental phoneme.

• **segmentalization**. **segmentalize**. **segmentally**.

segregatory coordination

Coordination in which the two elements could be separated and still make sense. Contrasted with COMBINATORY *coordination*.

Henry and Margaret had dinner is an example of *segregatory coordination*, since we can reasonably say *Henry had dinner* and *Margaret had dinner*.

selectional restriction

A limitation on what words can go with a particular word.

As propounded in early Generative Grammar, *selectional restrictions/rules/features* were said to relate to the syntactic frame in which a word could appear. Such selectional restrictions were contrasted with what were called 'subcategorization features'. But the latter were strictly syntactic, such as that a transitive verb needs a noun phrase object (hence the deviance of **John found sad*), whereas selectional features in fact were partly based on semantic criteria, e.g. that the verb *find* normally needs an animate subject. Chomsky's famous sentence *Colourless green ideas sleep furiously*, presented as an example of 'failure to observe a selectional rule', is clearly deviant in a different way from both *John found sad* (which breaks a category rule) and from 'clearcut cases of violation of purely syntactic rules, for example *sincerity frighten may boy the*'.

In later literature, selectional restrictions are sometimes called *semantic restrictions*. They are also sometimes equated with co-occurrence restrictions, though the latter tend to be more narrowly based on syntax. Since many restrictions are partly syntactic and partly semantic (and the term *subcategorization* is not in general use), there is considerable overlap and looseness in the use of the different terms for describing restrictions.

semantic

Relating to meaning.

> 1988 W. A. LADUSAW The principal descriptive goal of semantic theory is an account of the semantic structure of a language, the properties and relations which hold of the expressions of a language in virtue of what they mean. On analogy with syntactic theory, it is an account of part of the native speaker's linguistic competence, namely, that knowledge underlying 'semantic competence'. As such, it presumes an account of the syntax of a language and predicts judgements of semantic relations between its expressions based upon proposals about what they mean. Chief among the relations to be accounted for are paraphrase (or semantic equivalence) and semantic consequence (or entailment).

Terms such as *semantic interpretation* and *semantic representations* belong to various theories of Generative Grammar. In early versions of this, the *semantic component* had a technical status along with the syntactic and phonological components. *Semantic meaning* is sometimes contrasted with *grammatical meaning*.

semantic feature

One of a set of meanings into which a word can be analysed.

The term is part of the vocabulary of COMPONENTIAL ANALYSIS, where [-male] and [-adult] could be semantic features of the word *girl*. Note that semantic features are inherent in the word, in contrast to semantic restrictions (also called *selectional features*, *rules*, or *restrictions*), which require a semantic feature to be present in a collocating word.

semantic field

See FIELD.

semantic restriction

A limitation on what other words can go with any particular word, based on the meaning of the latter.

Semantic restrictions (sometimes also called *selectional restrictions*) are part of the framework of Generative Grammar. But the concept is of general relevance.

For example, **The student congregated* is deviant because an inherent feature of the verb *congregate* is the requirement of a plural subject. (Note that this requirement is different from a strictly grammatical rule, such as the concord rule requiring both subject and verb to be plural: e.g. *They are congregating*.) Other verbs may restrict the kinds of subject or object they allow, perhaps to 'human' or 'animate' (**The cupboard laughed*); some adjectives may require a concrete noun (**a hand-made oblong suggestion*).

Semantic restrictions may of course be broken in poetry or other imaginative literature, where the special effect is partly due to the violation of the rule, as in Dylan Thomas's *a grief ago*.

Compare COLLOCATIONAL *restriction*.

semantics

(The study or analysis of) the relationships between linguistic forms and meaning.

> 1912 E. WEEKLEY The convenient name semantics has been applied of late to the science of meanings, as distinguished from phonetics, the science of sound.
>
> 1964 E. A. NIDA While semantics deals with the relationship of symbols to referents, syntactics is concerned with the relationship of symbol to symbol.

Traditionally there is a division between syntax and word meaning, which is shown by the separation of information about language into grammar books and dictionaries. *Semantics* goes beyond 'word meaning' in viewing words as part of a structured system of interrelationships.

Later versions of Generative Grammar have given rise to GENERATIVE SEMANTICS and *interpretative semantics*. Other theories include *truth-conditional semantics, logical semantics*, and STRUCTURAL *semantics*.

See also FIELD.

The term is a borrowing of the French term *la Sémantique* (M. Bréal 1883). An older word (now disused) with much the same meaning was *semasiology*.

Compare PRAGMATICS.

semantic shift

A change in the meaning of a word taking place over time. (Also called *semantic change*.)

There is a general tendency for words to develop new meanings and to relinquish other meanings over time. Much of this change occurs not in isolation but in relation to other words whose meanings are changing in other ways. *Meat* once meant 'food in general' while *flesh* had a wider coverage than at present, taking in both living flesh and dead flesh as food. Individually considered, each word has contracted its field of reference, but taking them together it becomes clear that a certain reclassification has taken place. *Collide*, once used mainly of pairs of trains and ships in motion, has expanded its scope, merely as a result of technological change, so as to refer to motor vehicles and aircraft. With this momentum it has been able to achieve generalization not only to the encounter of almost any objects whose paths might cross (e.g. pedestrians, sub-atomic particles, etc.) but also to the meeting of a moving object with a static one (e.g. a car colliding with a tree).

sememe

1 The unit of meaning carried by a morpheme.

As for example the meaning of 'female, feminine' in the suffix *-ess* in *duchess, hostess, lioness.*

2 A minimal unit of meaning.

For example, the grammatical meaning 'past' or 'plural', which can of course be carried by different physical realizations.

> 1974 P. H. MATTHEWS In Bloomfield's formulation [d]/[t] would be one 'morpheme' (one phonetic form being a 'phonetic modification' of the other), and [n] [ən] would be a different 'morpheme'; they are related only in that both could be associated with the same 'sememe' or unit of meaning. But the notion of 'sememe' is decidedly problematic (particularly for a concept such as that of the 'Past Participle').

The term was originated by the Swedish linguist A. Noreen (1904).

semi-auxiliary (verb)

A two-part or three-part verb beginning with *be* or *have*, such as *be able to, be about to, be going to, have to.*

A number of verbs that seem to share some characteristics with auxiliaries cause grammarians problems of analysis. This category seems very similar to, but not quite the same as, PHRASAL AUXILIARY VERB.

semicolon See PUNCTUATION.

semi-modal

(*n. & adj.*) (A verb) that is, formally, partly like a modal verb and partly like a full (lexical) verb. (Also called *marginal modal* (*auxiliary*).)

Verbs included here are *dare, need, ought to,* and *used to,* with which both inversion and *do*-support occur (though not freely), and with the first two of which there is also a choice (again, not always) in the present tense third person singular between forms with and without *-s*:

> It was—dare I say it?—a success
> I used not to play/I didn't use to play
> I don't think he need take any action/I don't think he needs to take any action

Ought to and *used to* are formally past in tense, and *used to* is confined to past meaning.

semi-negative

(*n. & adj.*) (A word) that is almost negative in meaning, and grammatically often has the same effect as a negative word. (Also called *near negative.*)

A number of adverbials, e.g. *barely, hardly, little, scarcely*, and the determiners/pronouns *little* and *few* are so nearly negative that they function much like true negative words. Thus they take positive question tags:

> It's barely/scarcely possible, is it?
> Few people know this, do they?

Similarly, fronted semi-negative adverbs require subject–operator inversion in the same way that some true negatives do:

Hardly had they arrived when the lights went out
Only then did we realize our mistake

Compare with true negatives:

Nowhere have I seen such incompetence
Never did I expect to see this day arrive

Compare BROAD NEGATIVE.

semiology

The study of linguistic signs and symbols.

The same as SEMIOTICS. The term was coined (as French *sémiologie*) by de Saussure (1916).

semiotic

(*adj.*) Relating to the use of signs as a form of communication.

(*n.*) The same as SEMIOTICS, which is now the usual term.

1973 R. JAKOBSON The subject matter of semiotic is the communication of any messages whatever, whereas the field of linguistics is confined to the communication of verbal messages.

semiotics

The study of sign language.

Semiotics is sometimes restricted to non-linguistic communication, including possibly 'body language' and the kinds of visual sign and symbol that 'translate' into words, e.g. traffic signs and signals, company logos and such old established signs as a barber's pole or the three balls of a pawnbroker's sign. However *semiotics* (formerly also *semiotic*) may also include language: see the quotation at SEMIOTIC.

semi-passive

A class of passives which have both verbal and adjectival properties.

Semi-passives are like true passives in having active counterparts, e.g.

I *was impressed* by his fluency (*compare* His fluency impressed me)

But they are like adjectives in a number of ways: they can be coordinated with an adjective; the participle can be modified by adverbs such as *more*, *most*, *quite*, etc.; and the verb *be* can be replaced by lexical copular verbs such as *look*, *seem*, etc.

The whole family *were upset* and angry
They *were worried* about Tom's disappearance
Sarah didn't seem *interested* in his explanation

Compare CENTRAL *passive*, PSEUDO-PASSIVE.

semi-vowel

Phonetics. A sound which is phonetically vowel-like because it is a glide, but phonologically consonant-like in being marginal to a syllable.

In English the phonemes /j/ as in *you, use, view*, and /w/ as in *way, suave, choir* are semi-vowels. Traditionally they are classified as consonants because they function marginally; it is noticeable too that, like consonants, they are preceded by *a*, not *an* (e.g. *a youth*, *a window*).

sense

Semantics. Meaning.

The word *sense* is used in a wide variety of ways, just as *meaning* and *reference* are. In semantic theory it is used to describe verbal (lexical) meaning, derived partly from the meaning of other words (*sense relations*), in contrast to the relationship of a word to the outside world, which is REFERENCE.

sense group

Phonetics. A group of words united by sense.

Sense group is a term used in describing intonation patterns. It is roughly the same as a TONE UNIT (or *tone group*). The term emphasizes the fact that intonation is affected by meaning.

senser

The role of the logical subject of a mental process or perception verb.

This is not a very generally used term, but '*senser* + *phenomenon*' can replace '*actor* and *goal*' etc. in sentences such as

> I don't like your attitude (*I* = senser; *your attitude* = phenomenon)
> It upsets me that you won't even try (*It* + *that you won't even try* = phenomenon; *me* = senser)

EXPERIENCER is probably a more usual term.

sentence

The largest unit of language structure treated in traditional grammar; usually having a subject and predicate, and (when written) beginning with a capital letter and ending with a full stop.

(*a*) Traditional definitions of the sentence are often in notional terms, e.g. 'a set of words expressing a complete thought', but this is too vague to be useful. However, attempts at rigorous structural definitions (as above) are not entirely satisfactory either.

(*b*) Not all sentences have a subject and predicate (e.g. imperatives usually lack an expressed subject). At the same time more than one 'grammatically complete sentence' can be run together in writing with only one full stop, so that grammatical and orthographic sentences may not correspond (e.g. *I came, I saw, I conquered*). As for spoken language, it is often impossible to say where one sentence ends and another begins.

(*c*) Another problem affects the definition of a sentence as 'the largest unit of language structure', since pro-forms, connectors, and other cohesive devices straddle sentence boundaries, even if this is largely ignored in traditional grammar. We also have to recognize the theoretical possibility of a sentence of infinite length containing an infinite number of clauses. For this reason modern grammarians often prefer to analyse syntactic structure in terms of the clause, even though reference and substitution often operate over stretches of discourse larger than the sentence.

(*d*) Sentences are categorized in modern grammar, as in traditional grammar, into *simple*, *compound*, and *complex* sentences on the basis of the number and type of clauses they contain.

Compare ABBREVIATED, CLAUSE, FULL (3), FUNCTION.

sentence adjunct

An adjunct that is not structurally essential and relates to the sentence as a whole. This is a rather specialized subdivision of ADJUNCT (3). Many adjuncts belong to the predication, are sometimes obligatory, and normally come in end position (PREDICATION *adjuncts*). *Sentence adjuncts* are more marginal, are optional, and can come in initial position; they relate to the sentence as a whole. Contrast:

She lives *in Oxford* (predication adjunct, obligatory here)
In Oxford, you can visit the colleges (sentence adjunct)

sentence adverb

The same as SENTENCE ADVERBIAL

1932 C. T. ONIONS *Else*, *only* (= 'but'), *so*, *accordingly*, *hence*, *also*, *too*, *likewise*, *moreover*, though some of them frequently come at the beginning of a sentence, are not Conjunctions at all, but Adverbs. They qualify the sentence as a whole rather than any particular part of it, and may therefore be called Sentence Adverbs. Other Adverbs which may be used thus are *truly*, *certainly*, *assuredly*, *verily*, *undoubtedly*: e.g. 'This is certainly false' (= 'It is certain that this is false').

sentence adverbial

(Also called SENTENCE ADVERB.)

1 A term covering CONJUNCTS and DISJUNCTS.

2 Another term for DISJUNCT only; distinguished from CONJUNCT, which may then be labelled CONNECTOR.

sentence element See ELEMENT.

sentence modifier See MODIFIER 1.

sentence stress See STRESS.

sentence structure See STRUCTURE.

sentence type

Any of the major categories into which sentences (or clauses) are analysed. Traditionally sentences are divided into types such as 'statement', 'question', and so on. More precisely, they can be categorized by form and by function. Commonly adopted terms for these distinctions are:

formal (syntactic): DECLARATIVE, INTERROGATIVE, EXCLAMATIVE, IMPERATIVE.
functional: STATEMENT, QUESTION, EXCLAMATION, COMMAND (or DIRECTIVE).

But this terminology is not universal.

Compare MINOR SENTENCE.

sentential

Being or pertaining to a sentence.

•• **sentential relative clause**: see RELATIVE.

separability

The possibility of dividing a phrase by inserting an intervening word or words. This is a feature of many verb + adverb combinations (i.e. phrasal verbs in the narrow sense):

We looked up some old friends
We looked them up

Compare DISCONTINUITY.

sequence of tense rule See BACKSHIFT.

sequent

Of (the tense system of) the verb used after a past reporting verb.

Some of the tense distinctions normally available are NEUTRALIZED after a verb such as *said* or *thought*, with the result that the following (or *sequent*) verb has fewer options. For example a sequent past perfect may represent a present perfect, a past perfect, or a simple past:

They said he had left (i.e. they said either 'He has left', or 'He had left (before we got here)', or 'He left ten minutes ago')

set expression See FIXED *phrase*.

sex-neutral See EPICENE.

-*s* form

The third person singular ending of the present tense of lexical verbs.

e.g.

looks, sees, wishes

shift

(*n.*)

1 A phonetic change.

Usually with a defining word; see GREAT VOWEL SHIFT, STRESS *shift*.

•• **sound shift**: see SOUND LAW.

2 A change of meaning. See SEMANTIC SHIFT.

3 A change from one grammatical category to another.

Usually with a defining word; see BACKSHIFT, FUNCTIONAL *shift*, RANK-SHIFT.

> 1966 G. M. LEECH *Embedding* is a shift in rank whereby a group acts as a word or a word acts as a morpheme, etc.

(*v.*) Change phonetically, semantically, or grammatically.

> 1962 B. M. H. STRANG A form may have inherent rank as morpheme, word, phrase or clause, but may in special circumstances be shifted to any other rank.

shifting stress See STRESS.

short form

The same as CONTRACTION (2); a contracted form.

short vowel See LENGTH.

shwa See SCHWA.

sibilance

Phonetics. The quality of being SIBILANT.

sibilant

Phonetics. (*n. & adj.*) (A speech sound) made with a hissing effect.

Sibilant describes an auditory quality: a hissing perceived by the listener. In English four fricative phonemes are sibilants:

/s/ as in *sing, ice, hiss*
/z/ as in *zoo, rise, dessert*
/ʃ/ as in *ship, chute, issue, ocean*
/ʒ/ as in *genre, mirage, vision, leisure*

plus the AFFRICATES /tʃ/ and /dʒ/.

They contrast with *non-sibilant* fricatives.

sign

Semantics. That which conventionally stands for or signifies some other thing, of which linguistic units such as words are examples, but not the only examples.

The word *sign* and related terms are used in specialized ways in some linguistic discussion:

> 1977 J. LYONS The meaning of linguistic expressions is commonly described in terms of the notion of signification; that is to say, words and other expressions are held to be signs which, in some sense, signify, or stand for, other things. What these other things are . . has long been a matter of controversy. It is convenient to have a neutral technical term for whatever it is that a sign stands for: and we will use the Latin term significatum, as a number of authors have done, for this purpose.

De Saussure (1916) observed that *sign* in current usage generally designated only the acoustic impression, but proposed to keep the word for the combination of this and the associated concept, replacing 'concept' with *signifié* and 'acoustic impression' with *signifiant*. Other words for *signifié* are *significatum* and *signified*; and for *signifiant*, *significans* and *signifier*.

silent

Designating a letter in the written form of a word which is not sounded in speech.

Given the vagaries of English spelling, many letters could be said to be silent in certain contexts. The term however tends to be applied particularly to *silent e*, as in *done, infinite, corpse, have* (although in many cases, such as *hope, rate* as compared with *hop*, *rat*, the *e* in fact indicates the pronunciation of the preceding vowel—it is children's 'magic *e*') and to such 'lost consonants' as are seen in *(g)nat*, *(k)nife*, *ha(l)f*, and *croche(t)*.

simple

1 Of a word: not COMPOUND or COMPLEX.

Steam may be described as *simple* in contrast to *steamboat* (a compound) or *steaminess* (a complex word).

•• **simple preposition**: a single-word preposition (e.g. *from*, *before*) in contrast to a complex preposition (e.g. *out of*, *in front of*).

2 Of a sentence, its structure: (traditionally, and still generally) a single independent clause containing a single finite verb phrase.

e.g.

> Britain's role in Europe is important

Simple sentences contrast with *compound* and *complex sentences*.

However, in some analyses, a sentence containing another finite clause may still be simple provided that the sentence does not have a clause functioning as one of its elements (i.e. as Subject, Object, Complement, or Adverbial). By this definition a sentence containing a postmodifying relative clause is still a simple sentence, e.g.

> This presents a choice which will affect every aspect of your life and future

The relative clause is part of the noun phrase (which functions as object), so this sentence has a *simple* SVO structure.

3 Of a tense: consisting (in the positive) of a single word form.

English has two simple tenses: the *present simple* (e.g. *look*, *looks*) and the *past simple* (e.g. *looked*).

In traditional grammar, the *shall/will* future is sometimes called the *simple future*, but in more modern grammar this is analysed as a complex verb phrase.

4 Of a verb phrase: consisting of a single verb form without ellipsis.

e.g.:

'*Look*!' he *said*. 'I *know*.'

singular

(*n.* & *adj.*) (Designating) a word or form that denotes or refers to a single person or thing.

(*a*) *Singular* contrasts mainly with PLURAL in the description of nouns, pronouns, and verb forms. e.g.

The *girl/She is/looks* confident (singular)

versus

The *girls/They are/look* confident (plural)

(*b*) Uncount nouns are sometimes described as singular because they take singular verbs. But this is misleading, since singular count nouns and uncount nouns do not share all the same determiners (e.g. *a/one* roll but *some/much* bread).

(*c*) Invariable nouns of plural meaning lacking an *-s* but taking a plural verb (e.g. *police*) are sometimes described as singular nouns.

(*d*) The term '*singular noun*' is also sometimes applied to a noun which, in a particular meaning, can be used with *a/an* (e.g. *What a pity!*), but which has no plural form.

Compare NON-SINGULAR.

singulare tantum (Pronounced /sɪŋg(j)ʊˈlɑːreɪ ˈtæntəm/.) (Plural **singularia tantum**.)

A word which has only a singular form.

A little used term which is sometimes applied to mass (or uncount) nouns. Compare PLURALE TANTUM.

situation

The extralinguistic context of language.

See also CONTEXT *of situation*.

situational

Relating to or determined by the situation.

•• **situational context**: see EXTRALINGUISTIC.

situational meaning: see INTERPERSONAL.

situational recoverability: see RECOVERABILITY.

slang

Words, phrases, and uses that are regarded as very informal and are often restricted to special contexts or are peculiar to a specified profession, class, etc. (e.g. *racing slang*, *schoolboy slang*).

slot

A position within a structure that can be filled by any of a number of interchangeable items (a FILLER).

> 1973 R. QUIRK & S. GREENBAUM Existential *there* . . may be regarded as an empty 'slot-filler'.

slot-and-filler

Designating a method of analysing sentence structure in which various functional slots are first identified, and the words and phrases that can fill them (fillers) are then further analysed. For example in

> I can resist everything except temptation

there are three slots:

> Subject (I)
> Verb (can resist)
> Object (everything except temptation)

The system underlies present-day analysis into clause ELEMENTS. It also links up to some extent with the functional (rather than notional) methods used in defining parts of speech. Thus the subject slot could be filled by a vast variety of noun phrases (e.g. *you*, *the neighbours*, *everyone I know*, etc.), but all would contain a noun or pronoun as head. Similarly anything that can fill the second slot must be a verb.

Compare IMMEDIATE CONSTITUENT.

small clause

(In Government-Binding Theory.) A structure which resembles a clause in that it has words or phrases that are interpretable semantically as a subject + predicate, but which contains neither a finite verb nor a *to*-infinitive.

For example:

> We think *the scheme preposterous*
> I heard *the bureaucrat extol it*

In standard grammar such structures are not handled as clauses, but are commonly dealt with in terms of verb complementation.

social dialect See DIALECT.

social distancing

The explanation given for a use of the past tense which is attributable neither to the event being in the past nor to hypothetical meaning.

Past tenses are sometimes used where a present tense could as well be used, e.g.

Did you want to see me?
I was wondering whether you could spare me a minute?
We were hoping you would help
I wouldn't have thought so

Sometimes politeness is the motivation: it may be easier for the addressee to say no than to a blunter request (e.g. *We are hoping you will help*); but at other times the 'distancing' may sound over-formal or cold.

social meaning

The same as INTERPERSONAL meaning.

sociolinguistics

The study of language in relation to social factors.

Sociolinguistics is concerned with the interrelationship between language and region, language and national identity, language and peer group pressure, language and the age, sex, and social class of the user, and so on.

• **sociolinguist. sociolinguistic.**

1978 W. LABOV The sociolinguistic behaviour of women is quite different from that of men because they respond to the commonly held normative values in a different way.

soft

(In popular usage.) Designating the letters *c* and *g* when they represent the sounds /s/ and /dʒ/ respectively, normally when preceding the vowel letters *i*, *e*, and *y*. Contrasted with *hard* (when they represent the plosives /k/ and /g/).

The contrast between soft and hard values in these consonants belongs originally to the French stratum of the English vocabulary, and goes back in turn to the Romanic palatalization of the Latin velar plosives /k/ and /g/ when these were followed by front vowels. (A similar palatalization had also occurred in Old English, but the resulting pairs of sounds were differentiated by contrasting consonant symbols in Middle English spelling.)

soft palate See VELUM.

solecism

A mistake of grammar or idiom; a blunder in the manner of speaking or writing.

This is a general term. Grammarians are more likely to speak of error, incorrect usage, hypercorrection etc.

• **solecistic.**

sonorant (Pronounced /ˈsɒnər(ə)nt/.)

Phonetics. (*n. & adj.*) (A sound) produced with the vocal organs so positioned that spontaneous voicing is possible; a vowel, a glide, or a liquid or nasal consonant. Contrasted with OBSTRUENT.

> 1968 N. CHOMSKY & M. HALLE Vowels, glides, nasal consonants and liquids are sonorant.

This type of definition has been criticized on the grounds that voicing is essential:

> 1982 P. LADEFOGED Our definition makes voicing a prerequisite for sonorants . . . Chomsky and Halle's physiologically based definition leads to glottal stops and voiceless vowels being classified as sonorants, which is counterintuitive to say the least.

sound change

(In historical linguistics). Change in pronunciation over time.

Changes in pronunciation (as in syntax and vocabulary) over time are part of the dynamic of a living language. The phonological structure of Old English is so different from modern English that it is like a foreign language. Middle English exhibits quite different sound patterns from our own. Shakespeare's English looks deceptively more like ours than the contemporary descriptions show it really to have been because we have preserved the spelling of his time relatively unaltered. And many of the apparently arbitrary spellings of modern English are explainable in terms of earlier pronunciations.

Changing sound patterns are discernible in our own time. Daniel Jones in his *Outline of English Phonetics* (1918) included a diphthong /ɔə/ heard in such words as *coarse*, *score*, and *four* and distinct from the monophthong /ɔː/ used in *cork*, *short*, and *fork*, though he noted that many RP speakers used /ɔː/ for both. Present-day phonetic descriptions of English (RP) do not mention the distinction, since virtually no RP speakers make it now. The vowel /ɔː/ is also frequently heard in words such as *tour* (traditionally /tʊə/), but it is now virtually never heard, as it once was, in words such as *off* and *cloth* (a pronunciation which was old-fashioned even in Jones's day).

Another observable kind of sound change is the shifting of stress patterns within the word. One fairly systematic change in British English is a tendency in polysyllabic words for stress to move from earlier or later in the word to the third syllable from the end: hence changes such as

ˈcontroversy → conˈtroversy
ˈprimarily → priˈmarily
conˈtribute → ˈcontribute
soˈnorous → ˈsonorous
artiˈsan → ˈartisan
promeˈnade → ˈpromenade

all of which have been observed changing during the past century or so.

sound law

(In comparative and historical linguistics.) A regularity among sound changes. (Also called *sound shift*.)

Structural correspondences between semantically similar words in related languages have been noted for centuries, but first began to be analysed rigorously in the late eighteenth and early nineteenth centuries.

Grimm's law, proposed by the German scholar Jakob Grimm in the 1820s, suggested that a sound shift had taken place in a prehistoric period of Germanic, which accounted for various consistent correspondences between different languages derived from Indo-European. For example Latin voiceless plosives (/p/, /t/, /k/) regularly correspond to Germanic fricatives (/f/, /θ/, /h/) (e.g. *pater*: *father*, *tenuis*: *thin*, *cornu*: *horn*), while Latin voiced plosives (/b/, /d/, /g/) regularly correspond to Germanic voiceless plosives (/p/, /t/, /k/) (e.g. *labia*: *lip*, *decem*: *ten*, *ager*: *acre*).

Similar correspondences can be seen between English and related Germanic languages: for example, where Old English had /ɑː/ and Modern English often has /əʊ/, Dutch has a vowel spelt *ee* and German has a vowel spelt *ei* (e.g. *dole*, *ghost*, *stone*: Dutch *deel*, *geest*, *steen*, German *Teil*, *Geist*, *Stein*).

And the same kinds of correspondence can be observed between dialects of English that have drifted apart over history: for example, where southern dialects have /aʊ/, northern and Scottish traditional dialects have /uː/ (e.g. *now*: '*noo*'; *house*: '*hoose*').

Compare GREAT VOWEL SHIFT.

sound quality See QUALITY (1).

sound symbolism

A (fancied) representative relationship between the sounds making up a word and its meaning.

Various kinds of sound and meaning correlations are said to exist; specialized terms include ICONICITY, ONOMATOPOEIA, PHONAESTHESIA, and SYNAESTHESIA.

Usage is inconsistent. *Sound symbolism* can be used as a cover term for all such phenomena, but elsewhere in the literature the term seems often to exclude obvious onomatopoeia such as *cuckoo* and *cock-a-doodle-doo*.

sound system

The phonemic system of a language.

source

1 The place etc. from which something originates; used in classifying the meaning of adverbials and prepositions of space. Often contrasted with GOAL.

1975 D. C. BENNETT Three directional cases are posited: 'source', 'path' and 'goal' . . . A sentence such as *We went from Waterloo Bridge along the Embankment to Westminster* is considered to contain three directional expressions in its semantic representation: a source expression, a path expression and a goal expression. The source expression, realized in surface structure as *from Waterloo Bridge*, specifies the starting-point of the change of position described by the sentence.

2 *Linguistics*. In a model of communication, the sender of a 'message'.

The person who receives (or is meant to receive) the message is the *destination*.

space

The three-dimensional expanse in which physical objects exist; place; used in describing the meaning of adjuncts and prepositions, and including both *position* and *direction*.

Many *adverbials* and *prepositions of space* have both meanings; e.g.

abroad, beneath, downstairs, outside, etc.

• **spatial**.

Compare TEMPORAL, TIME.

specific

Clearly defined, particular, referring to that thing etc. and no other.

(*a*) In one model, *specific* is particularly used in relation to article usage. A distinction is made between *specific* and GENERIC (or CLASSIFYING) meaning, a distinction that cuts across *definite* and *indefinite*. Thus *specific* meaning itself may be either *definite* or *indefinite*. E.g.

I met an interesting man on holiday (specific indefinite).
The man told me . . . (specific definite.)

(*b*) In another model, *specific* is contrasted with GENERAL, but the distinction made is very different. All determiners are either one or the other: *specific* refers to *the* in all its uses, to demonstratives (*this* etc.), and to possessives (*my* etc.); *general determiners* include *a*, *an* (whatever its meaning). Thus *specific* includes *definite*, and *general* equates roughly with *indefinite*.

A further difficulty is that, since the word *specific* is in everyday use, and indeed is sometimes glossed in dictionaries as 'definite', the word is often used loosely and the necessary distinction is frequently ignored or confused in popular grammar books.

speech See CONNECTED SPEECH, DIRECT SPEECH, INDIRECT SPEECH, REPORTED SPEECH.

speech act

A social or interpersonal act that takes place when an utterance is made.

The theory of *speech acts* was popularised by J. R. Searle in his book *Speech Acts* (1969), in which he expounded and expanded the theories of the philosopher J. L. Austin. The theory of speech acts propounds various important distinctions between

(*a*) *performative utterances*, which are used to perform some action, in contrast to *constative utterances*, which contain propositions that are either true or false. See PERFORMATIVE.

(*b*) ILLOCUTIONARY, LOCUTIONARY, and PERLOCUTIONARY acts.

speech chain

Phonetics. The series of links between speaker and listener.

The speech chain, beginning with the speaker's brain and ending with the listener's brain, is of considerable interest to phoneticians. What happens in the brains of listener and speaker, the two ends of the chain, are the most difficult parts to understand, but considerable progress has been made with the intermediate stages. See ARTICULATORY, ACOUSTIC, and AUDITORY *phonetics*.

speech community

Any social or geographical group sharing roughly the same language.

speech organ

Any part of the mouth, nose, throat, etc. involved in the production of speech sounds. (Also called *organ of speech*.)

Hence the lips, alveolar ridge, soft palate, larynx, and so on are all referred to as *speech organs*. The movable organs are sometimes distinguished as ARTICULATORS.

[See diagram p. 447]

speech sound

An elementary sound occurring in a language, considered phonetically without regard to the oppositions and combinations in which it may occur (which are the concern of phonology).

spelling pronunciation

The pronunciation of a word according to its written form.

Such pronunciations are considered 'incorrect' by many speakers, but gain wide acceptance through being closer to the spelling and may eventually supersede the traditional pronunciation.

Thus the traditional pronunciations (in RP) of *ate*, *forehead*, and *often*, /et/, /'fɒrɪd/, and /'ɒf(ə)n/ are probably less often heard now than the spelling pronunciations /eɪt/, /'fɔːhed/, and /'ɒft(ə)n/. Spelling pronunciations that are now fully established include /tʊ'wɔːdz/ (replacing /tɔːdz/) for *towards*, /'weɪs(t)kəʊt/ (replacing /'weskɪt/) for *waistcoat*, and *humour*, *hotel*, *hospital*, and *herb* with initial /h/ (replacing the traditional *h*-less forms) (*herb* is still /ɜrb/ in North America).

spirant

Phonetics. (*n. & adj.*) The same as FRICATIVE.

split infinitive

A *to*-infinitive with a word or words separating the *to* from the actual infinitive, regarded as incorrect usage.

The prescriptive grammarians' objection to the split infinitive is based on the fact that in Latin the infinitive is a single indivisible word.

Avoidance may at times be wise, since a split infinitive can be stylistically awkward, and anyway may distract a prescriptively minded reader from the message. But avoidance also has its pitfalls, since the word or words that 'split' the infinitive, when moved, may become attached to another part of the sentence, causing its meaning to change; and avoidance can also result in stylistic awkwardness. Split infinitives may be the best option in:

> According to the oil company it is essential to drill on the National Trust land to fully exploit a new deeper oilfield which has been found at the farm (substituting *fully to exploit* attaches *fully* to *to drill on the National Trust land*, while substituting *to exploit fully* separates the verb from its object)
> He has promised to be the first president to really unite the country (?*really to unite*, **to unite really*, **to unite the country really*).

spread lips See NEUTRAL.

standard

(*n. & adj.*) (Of, pertaining to, or designating) the variety of a language (or occasionally, of pronunciation) with the most prestige.
See STANDARD ENGLISH.

Standard English

The variety of English employed by educated users, e.g. those in the professions, the media, and so on; the English defined in dictionaries, grammars, and usage guides.

Standard English is a wider term than *Received Pronunciation* (*RP*) since

> (*i*) it primarily refers to a system of grammar and lexis (Standard English can be spoken in a variety of different accents—Northern, Scottish, American, etc.);
> (*ii*) it includes written as well as spoken English;
> (*iii*) it is an English norm used (with relatively minor regional variations) worldwide.

Standard English enjoys greater prestige than dialects and non-standard varieties: non-standard varieties are felt to be the province of the less educated. But linguists insist that standard English owes its position to political and social causes, and that other varieties are not linguistically substandard.

> 1940 C. C. FRIES 'Standard' English . . is, historically, a local dialect, which was used to carry on the major affairs of English life and which gained thereby a social prestige.

In addition to the basic 'common core' standard, many national standards are recognized, including American English (Am.E) and British English (Br.E), which have few grammatical differences but an appreciable degree of lexical divergence.

Standard Theory (Also **Standard Model**.)

The model of Generative Grammar presented by Noam Chomsky in *Aspects of the Theory of Syntax* (1965). (Also called *Aspects Theory* or *Aspects Model.*)

This model introduced the concepts of *deep* and *surface structure* and also posited differences between a speaker's *competence* and *performance*. It has since been developed and changed by both Chomsky himself and other linguists. Later models have included *Extended Standard Theory*, *Revised Extended Standard Theory*, and *Government-Binding Theory*, plus more radically different alternatives, such as *Generative Semantics*.

statal

Of a passive verbal form: expressing a state rather than a dynamic action. Contrasted with ACTIONAL.

Some passive sentences are ambiguous. Thus *The opera house was finished in 1980* probably, out of context, has an ACTIONAL meaning, namely 'the building was completed (late or on time), and then (perhaps) officially opened in 1980'. But it could have a *statal* meaning, i.e. 'the building was already finished when I went there in 1980'. In an *actional passive*, *be* + past participle clearly forms a passive verb; in a *statal passive*, *be* is arguably a copular verb, the *-en* form functions as an adjective, and the whole sentence is a PSEUDO-PASSIVE.

Compare STATIVE.

state

(Expressing) a relatively permanent and unalterable situation. Contrasted with *action*, *event*, *happening*, etc.

State is used of verbal meaning. For example, all the following describe states:

They *have lived* here for twenty years
They *own* their own house
It *has* four bedrooms
The garden *is* huge
I *dislike* bats

whereas the following describe actions or events (whether single or repeated):

We'*ve* just *bought* a flat
They *have moved* ten times in fifteen years
They *are buying* a house

•• **state verb**: (a popular term for) STATIVE *verb* (as illustrated above).

State verb versus ACTION VERB makes broadly the same contrast as *stative* versus *dynamic*. Theoretically therefore state verbs are not used in the progressive; in practice some state/stative verbs are used with dynamic meaning (*We're having a party*, *You're being difficult*). Notice also that some state verbs may denote a brief temporary state (*I think I'll go to bed*), but this is still in contrast to a temporary action (*I'm thinking about it*).

statement

A sentence or utterance that states or declares.

Commonly the word *statement* is used as one of the formal categories by which sentences are analysed, and it then contrasts syntactically with question, exclamation, etc.

Where sentence types are distinguished by semantic or functional, as well as formal criteria, *statement* is often reserved as the semantic label, in contrast to the formal label DECLARATIVE.

> 1984 R. HUDDLESTON Precisely because they are semantic categories, the criteria that distinguish statement, question and directive from each other are of a quite different nature from those that distinguish the syntactic clause type categories. In performing the illocutionary act of stating, I express some proposition and commit myself to its truth: I tell my addressee(s) that such and such is the case. Statements, in the 'product' sense, are assessable as true or false: questions and directives are not.

static

The same as STATIVE.

stative

Expressing a state or condition. Contrasted with DYNAMIC (I).

The term is mainly used, like *dynamic*, in the classification of verbs, but like that term can also be applied to other word classes.

> 1990 S. GREENBAUM & R. QUIRK A stative adjective such as *tall* cannot be used with the progressive aspect or with the imperative: **He's being tall*, **Be tall.*

stem

Morphology. An element in word formation.

The terms *stem*, *root*, and *base* are sometimes used interchangeably, and sometimes in contrasting (and conflicting) ways.

For some linguists *stem* is the primitive minimal unit of morphology, rather than BASE. Such a stem (possibly the same as a morpheme) may be free (as in *self*-ish) or bound (as in *jeal*-ous). More complicated words may then be analysed in terms of stems and BASES. On this model, *pole* is the stem of *polar*, *polarize*, and *depolarize*; but in the context of *depolarize* (where no such word as **depolar* exists and the word is formed by adding the prefix *de-* to *polarize*), *polarize* is not a primitive stem but a base. For other linguists, the two terms are reversed.

Other linguists avoid the term BASE in favour of terms such as *compound* and *complex* stems. Thus *self-conscious*, consisting of two *simple stems* (*self-* +

conscious) can be described as a *compound stem* for the word *unselfconscious*. This in turn is a *complex stem* for *unselfconsciousness*.

In Latin and Greek grammar, and in historical linguistics, the *stem* is the part of the word to which inflectional endings are attached, and which is usually an extension (by means of a *thematic vowel* or a formative suffix) of the *root*; for example, OE *lufode* 'loved' has the root *luf-* but the stem *lufo-* where the root is extended by a thematic vowel *-o-* (= Latin *-a-* in *amabam*).

Compare BASE, ROOT.

stop

Phonetics. A consonant in which there is a complete closure somewhere in the vocal organs. (Also *stop consonant*.)

The term by itself normally means a PLOSIVE. It can, however, be extended to NASALS, in which case plosives can be distinguished as *oral stops* and nasals as *nasal stops*.

strand

Cause (a word or words) to become grammatically isolated. Usually as the participle **stranded**.

Ellipsis frequently leaves an operator stranded, e.g.

She promised to telephone us last night, but she *didn't*

Similarly, a subject may be stranded in a comparison, e.g.

You play the flute so much better than *George*

Stranded prepositions are also described as *deferred* prepositions, e.g.

What ever were you thinking *of*?
The boiler needs seeing *to*
Here's the book I was telling you *about*

• **stranding**.

See PIED PIPING.

stratificational grammar

A model of grammar, developed in the 1960s by the American linguist Sydney M. Lamb, in which the various levels are known as *strata*.

stress

Phonetics.

(*n*.) Force or energy used in the articulation of a syllable.

(*a*) The terms *stress* and *accent* are often used interchangeably, but some phoneticians use the terms more precisely, relating stress to the energy involved in the production of speech.

1982 P. LADEFOGED Phonetic literature is full of vague remarks about the nature of stress; but the data summarized in Ladefoged (1967) shows conclusively that a primary or secondary stress involves a gesture of the respiratory muscles which can be quantified in terms of the amount of work done on the air in the lungs.

Acoustically, stress is perceived as involving greater loudness, but this is oversimple as an explanation. Stress involves greater force than the ordinary syllable pulse (or chest pulse).

(*b*) *Lexical stress* (also called *word stress* or *word accent*) refers to the *stress* (or *accentual*) patterns of words. In English these are for the most part fixed for each word, though the stress occurs on different syllables in different words, e.g. *'yesterday*, *to'morrow*, *under'stand.* (But see SOUND CHANGE.)

Some word pairs, involving two different parts of speech, are distinguished by stress. Contrast

'Exports (*noun*) rose in the second quarter. But we still need to ex'port (*verb*) more.

Compare also

They have a 'green 'house, but not a 'greenhouse.

Some words have both *primary* and *secondary stress.* Primary stress potentially allows a pitch change (which would normally accompany the word if said in isolation). Secondary stress involves less energy and is heard as less loud.

microcomputer /'maɪkrəʊkəmˌpju:tə/ (primary, secondary)
anti-aircraft /ˌæntɪ'eəkrɑ:ft/ (secondary, primary)

Tertiary stress is recognized by some phoneticians. Many syllables are unstressed.

(*c*) *Shifting stress* (*stress shift* or *variable stress*) is a phenomenon of connected speech. Words containing secondary stress may change their stress patterns, as a way of 'balancing' a phrase or sentence, e.g.

The ˌprin'cess but *the ˌPrincess 'Royal*
ˌnumber thir'teen but *ˌthirteen 'people*

(*d*) *Sentence stress* refers to the way in which some words in an utterance are stressed and others not. In general, lexical words (nouns, verbs, etc.) are stressed, and form words (articles, prepositions, etc.) are not. Strictly speaking, this kind of stress is not a characteristic of the sentence but of the TONE UNIT. See CONTRASTIVE STRESS.

(*e*) *Tonic stress* is stress on the NUCLEUS (also called *nuclear stress*).

(*v.*) Place the (main) stress on. [See table p. 445]

stress-timed

Phonetics. Of a language: having the stressed syllables occurring at regular intervals, irrespective of how many unstressed syllables there may be.

English is predominantly stress-timed, in contrast to *syllable-timed* languages (such as French) in which the syllables occur at more or less regular intervals. Thus in the sentence

'Both of them are ˎmine

the unstressed syllables (*of them are*) are compressed, with vowel weakening (/əv ðəm ə/), while the monosyllable *mine* takes roughly as much time as the preceding *Both of them are*. This does not mean that all sequences containing one stress are of absolutely equal length, but the rhythms of stress-timed and syllable-timed languages are noticeably different.

• **stress-timing**.

stricture

Phonetics. A degree of constriction somewhere in the vocal tract in the course of the articulation of a speech sound.

> 1988 J. C. CATFORD Articulations can be described and classified in terms of their manner of articulation, or 'type of stricture' (in this case 'fricative'), and their place of articulation.

string

A sequence of linguistic elements and symbols.

The term is particularly used in Generative Grammar. For example, a *terminal string* is the final representation of a sentence, after various rules have been applied, which leads to the production of the utterance itself.

The term was introduced into linguistics from mathematics and computing, probably by Chomsky (1955a).

strong

1 Having some prominence of phonetic quality. Contrasted with WEAK (1).

•• **strong form**: the form of a FORM WORD that contains a strong vowel.

Many FORM WORDS (or GRAMMATICAL *words*) have two pronunciations: a *strong form* and a *weak form*. The strong form, containing a *strong vowel*, is used when the word is spoken in isolation or occurs in a prominent position (e.g. at the end of a sentence), or is stressed for emphasis. The weak form, containing a weak vowel, is its usual pronunciation. Thus the articles *a/an* and *the* usually have weak forms (/ə/ and /ðə/ before a consonant sound, /ən/ and /ðɪ/ before a vowel sound); their strong forms are /eɪ/, /æn/, and /ðiː/. A similar alternation is found with prepositions, e.g.

> He'll have got to (/tə/) London by now.
>
> Where's he ˎgot to (/tuː/)? (strong form, because in final position, although unstressed)

and with auxiliary verbs, e.g.

> You ought to have (/əv/) ˎdone it.
>
> But I ˋhave (/hæv/)

and also with a number of other short words, including personal pronouns, *that* as a conjunction or relative pronoun, *Saint*, *Sir*, existential *there*, and *and*, *but*, *as*, *than*.

strong vowel: a stressed vowel or any instance of a vowel that retains the same quality in unstressed position as it has when stressed (contrasted with WEAK *vowel*).

All vowels in stressed syllables are clearly identifiable, and therefore strong. In unstressed syllables vowels often become weak, but sometimes remain strong. Thus all the vowels in the following words, whether stressed or unstressed, are strong:

anarch /ˈænɑːk/, armrest /ˈɑːmrest/, fatstock /ˈfætstɒk/, obese /əʊˈbiːs/, photo /ˈfəʊtəʊ/, unlike /ʌnˈlaɪk/

A syllable, whether stressed or unstressed, containing a strong vowel is sometimes called a *strong syllable*.

2 (In historical linguistics and some traditional grammar.) Of a verb or its conjugation: forming the past tense and past participle with ABLAUT and without a suffix *-(e)d/-t* (though often with a suffix *-(e)n* in the past participle). Contrasted with WEAK (2).

The distinction between *strong* and *weak* verbs has been obscured since Old English times by sound changes which, in some originally weak verbs, have introduced a variation in stem vowel between present tense and past tense (and participle), or have caused the disappearance of the suffix, or both, e.g.

keep, kept (OE cep-an, cep-te)
spread, spread (OE spræd-an, spræd-de)
feed, fed (OE fed-an, fed-de)

Hence English verbs are more helpfully divided into regular and irregular verbs.

The terms *strong* and *weak* are a translation of the German *stark* and *schwach*, introduced by Jakob Grimm. A 'strong' verb had no need of the 'help' of a suffix to express tense.

Compare FORTIS.

structural

Relating to (the analysis of) language structure or organization.

In a sense all grammarians take a structural approach to language, since the object of grammatical description must be to show systematic relationships and rules.

The term is however associated (i) (particularly in Europe) with the work of de Saussure (see STRUCTURALISM); and (ii) (more narrowly) with the American linguistic school of Bloomfield and his followers which flourished in the 1930s, 40s, and 50s. A characteristic of the BLOOMFIELDIAN approach was to emphasize formal rather than semantic features of language, so that the term *structural* often carries this sense.

1952 C. C. FRIES One of the basic assumptions of our approach here to the grammatical analysis of sentences is that all the structural signals in English are strictly formal matters that can be described in physical terms of forms, correlations of these forms, and arrangements of order.

•• **structural ambiguity**: see AMBIGUITY.

structural linguistics: see STRUCTURALISM.

structural recoverability: see RECOVERABILITY.

structural semantics: a theory of lexical structure embracing semantic (lexical) fields, the theory of paradigmatic and syntagmatic relations, and such sense relations as opposition and contrast, hyponymy, part-whole relations (e.g. of *leg* to *body*), and some version of componential analysis.

structural word: the same as FORM WORD.

structuralism

A theory in which language is considered primarily as a system of structures. (Also called *structural linguistics*.)

Narrowly, *structuralism* is taken to mean the theories and methods associated with Bloomfield and his followers, which emphasized form rather than meaning. Great attention was paid to what were held to be scientific 'discovery' procedures by which the phonemes and morphemes of any language could be discovered without reference to meaning. It is this narrow structuralism that was attacked by the generative grammarians, though their debt to it was greater than sometimes admitted.

In the European tradition, structuralism is altogether older and broader. De Saussure is generally regarded as the father of modern structural linguistics, even though some of his principles can be found in earlier writers.

> 1977 J. LYONS What must be emphasized . . is that there is, in principle, no conflict between generative grammar and Saussurean structuralism . . . Saussurean structuralists, unlike many of the postBloomfieldians, (for whom 'structural semantics' would have been almost a contradiction in terms), never held the view that semantics should be excluded from linguistics proper.

The terms *structuralism* and *structural linguistics* are broadly synonymous. Where they are differentiated today, the former tends to be confined to the Bloomfieldian type of sentence analysis, while the latter may then refer to a more generative type of approach to linguistics.

> 1972 D. LODGE Structural linguistics goes beyond the description of any particular language to pursue the 'deep structures' that are common to all languages.

• **structuralist**: (of or characteristic of) an adherent of structuralism.

> 1970 J. LYONS Examples . . were used by Boas to support the view that every language has its own unique grammatical structure and that it is the task of the linguist to discover for each language the categories of description appropriate to it. This view may be called 'structuralist' (in one of the main senses of a rather fashionable term).

structure

The mutual relation of the constituent parts or elements of a language as they determine its character; organization, framework.

(*a*) The term is used in a very general way (though often more formally than *construction*). The analysis of *clause structure* (or *sentence structure*) is essential to any grammar: see CLAUSE.

It is equally possible to talk of *morphological structure* (in word formation); *phonological structure* (the system of phonemes and their interrelationships etc.); *syllable structure* (what combinations of consonants and vowels are possible); *semantic structure* (lexical fields etc.), and *information structure* (the arrangement of 'given' and 'new' information, etc.).

(*b*) *Structure* may be distinguished from *system*.

1964 R. H. ROBINS *Structure* and *system*, and their derivatives, are often used almost interchangeably, but it is useful to employ *structure* . . specifically with reference to groupings of syntagmatically related elements, and *system* with reference to classes of paradigmatically related elements.

In Systemic-Functional Grammar, *structure* is one of the four CATEGORIES (together with *unit*, *class*, and *system*) which contrast with SCALES.

Compare DEEP STRUCTURE, SURFACE STRUCTURE, SYSTEM.

•• **structure word**: the same as FORM WORD.

style disjunct

An adverbial that comments on the way in which the statement to which it is attached is being made, in contrast to a CONTENT DISJUNCT.

e.g.

Seriously, I don't want to get involved
Not to put too fine a point on it, she lied
Going by what the locals told me, little foreign aid reaches its destination

suasive verb

A verb having the meaning of 'persuade' which is followed by a *that*-clause.

Suasive verbs, in one classification, contrast both semantically and syntactically with *factual verbs* (FACTUAL (2)). Unlike factual verbs, which take an indicative in the subordinate clause, suasive verbs permit a choice, e.g.

She demanded that { he returned the money / he should return the money / he return the money }

subject

1 That part of the sentence that usually comes first and of which the rest of the sentence is predicated.

The division of a sentence (or clause) into two parts, subject and PREDICATE, is a long established one, derived originally from Logic and Philosophy (Latin *subjectum* translating Aristotle's *to hupokeimenon*, which primarily means 'the material of which things are made', hence 'the subject of an attribute or of a predicate').

2 (In modern analysis.) One of the five possible major formal constituents of clause structure, abbreviated *S*, (the predicate being analysed in detail into Verb, Object, Complement, and Adverbial); normally, in a declarative sentence, preceding and 'governing' the verb (which must agree with it in number and person).

A subject is normally essential in English sentence structure—so much so that a dummy subject must sometimes be introduced (e.g. *It is raining*). Subjects are, however, usually missing from imperative sentences (e.g. *Listen!*) and may be ellipted in an informal context (e.g. *See you soon*).

Although the grammatical subject is typically a noun phrase, it can also be realized by a clause or even a prepositional phrase, e.g.

That you could do such a thing really shocks me
After nine o'clock would be more convenient

In traditional grammar the subject (in senses 1 and 2) was often defined as the 'doer' of the verbal action, but this definition often fails to reflect the real meaning, e.g.

It is raining (dummy subject)
The match has been cancelled (a subject which does nothing!)

Hence the more syntactic way of defining the term is now general.

3 An element which is not grammatically the subject but may be regarded as the 'real' subject from some other point of view.

To overcome the ambiguity of the word *subject*, traditional grammar sometimes qualified the word. Thus in addition to a *grammatical subject* there might be a *logical subject*, particularly with a passive verb. Thus in

The building was designed by my favourite architect

the grammatical subject is *The building*, but the logical subject (the 'doer', or in present-day terminology the AGENT) is *my favourite architect*.

Traditional grammar also sometimes introduced a PSYCHOLOGICAL SUBJECT, roughly equivalent to the present-day THEME or TOPIC, e.g.

That question I cannot answer (grammatically the object)

subject-attachment rule See ATTACHMENT RULE.

subject case

The case taken by pronouns when in (grammatical) subject position. (Also called SUBJECTIVE *case*.)

Only six distinct subject pronouns occur in modern English: *I, she, he, we, they, who*.

Compare NOMINATIVE.

subject complement See COMPLEMENT.

subjective

1 Designating (the case of) an element which is the subject.

2 Relating or referring to the subject.

•• **subjective complement**: see COMPLEMENT.

In *subjective genitive* the reference is to a 'deep' subject rather than to the subject of the actual sentence or clause. Thus the genitive has subjective meaning in

The Government's plans for privatization (cf. The Government is planning privatization)

Contrast OBJECTIVE *genitive*.

subject-operator inversion

Inversion of a subject and auxiliary verb. See INVERSION.

subject-raising See RAISING.

subject territory

The position of the grammatical subject before the verb. Contrasted with OBJECT TERRITORY.

subject-verb inversion See INVERSION.

subjunct

A member of a subclass of adverbs (adverbials), contrasting with ADJUNCT, CONJUNCT, and DISJUNCT.

This category (in the sense given below) was introduced by Quirk et al. in 1985, the adverbs thus categorized having in the authors' 1972 Grammar been part of the adjunct class.

1985 R. QUIRK et al. We apply the term SUBJUNCTS to adverbials which have, to a greater or lesser degree, a subordinate role . . in comparison with other clause elements.

Subjuncts are distinguished syntactically from adjuncts in several ways, perhaps most obviously by the fact that whereas adjuncts operate as Adverbial in clause structure on a level with other clause elements, subjuncts do not. In the following, contrast *financially*, operating as an adjunct within clause structure in

He helped them financially

which allows

It was financially that he helped them
How did he help them? Financially

with *technically* (meaning 'from a narrow point of view'), operating as a subjunct in

Technically, he helped them

which does not yield

*It was technically that he helped them
How did he help them? *Technically

In general, subjuncts are either outside clause structure, expressing a 'viewpoint' (e.g. *Politically, the idea is suicidal*) or as a 'courtesy' marker (*Kindly be seated*), or they are more narrowly linked to a single word or phrase (e.g. *really odd, hardly possible, too dreadful*).

See EMPHASIZER, INTENSIFIER, and FOCUSING ADVERB.

2 (In Jespersen's terminology.) A word or group of words of the third rank of importance in a phrase or sentence.

> 1914 O. JESPERSEN The adjunct in *perfect simplicity* is a shifted subjunct of the adjective contained in the substantive *simplicity*, cf. *perfectly simple*.

subjunctive

(*n. & adj.*) (A verbal form or MOOD) expressing hypothesis or non-factuality. Contrasted particularly with INDICATIVE.

Since Modern English (unlike, say, French) has few distinct verb forms that differentiate subjunctive from indicative, the status of the subjunctive can be challenged. Its disappearance has long been forecast:

> 1860 G. P. MARSH The subjunctive is evidently passing out of use, and there is good reason to suppose that it will soon become obsolete altogether.

But it survives, and in one area (see below) its use seems to be on the increase.

Traditionally, the uses of ordinary indicative tenses to express hypothesis etc., like, for example, the use of a past tense to refer to a present or future condition (e.g. *If you came tomorrow* . . .), have been described as examples of subjunctive mood or tense—perhaps because in translation such a usage might need a subjunctive form in another language. Modern grammar considers this to be quite unjustified, and restricts the use of the term *subjunctive* to two distinct tenses (as follows).

•• **present subjunctive**: a finite verb form identical with the base of the verb.

The so-called *present subjunctive* is, formally, exactly the same as the present indicative tense, except in the third person singular which lacks *-s*, and in the verb *be* where the subjunctive is *be* (instead of *is*, *am*, or *are*). Functionally, it can be used with reference both to the present and to the past. It is used in three ways:

(*a*) The *mandative subjunctive* is used in subordinate clauses following an expression of command, suggestion, or possibility, e.g.

> I recommended he write and apologize
> She requested that she not be disturbed

This subjunctive has made a considerable comeback in British English in recent years, probably under American influence.

(*b*) Rather formally, the present subjunctive can be used in subordinate clauses of condition and concession, but not with past reference, e.g.

> If that be the case, our position is indefensible

(*c*) The *formulaic* or *optative subjunctive* is used in independent clauses, mainly in set expressions, e.g.

> God save the Queen

Some such clauses have unusual word order, e.g.

Perish the thought!
Come hell and high water

•• **past subjunctive**: the word *were*, used as the 'past' tense of the verb *be* for all persons.

The so-called *past subjunctive* (also called the *were*-subjunctive) is used in clauses of hypothetical condition. It differs from the past indicative of *be* only in the first and third person singular, which popularly replace it. The reference is to present (or future) time, e.g.

If I were you, I'd own up (If I was you . . .)
If only my grandfather were alive today (If only my grandfather was . . .)
If she were to come tomorrow . . . (If she was to . . .)

The subjunctive was so named because it was regarded as specially appropriate to 'subjoined' or subordinate clauses.

submodification

Modification of a modifier, as in *a very unusual house*, where *very* is a SUBMODIFIER of *unusual*; contrast *a large unusual house*, where both *large* and *unusual* directly modify *house*.

submodifier

An adverb used in front of an adjective or adverb to modify its meaning, e.g.

very unusual, a *quite* extraordinarily confusing letter

Such words are called SUBJUNCTS in another classification, but *subjunct* is a much wider class.

subordinate

(*adj.*) (Pronounced /səˈbɔːdɪnət/.) Grammatically dependent. Contrasted variously with COORDINATE, INDEPENDENT, and SUPERORDINATE.

•• **subordinate clause**: a clause that forms part of a sentence and is dependent on (or forms part of) another clause, phrase, or sentence element.

Traditional grammar recognized three types of subordinate clause:

(i) ADVERB CLAUSE (ADVERBIAL *clause*), e.g.
I was surprised, *because it was so unexpected*
(ii) NOMINAL CLAUSE (NOUN CLAUSE), e.g.
It was odd *that he didn't telephone*
This includes clauses often separated as NOMINAL RELATIVE CLAUSES, e.g.
I was surprised by *what you said*
(iii) RELATIVE CLAUSE (ADJECTIVE CLAUSE, ADJECTIVAL CLAUSE), e.g.
The news *(that) you gave us* is very odd

Modern grammar often includes some non-finite and verbless structures as clauses, e.g.

Knowing them, I can well believe it

With you here, things will be easier

Another kind of clause sometimes classified as subordinate is the COMMENT clause, e.g.

It's not easy, *you see.*

Compare COMPARATIVE CLAUSE.

(*v.*) (Pronounced /səˈbɔːdɪneɪt/.) Make grammatically subordinate, usually by means of a subordinating conjunction.

•• **subordinating conjunction**: the same as SUBORDINATOR.

subordinating correlative: see CORRELATIVE.

subordination

The joining of a subordinate clause to a higher linguistic unit.

Subordination is often formally indicated by the use of a SUBORDINATOR, particularly in adverbial clauses. There may be no marker in some nominal clauses, nor in comment clauses, e.g.

I thought *(that) I had told you*
It's not easy, *you know*

Wh-words introduce some clauses of condition and concession:

Whatever happens, don't panic

Non-finite and verbless clauses may be introduced by a subordinator or zero, or sometimes by a preposition, e.g.

On hearing this, she rushed to the bank
With the money under her belt, she felt better

See also MULTIPLE *subordination*.

subordinator

A conjunction introducing a subordinate clause.

Most subordinators are single-word conjunctions, e.g.

although, because, before, since, whereas

but there are also multi-word subordinators

in order that, provided (that), as long as, in case

See also MARGINAL *subordinator*.

substance See FORM (4).

substandard See STANDARD.

substantive

(In older grammar: now disused.)

I (*n. & adj.*) (*a*) (A word) denoting a substance.

•• **noun substantive**: a noun.

This was frequently abbreviated to *substantive*. See the note at ADJECTIVE.

(*b*) (By extension.) Nounlike.

1824 L. MURRAY Some writers are of opinion, that the pronouns should be classed into substantive and adjective pronouns.

The above is presumably a reference to the difference between pronouns and determiners in today's terminology.

2 Expressing existence.

•• **substantive verb** (or **verb substantive**): the verb *be*.

• **substantival**.

substitute

(*n. & adj.*) (A word) that is used to replace another word or words.

(*v.*) Use (a word) as a replacement (for another word or words).

This is a grammatical device, related to ellipsis, and does not refer to the use of synonyms. A *substitute word* is now often called a PRO-FORM.

• **substitutability**. **substitutable**.

1968 J. LYONS The notion of distribution, which is based on substitutability, is simply not applicable to sentences.

See SUBSTITUTION.

substitution

The use of pro-forms in place of the repetition of a linguistic unit.

Substitution differs from CO-REFERENCE. In substitution, pro-forms replace other words, e.g.

I like your golf umbrella. Where can I get *one* like it?

In co-reference, the pro-form refers to the very same referent, e.g.

I like your golf umbrella. May I borrow *it*?

Substitution also differs not only in being concerned with the referents of a noun phrase, but also in using pro-forms to replace predications, clauses, and so on. E.g.

If you'll contribute £20, I'll *do so* too/I'll *do the same*

suffix

Morphology. (*n.*) An affix added at the end of a word or base to form a new word. Contrasted with PREFIX.

Suffixes, unlike prefixes, usually have a grammatical effect on the word or base to which they are added. They are broadly of two kinds:

inflectional suffix: a suffix used to form an inflection.

e.g. *look* + *-s* or *-ed*; *kind* + *-er* or *-est*.

derivational suffix: a suffix used to form a derivative from a base.

e.g.

(noun) book*let* (M19), kind*ness* (ME), play*er* (OE)

(adjective) connect*ive* (M17), care*less* (OE), hope*ful* (ME), manage*able* (L16)
(verb) idol*ize* (L16), wid*en* (E17)
(adverb) pretti*ly* (LME)

(*v.*) Add as a suffix.

• **suffixation**: the adding of a suffix.

summation plural

A type of noun which exists only in the plural and denotes something (either a tool, an instrument, or an article of dress) consisting of two equal parts joined together. (Also called *binary noun*.)

e.g.

binoculars, culottes, leggings, pliers, secateurs

See PLURALE TANTUM.

superlative

(*adj.*) Of a gradable adjective or adverb form, whether inflected (essentially, by the addition of *-est* to the positive form) or periphrastic (by the use of *most*): expressing the highest degree of the quality or attribute expressed by the positive degree word.

e.g.

best, happiest, soonest, most beneficial, most energetically

•• **superlative degree**: the highest degree of comparison, above POSITIVE and COMPARATIVE.

(*n.*) (An adjective or adverb that is in) the superlative degree.

superordinate

(*n. & adj.*) (A linguistic unit) operating at some higher level than another, which is subordinate to it.

(*a*) The term is particularly applied to clauses. In popular grammar clauses are either main or subordinate. But in some analyses, the main clause is said to 'contain' the subordinate clause and therefore to be *superordinate* to it.

A possible advantage of introducing this extra term is that in some complex sentences one clause that is subordinate (in traditional terms) may at the same time be superordinate to another clause. For example, in *I'm sure you'll enjoy it when you get there*, *you'll enjoy it* is subordinate to *I'm sure* but superordinate to *when you get there*.

(*b*) In Semantics *superordinate* is applied to a word of more general meaning than, and therefore implied by or able to replace, other more specific terms; a HYPERNYM. Thus *animal* is superordinate to (is a superordinate of) *tiger* and *kangaroo*.

Compare HYPONYM.

superordination

Semantics. The relationship of being superordinate to another linguistic unit. Contrasted with HYPONYMY.

The term is used in describing the hierarchical structure of some sets of lexical items:

> 1977 J. LYONS Let us say, then, that 'cow' is a hyponym of 'animal' .. and so on ... The obvious Greek-based correlative term for the converse relation, 'hyperonymy' .. is unfortunately too similar in form to 'hyponymy' and likely to cause confusion. We will use instead superordination, which, unlike 'subordination', is not widely employed as a technical term in linguistics with a conflicting sense.

supplementive clause

A non-finite or verbless clause without a subordinator.

This term is not widely used.

> 1985 R. QUIRK et al. Adverbial participle and verbless clauses without a subordinator are *supplementive clauses* ... The formal inexplicitness of supplementive clauses allows considerable flexibility in what we may wish them to convey.

suppletion

The occurrence of an unrelated form so as to supply a gap in a conjugation.

Went as the past of *go* and *was* and *were* as the past of *be* are obvious examples of suppletion.

- **suppletive**: designating a form used in suppletion.

suprasegmental

Phonetics. (*n. & adj.*) (Designating) a feature of intonation extending beyond the phoneme. Contrasted with SEGMENTAL.

Features of intonation such as pitch, stress, and juncture are suprasegmental.

> 1975 J. J. OHALA How we perceive duration, pitch, and intensity is an area that seldom receives serious attention in the literature on suprasegmentals.

- **suprasegmentally**.

Compare PROSODIC.

surface structure

1 (In Transformational-Generative Grammar.) The last, 'uppermost', structural representation of an utterance.

2 Loosely, the actual final utterance; the words spoken or written.

In both senses, *surface structure* contrasts with two similar senses of DEEP STRUCTURE.

switching See CODE-*switching*.

syllabic

Relating to or constituting a syllable.

In some phonetic analyses, *syllabic* and *non-syllabic* are contrasted features, particularly in relation to those consonants which can be pronounced as separate syllables.

Syllabic consonants in English are /m/ as in the pronunciation of *mm*, /n/ as in butt*on*, and /l/ as in app*le*. Some phoneticians describe these sounds as actually having an extremely weak /ə/ in front of them. The diacritic is [̩].

In rhotic accents /r/ also sometimes has a syllabic function, for example in words such as *metre*, where the final syllable in a non-rhotic accent would be /ə/.

syllabification (Also **syllabication**.)

The division of a word into syllables.

Phonetic syllabification and orthographic syllabification do not necessarily correspond. For example, the word *syllable* itself is phonetically a three-syllable word, but when written across two lines it could only reasonably be split at one place, i.e. as *syll-able*.

syllable

A unit of pronunciation forming the whole or part of a word, and having one vowel (or syllabic consonant) phoneme, often with one or more consonants before or after.

Definition of the syllable in universally valid phonetic terms has proved difficult, whether based on the auditory feature of prominence or on the articulatory feature of 'pulse'.

In the prominence theory, some sounds which are more prominent than others form the core of a syllable, with the less prominent sounds at the syllabic boundaries. This theory produces two problems. Firstly, sounds such as /s/ can sound prominent, and this would make a word such as *instance* a four-syllable word, which is counter-intuitive to native speakers. Secondly, it is not always clear where the boundaries of a syllable should be.

In the pulse theory, it is claimed that the number of syllables correspond to the number of chest 'pulses', with vowel sounds again being central to the syllable. The problem here is that if there are adjoining vowels with one that is only weakly stressed (as in *knowing* /ˈnəʊɪŋ/, it is not certain that two chest pulses can be measured, though linguistically the word has two parts.

The solution is often to define syllable in phonological terms (i.e. for a particular language). In English a syllable contains one, and only one, vowel sound. This may be a pure vowel (long or short), a diphthong, or a SYLLABIC consonant. Some syllables consist of such a sound alone (as in the single-syllable words *eye* /aɪ/, *ah* /ɑː/). Consonants are marginal to a syllable: English syllable structure can be quite complicated with as many as three consonants before the vowel and up to four after it (symbolized (CCC) V (CCCC)). See CONSONANT CLUSTER.

Syllables, rather than individual phonemes, carry stress and pitch patterns.
•• **open syllable**: see OPEN (3).

syllable-timed

Of a language: having each syllable pronounced with roughly the same duration.
French is considered to be syllable-timed, whereas English is STRESS-TIMED, but these are tendencies rather than absolute distinctions.

syllepsis (Plural **syllepses.**)

The use of a word or phrase in two ways in the same clause or sentence, when it can only properly relate to one way.
(*a*) When a word is used at one and the same time in two different senses, this is a figure of speech, sometimes also called *zeugma*, e.g. *She caught the train and a bad cold.*
(*b*) *Syllepsis* is grammatical when a particular word form is used in connection with two others, but can only agree with one of them, e.g.

Neither John nor I *agree*
John and Mary each want *their* own way

The term is little used in modern grammar. Compare AGREEMENT.

synaesthesia See SOUND SYMBOLISM.

synchronic

Linguistics. Concerned with the linguistic phenomena of a single period of time (especially the present) as a unified system. Contrasted with DIACHRONIC.
Synchronic linguistics can theoretically be concerned with a language at any point in time, but the term is often shorthand for the study of the language here and now.
The importance of distinguishing synchronic and diachronic observations is obvious. What is syntactically or grammatically acceptable in one period may not be in another. Similarly, the suggestion that the real meaning of a word is its original meaning is based on a failure to recognize the changing nature of language.
The term was coined by F. de Saussure in his posthumously published work *Cours de linguistique générale* (1916).
• **synchronically. synchrony**: synchronic method or treatment.
Compare DESCRIPTIVE (1).

syncopate

(In historical linguistics.) Shorten (a word) by dropping interior sounds, particularly vowels, usually so that a syllable is lost.

e.g. the loss of the medial syllable in *Gloucester* /ˈglɒstə/, *Salisbury* /ˈsɔːlzbərɪ/ or the loss of the /ə/ before *r* in words such as *every*, *secretary*.

In phonological terms, the preferred verb now is ELIDE and the phenomenon is dealt with under ELISION.

• **syncopation, syncope**: contraction of a word by the omission of sounds.

Compare HAPLOLOGY.

syncretism

The realization of two or more inflectional categories in a single form.

The term was originally used in historical linguistics to refer to the merger of inflectional categories by the transfer of the functions of one category to the form used for the other. Thus in early Middle English the functions of the dative case of nouns (marking the indirect object, the complement of certain prepositions, etc.) were transferred to the accusative; this had already occurred in preliterary times with several personal pronouns (e.g. *me* = 'me', 'to me', *us* = 'us', 'to us', etc.). Thus *bedd* 'bed', *lif* 'life', used as objects of a verb, were distinct in Old English from prepositional uses such as *on bedde* 'in bed' and *on life* (in which the dative form has survived fossilized in Modern English *alive*)—a distinction that had disappeared by 1500. In early Modern English the functions of the subjective form of the second person pronoun (*ye*) were transferred to the objective form (*you*), but this did not occur in any other personal pronouns.

Syncretism is now used in synchronic descriptions.

> 1974 P. H. MATTHEWS In *He came* and *He has come* we distinguish a Past Tense *came* and a Past Participle *come* . . . In *He tried* and *He has tried* the first '*tried*' must again be Past Tense and the second again the Past Participle; for TRY (as for most English verbs) the two forms are identical both in spelling and in phonetics. The term 'syncretism' (in origin a term in diachronic linguistics) is often applied synchronically to this situation.

syndetic

Of coordination: indicated by means of conjunctions. Contrasted with ASYNDETIC.

Syndetic coordination is the more usual form of coordination.

synonym

A word or phrase that means the same or almost the same as another in the same language.

• **synonymous**: in a relationship of synonymy *with* another word. **synonymy**: the fact of having (almost) the same meaning.

Strictly speaking there are few, if any, 'true' synonyms, that is, words that are completely and always interchangeable. But pairs of words such as *close* and *shut* are sufficiently alike to rank as synonyms, even though one cannot be substituted for the other in, for example, *I'm going to close my bank account*, *The meeting closed with a vote of thanks*, or *The water supply was shut off*.

In discussion of the semantics of a linguistic unit longer than the word, *synonymy* tends to refer to the identity of denotational or referential meaning, even though the emphasis or affective meaning may be different.
Compare ANTONYM.

synsemantic

Semantics. (*n. & adj.*) (A word or phrase) that has meaning only in a context.
See AUTOSEMANTIC.

syntactic

Of or relating to SYNTAX.
In grammar generally, no special need is felt to use this adjective. A *syntactic class* is usually called a *word class* or simply a *class*; a *syntactic function* can simply be referred to as a *function*.
In Generative Grammar, however, the *syntactic component* contrasts formally with the *semantic* and *phonological* components in the general framework. *Syntactic structures* are generated, there are *syntactic categories* (classes), *syntactic rules*, *syntactic relationships*, *syntactic specifications*, and *syntactic features*.

syntactics

Syntax, particularly when contrasted with *pragmatics* and *semantics* as a subdivision of SEMIOTICS; the formal relationship of signs to each other.

> 1964 E. A. NIDA While semantics deals with the relationship of symbols to referents, syntactics is concerned with the relationship of symbol to symbol; for the meaning of expressions is not to be found merely in adding up symbols, but also in determining their arrangements, including order and hierarchical structuring. For example, the constituents *black* and *bird*, when occurring in juxtaposition, may have two quite different meanings.

syntagm (Pronounced /ˈsɪntæm/.) (Also **syntagma**, pronounced /sɪnˈtægmə/.)

(A set of linguistic forms in) a serial relationship. Contrasted with a PARADIGM.

Syntagms, the constitutents of a larger unit, are in an *and*-relationship (a chain relationship) with other forms (see CHAIN).

> 1959 W. BASKIN [translating F. de Saussure, coiner of the term] In discourse, . . words . . . are arranged in sequence on the chain of speaking. Combinations supported by linearity are *syntagms*. The syntagm is always composed of two or more consecutive units.

Syntagms operate at all levels of linguistic analysis. Thus phonemes join into syntagms to produce words (e.g. /b/ + /ʊ/ + /k/ = book); morphemes too join into syntagms to produce words (*book* + *ish* = *bookish*). And words in a chain relationship form clauses.

syntagmatic

Of or pertaining to a syntagm or syntagms.

Syntagmatic relationships are contrasted with PARADIGMATIC ones. Thus in *All power corrupts*, *all* + *power* + *corrupts* are in a syntagmatic relationship with each other; the constitutents can be analysed in terms of syntactic function, in this example a subject (consisting of a determiner and noun) plus a verb. Words that could grammatically substitute for each of the three words in the example are in a paradigmatic relationship with that particular word: thus *some*, *no*, *more*, etc. form a paradigm with *all*; a great many nouns are in a paradigmatic relationship with *power* (since we could substitute *control*, *strength*, *weakness*, *conversation*, *art*, or any of a large set of uncount nouns for *power*); and, for the same reason, an indefinite number of intransitive verbs are in a paradigmatic relationship with *corrupts*, as indeed are several other tenses of that verb (e.g. *has corrupted*, *will corrupt*).

• **syntagmatically**: in the manner of a syntagm.

1961 Y. OLSSON Both collocation and colligation operate syntagmatically, that is, along the line one-after-another.

syntax

The arrangement of words in sentences, and the codified rules explaining this system.

Traditionally *syntax* (the structure of sentences) is one of the two main branches of language study, alongside MORPHOLOGY (the structure of words). The extension of linguistic studies to cover phonetics, phonology, semantics, discourse, etc., and the recognition of the complexity of the relationship between syntax and morphology, might make this twofold division appear oversimple. Nevertheless, syntax is still basic to language study.

synthetic

1 Designating a language in which a word usually contains more than one morpheme. Contrasted with ANALYTIC.

English is an example, though not an extreme one, of a synthetic language.

Synthetic here is used in its etymological sense 'put together'. It does not mean 'artificial'.

2 *Semantics.* Designating a proposition or sentence that is true (or false) by virtue of extralinguistic facts or circumstances, in contrast to an ANALYTIC one.

system

1 (Generally.) A network of parts in an orderly arrangement; a regular set of relationships.

Thus the English language itself is a system, consisting of syntactic, phonological, and semantic systems, which in turn contain other systems (e.g. the vowel and consonant systems).

2 (Specifically.) A group of terms or categories, particularly in a closed, paradigmatic relationship.

1953 R. H. ROBINS Professor J. R. Firth has recently suggested that the terms 'Structure' and 'System' be kept apart in the technical vocabulary of linguistic description. 'Structure' might be used to refer to unidimensional, linear abstractions at various levels from utterances or parts of utterances . . . When . . categories have been devised by means of which the utterances of the language can be successfully described and analysed, closed systems are formed of these categories.

Closed word classes are also often described as systems; e.g. the article system, the pronoun system (see CLOSED (1)). It is not usual to describe open word classes as systems.

In Scale-and-Category Grammar, *system* has a special place as one of the three (or four) categories that contrast with scales in the organization of the grammar.

1985 G. D. MORLEY The fourth category, system, accounts for the range of choices (classes) which are available within a unit, and any given range of possible options is known as a set of 'terms'. Thus, for example, the relations between the terms from the system of mood . . . may be set out as follows.

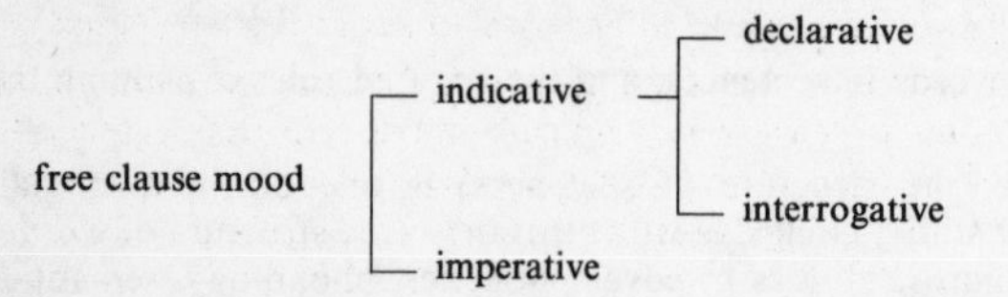

Other systems included in this analysis are simple versus non-simple sentences (which in turn have the subsystems complex, compound, and compound complex); number; person; voice; and so on.

systemic grammar

A later form of Scale-and-Category Grammar. (Also called **systemic-functional grammar**.)

Scale-and-Category Grammar, like early Transformational Grammar, was much concerned with syntax and structure, but as the theory developed the meaning and social functions of language became increasingly important.

1985 G. D. MORLEY During the latter half of the 1960s Halliday's work became increasingly influenced by ideas on the functional nature of language . . and a multifunctional semantic dimension was not merely added to systemic theory but became central to it . . . At the time of this reorientation, the theory became known as systemic functional grammar, or systemic grammar for short. (Many linguists, indeed, now use the name 'systemic grammar' in referring to all work in the Hallidayan mould since 1961.)

T

tag

A short phrase or clause added on to an already complete utterance.

Different types of tag are distinguished, for example:

(*a*) A noun phrase referring back to an earlier pronoun, for emphasis, e.g.

They use some confusing terms, *these grammarians.*

or to add an exclamatory comment, e.g.

He's won another prize, *clever man*!

or—perhaps more colloquially—with a verb, e.g.

They baffle you, *do those long words*
That was the week, *that was*

(*b*) Phrases such as *etc.*, *et cetera*, *and so on*, used to avoid further listing.

(*c*) Short questions (QUESTION TAGS). These usually take the form of an auxiliary verb + pronoun which are syntactically related to the sentence to which they are added. Sentences of this type are typical TAG QUESTIONS. But other types of question tag are found, e.g.

It cost £1000, *did you say?*
That's not exactly cheap, *would you think?*
Tell him to take it back, *why don't you?*
Or suppose I tell him, *shall I?*

Some grammarians include short comments and answers added by another speaker, when these too consist of a subject noun or pronoun and an auxiliary verb relating to the verb of the previous utterance. For example:

It costs £1000. *Does it?/It doesn't!/So does the small one*
That's not cheap. *Nor are the other ones*
Shall I tell him? *Yes, do*

tag question

An interrogative sentence formed with a question-tag.

1990 J. ALGEO There are at least five uses of tag questions, some of them characteristically British, showing a progressive decline in politeness and in the degree to which they draw the addressed person into the conversation.

Algeo's five types of tag question are:

You haven't still got that map I lent you, have you? (informational)
But we wouldn't be able to use it to find the post office, would we? (confirmatory)
Took you a while to realize that, didn't it? (punctuational: i.e. similar in function to an exclamation mark)

Well, I haven't seen it before, have I? (peremptory: i.e. fending off an unwelcome remark)
You didn't exactly give me a chance to show you it, did you? (aggressive: i.e. expressing a hostile reaction)

See QUESTION-TAG.

tail

Phonetics. That part of a TONE UNIT that comes after the nucleus.

The nuclear syllable of a tone group, the syllable on which the main pitch change occurs (the *tonic syllable*), may be the final syllable of an utterance, in which case there is no tail. Often however there is a short tail, consisting of one or two unstressed syllables, as in

Have you for´gotten?
Of course, I re˯member it

But a tail can contain stressed words (but without pitch change), e.g.

Well, `say something, ˌthen

tap See FLAP.

tautology

The saying of the same thing over again in different words (particularly as a fault in style).

Language necessarily contains some REDUNDANCY, and many speakers and writers repeat themselves for emphasis. Tautology, by contrast, is usually an unnecessary, and probably unconscious, repetition, as in *an unmarried young bachelor*, *one after the other in succession*, or *They shut and closed the door*.

taxis (Plural **taxes**.)

Order or arrangement of words.

An outdated term; but see HYPOTAXIS and PARATAXIS.

taxonomic linguistics See LINGUISTICS.

temporal

Relating to time.

Temporal is particularly used in relation to the meaning of some adjuncts, conjuncts, and prepositions (e.g. *then*, *meanwhile*, *before* etc.).

Temporal meaning is often contrasted with SPATIAL meaning, but many words have both meanings (e.g. *in* a box, *in* July).

Compare TIME.

temporary

Relating to an activity, attribute, etc., that lasts for a limited time. Contrasted with *permanent*.

Temporary is particularly used in relation to verbal meaning. Progressive aspect is often said to convey *temporary meaning* (e.g. *She's working in a bank*) in contrast to more permanent meaning (e.g. *She works in a bank*).

Similarly *temporary* and *permanent* qualities often, though not necessarily, correlate with the predicative and attributive uses of adjectives. Contrast:

> It's a long *involved* story (an inherent quality)
> Ask the people *involved* (temporarily, on this occasion)

Compare DURATION (2).

tense

(*n.*) A form taken by a verb to indicate the time at which the action or state is viewed as occurring; the quality of a verb expressed by this.

Traditionally, *tense* is defined in terms of time. But labels such as PAST, PRESENT, and FUTURE tense are misleading, since the relationship between the tenses is more complicated than the labels suggest. Past and present tenses can be used in some circumstances to refer to future time (e.g. *If he came tomorrow . . .*, *If he comes tomorrow . . .*), present tenses can refer to the past (as in newspaper headlines, e.g. *Minister resigns*, and in colloquial narrative, e.g. *So she comes up to me and says . . .*), and so on.

> 1975 T. F. MITCHELL Some linguists argue quite plausibly that there is no coherent relation between tense and time and would rather relate tense distinctions to speech function and attitude.

Some linguists define *tense* narrowly by form, which gives English only two tenses: the *present tense*, which in lexical verbs is the same as the base (except for the *-s* ending in the third person singular); and the *past tense*, which in regular lexical verbs has the *-ed* inflection.

In terms of meaning, the present tense is then defined as the unmarked tense, which is timeless in the sense that it can embrace any time that does not exclude the speaker's time (hence its use for general truths) and any time that the speaker does not want to distance himself from. The past tense is then defined as the marked tense, marked for separation from the speaker's 'now', or to indicate the hypothetical nature of the statement, or to convey SOCIAL DISTANCING.

More generally, many verb phrase combinations that incorporate features of ASPECT, MOOD, and VOICE are treated as part of the tense system, giving such tenses as past subjunctive (e.g. *If I were you . . .*), present progressive passive (*Are you being served?*), and so on.

Compare COMPOUND TENSE.

tense

(*adj.*) *Phonetics*. Articulated with more effort than usual.

See LAX.

tensed

Of a verb form: having tense.

Tensed is sometimes favoured instead of the more usual FINITE. Thus the past and present tense forms (*wanted*, *wants*, *want*) are *tensed*, and the base form, *-ing* form and *-en* form (*want*, *wanting*, *wanted*) are *non-tensed* (rather than *non-finite*).

Similarly, clauses can be described as tensed or non-tensed, on the basis of the verb form they contain.

tentativeness

The quality of being said provisionally or experimentally; one of the meanings conveyed by MODALS or by the use of the ATTITUDINAL past tense.

e.g.

I might be able to help
I was hoping you would come tomorrow

terminal

Occurring at or forming the end of something; final.

Especially used in the the terminology of Generative Grammar, e.g.

•• **terminal element, terminal symbol**: any of the elements or symbols occurring in the terminal string.

terminal string: the last series of symbols from which a sentence is derived.

territory

Used with a qualifying word to indicate the part of a sentence influenced by what the qualifying word refers to, as in ASSERTIVE *territory* (and *non-assertive territory*) and in SUBJECT TERRITORY and OBJECT TERRITORY.

tertiary

1 *Phonetics & Phonology*. Designating a level of stress weaker than SECONDARY stress.

Tertiary stress is posited by some analysts, though many restrict themselves to *primary* and *secondary*. It is marked [ˌ], as in *neologism* /nɪˈɒləˌdʒɪz(ə)m/ or *five o'clock shadow* /ˌfaɪvəˌklɒkˈʃædəʊ/.

2 (*n. & adj.*) (In some older grammar.) (Designating the RANK or level of) a linguistic unit that is third in order of importance.

Thus in *terribly cold weather*, *terribly* is tertiary. See RANK.

text

A piece of written or spoken language.

Text is intended to be a neutral term for any stretch of language, including transcribed spoken language, viewed not so much as a grammatical unit but as in some way a semantic or pragmatic unit.

Compare UTTERANCE.

text linguistics

The study of a 'communicative' text, rather than grammatical sentences, as the basis for language analysis.

> 1977 T. REINHART The rapidly growing school of 'text-linguistics' . . . The general belief shared by these scholars is that the 'natural domain' of linguistic theory consists of discourses, or texts, rather than sentences. However, . . this belief is not what distinguishes text-linguistics from other discourse-oriented . . trends in linguistics . . . Text-linguistics differs from these approaches in its interpretation of the claim that texts are the natural domain of linguistics. For generative text-linguists, this means that the grammar must actually generate (all and only) possible well-formed texts of the language.

Text linguistics is sometimes rather similar to DISCOURSE studies: the aim being to observe such devices as coherence and cohesion over a unit larger than a sentence. On the other hand, text linguistics also embraces short texts, such as signs (e.g. *No Entry*), which discourse studies might not.

textual

Of meaning: as structured in the text itself.

In Systemic-Functional Grammar, meaning can be analysed into three types: the IDEATIONAL, the INTERPERSONAL, and the *textual*. The textual meaning is concerned with the 'clause as message', the way this is structured into theme and rheme, and the relationship of the clause to its context.

that-clause

Primarily, a nominal clause beginning with *that*, or where *that* could be inserted.

e.g.

> That you believe such nonsense amazes me
> I'm sorry (that) you believe such nonsense

See ZERO *that-clause*.

Although some relative clauses begin with *that*, e.g.

> What's all this nonsense *that you're repeating*?

relative clauses are not always included in this category. Appositional clauses beginning with *that* may or may not be included.

> The idea *that you believe this* distresses me

thematic

Of, pertaining to, constituting, or designating the THEME.

1980 E. K. BROWN & J. E. MILLER In imperative sentences the thematic element is normally the main verb.

• **thematically**.

theme

That part of clause structure which establishes the subject-matter or viewpoint of the clause, about which more important information (the RHEME) will be stated.

• **thematization**. **thematize**: convert (an element) into the theme of the clause or sentence.

The THEME and RHEME analysis is a way of looking at information structure, not at functional grammatical elements. The theme always comes at the beginning of a clause or sentence. Various grammatical devices are available in English to thematize different elements if the grammatical subject is not the theme. One is the use of the passive to thematize an object:

Trespassers will be prosecuted
Mistakes cannot afterwards be rectified

Compare also:

1985 R. QUIRK et al. *-ing* clauses with thematization: He's worth listening *to*.

where the object of a preposition (*It's not worth listening to him*) has been turned into the theme.

Thematizing is also called FRONTING or TOPICALIZATION. See TOPIC.

theoretical grammar See GRAMMAR.

theoretical linguistics See LINGUISTICS.

there-existential See EXISTENTIAL.

third person

(Denoting, or used in conjunction with a word indicating) the person or people spoken about, in contrast to the speaker or addressee.

Third person pronouns and determiners are *he, him, himself, his; she, her, herself, hers; they, them, themselves, their, theirs.*

The third person singular form of the present tense of primary and lexical verbs is marked in writing by *-s* or *-es* (e.g. *is, has, does, wants, fixes*).

Compare FIRST PERSON, SECOND PERSON.

three-part verb

A MULTI-WORD VERB with two particles. (Also called *three-part word.*)

This is another way of describing a phrasal-prepositional verb, e.g. *look down on* (= despise), *put up with* (= endure).

time

1 Designating an adjunct, adverbial, or preposition relating to time; TEMPORAL.

Time adjuncts, *adverbials*, *prepositions*, and so on, are the same as TEMPORAL *adjuncts* etc. Time adjuncts are often subdivided under such categories as DURATION and FREQUENCY.

•• **time clause**: an adverbial clause relating to time and introduced by a temporal (time) conjunction (such as *when*, *while*, *after*, *until*, *since*, *as long as*, and *once*) e.g.

Once we receive the money, we will send out the orders

Some time clauses are non-finite, e.g.

While smoking in bed, he had fallen asleep
He awoke *to find the bedclothes on fire*

or verbless

He should fix a smoke alarm *as soon as possible*

For the relationship between time and tense, see TENSE.

2 **time phase**: see PHASE.

timed See FOOT-TIMED, STRESS-TIMED, SYLLABLE-TIMED.

timeless

Not concerned with, or not limited by, time.

Some uses of the so-called PRESENT tense (e.g. *water freezes at O° Celsius*) are described as *timeless*.

tip

Phonetics. The point of the tongue, behind which are the blade and the front.

The tip is called the APEX in some descriptions.

English /t/, /d/, /θ/, and /ð/ are usually articulated with the tongue tip.

tmesis (Pronounced /tə'miːsɪs/. Plural **tmeses**.)

The separation of parts of a word by an intervening word or words.

This is not a very productive operation in English, being largely confined to the insertion of swearwords for greater emphasis, as in *I can't find it any-blooming-where*.

The phenomenon is now usually described by using INFIX.

to-infinitive

The infinitive preceded by *to*. Contrasted with the BARE INFINITIVE.

The *to*-infinitive (or a *to*-infinitive clause) is used:

after many catenative verbs: *I want to know*
as a nominal: *To know all is to forgive all*
as an adverbial clause: *Pull tab to open*
as a post-modifier: *a book to read, nothing to do*
as an adjective complement: *nice to know, hard to imagine.*

Compare SPLIT INFINITIVE.

tonal

Of or pertaining to TONE.

1973 R. QUIRK & S. GREENBAUM Contrastive stress often involves moving a tonal nucleus from its normal, unmarked position on to the contrasted item.

tone

Phonetics. A distinctive pitch.

In any complete utterance, however short, there is one particularly prominent syllable, prominent not only because it is stressed, but because it carries a change of pitch, usually a FALL or RISE (or a more complicated variant) but occasionally a LEVEL pitch. This prominent pitch is called the *nuclear* tone, and the syllable is the *tonic syllable.* It forms the nucleus of a TONE UNIT.

In some languages different tones (pitch patterns) on an identical syllable can produce words of totally different meanings. Such languages (e.g. Mandarin and other languages in South-East Asia and Africa) are *tone languages.* In non-tone languages such as English (and other European languages) objective word meanings are not affected by intonation, although different tones can convey different attitudes. Thus *All right* with differing intonation can convey grudging acquiescence, enthusiastic agreement, a question, sarcastic disagreement, and so on.

Tones are variously analysed, but a common analysis of standard English gives high and low FALLS, high and low RISES, RISE-FALL, FALL-RISE, and LEVEL.

What is loosely in everyday parlance called 'tone of voice' depends on various features, and not merely *tone* in the technical sense. Loudness, for example, and perhaps a generally higher pitch might contribute to an angry or excited 'tone of voice'.

tone unit

Phonetics. The basic unit of intonation. (Also called **tone group**.)

(*a*) Intonation can be analysed in different ways, but in the mainstream British tradition the *tone unit* or *tone group* is basic, rather like the clause or sentence in the analysis of syntactic structure.

A tone unit/group, by definition, must contain a *nuclear tone* (a NUCLEUS), that is a marked PITCH change. Optionally, it may contain a PRE-HEAD and/or a HEAD before the nucleus, and a final TAIL, e.g.

I've ˌjust ˎtold you

A tone unit/group can cover a complete clause or shortish sentence if said with no special emphasis, but in a more lively style, speakers use frequent changes of pitch, with a succession of short tone units/groups. This is a noticeable characteristic of many television announcers.

(*b*) In some analyses of intonation attempts are made to describe intonation over a greater length of discourse or 'sequence of tone groups'. Distinctions are made between a *major tone unit/group*, roughly corresponding to a sentence, and a *minor tone unit/group*, based more on a phrase or individual word. In these cases terms such as *breath group*, *sense group*, and *tone group* itself may be used with more specialized meanings.

tongue

Phonetics. The principal organ of speech.

The tongue is involved in some way in the production of most speech sounds, and therefore figures in articulatory descriptions.

Vowel articulations are described in terms of tongue HEIGHT, and whether the FRONT or BACK or CENTRE of the tongue is highest.

For consonants, the tongue is theoretically divided into various parts, so that whichever part is important in the articulation can be named. Sometimes the ROOT is named, then slightly further forward is the BACK, then the FRONT (but see the note below), then the BLADE and, furthest forward of all, the TIP (or APEX). The sides are called the RIMS. But note that this nomenclature is unfortunately not universal, so that sometimes the blade is called the front, and the area behind that (front in other systems) is called the centre or top.

[See diagram p. 448]

tonic

The same as NUCLEAR. See TONE.

- **tonicity**

> 1975 T. F. MITCHELL One may find in tonicity . . a means of recognizing the constituency of idioms, as for example that of *put up with* = tolerate, any of whose three constituents may be tonic in appropriate grammatical circumstances: cf *A. You'll have to put úp with it. B. It's not the kind of thing you can pút up with*. This is tonicity in a context of repetition or 'second mention' of a lexical item.

topic

That part of a sentence about which something is said. Contrasted with COMMENT.

The *topic* and *comment* distinction, like THEME and RHEME or GIVEN and NEW, is a way of analysing the information structure of a sentence.

> 1958 C. F. HOCKETT The speaker announces a topic and then says something about it . . . In English and the familiar languages of Europe, topics are usually also subjects and comments are predicates.

However, although the topic frequently coincides with the subject and the comment with the predicate, as in

The land / lies in the Green Belt

the topic can be some other grammatical element, e.g.

At Layhams Farm, it is now proposed to construct a double artificial ski slope (*place adverbial* as topic)
Recreational it may be, but no development could be more inappropriate (*adjective complement* as topic)
More building we do not want (*object* as topic)

The topic and comment distinction is very similar to the theme and rheme distinction, so that putting the topic (or theme) at the front is variously described as TOPICALIZATION, THEMATIZATION, and FRONTING.

topicalization

The making of an element into the topic of a sentence.

• **topicalize**.

trachea

Phonetics. The windpipe.

traditional grammar See GRAMMAR.

traditional orthography

Standard spelling.

The term (abbreviated *t.o.*) is used by spelling reformers who would like to introduce simpler and 'more logical' spelling.

transcribe

Write (spoken language) in phonetic symbols.

See TRANSCRIPTION.

transcription

Phonetics & Phonology. The representation of spoken language in phonetic symbols.

The aim of transcription is to indicate speech sounds consistently. This is partly to overcome the vagaries of spelling: many dictionaries give standard pronunciations of a word and common variants. But transcription also makes it possible to represent accurately the assimilations and elisions of actual speech, and (if required) the idiosyncracies of an individual's speech on a particular occasion.

The most widely used script (or NOTATION) is the International Phonetic Alphabet, usually with adaptations according to the level of accuracy required, and according to the particular purpose of the transcription. Transcriptions are primarily PHONETIC or PHONEMIC. A phonetic transcription aims to represent actual speech sounds objectively and accurately, according to articulatory and auditory criteria. A high degree of accuracy can be achieved with special additional symbols if necessary and diacritics indicating such things as aspiration or the nasalization of vowels. A very detailed transcription is a NARROW transcription; one with few details is BROAD. A phonemic transcription is the broadest of all. It uses only one symbol for each phoneme of a language, regardless of actual variations. It therefore looks simple, but requires a knowledge of the allophonic conventions if it is to be reproduced aloud with approximate authenticity. It is phonological rather than objective.

A further complication is that there is often more than one way of transcribing phonemically. Thus the English vowels heard in *heat* and *hit* may be transcribed as /iː/ and /i/, showing a distinction of length only, or as /i/ and /ɪ/, showing a distinction of vowel quality only, or as /iː/ and /ɪ/ showing both. The vowel in the words *oh*, *slow*, etc. used to be represented as [oʊ], but is now more usually shown as [əʊ].

[See example p. 448]

transferred negation See NEGATION.

transform

(*n.*) A level of structure—possibly somewhat abstract—derived by the application of a transformation. Also, occasionally, the same as TRANSFORMATION.

1962 B. M. H. STRANG It is sometimes assumed that all positive and affirmative sentences have negative and interrogative transforms. This is not quite true.

(*v.*) (Cause (a basic structure) to) change into a grammatical but less elementary structure.

transformation

A rule-governed operation that converts a basic structure into an acceptable but less elementary one.

1955b N. CHOMSKY A sentence *X* is related to a sentence *Y* if, under some transformation set up for the language, *X* is a transform of *Y* or *Y* is a transform of *X*.

transformational

Of or pertaining to a transform or transformation.

In the early theory of Transformational Grammar, *transformational rules* operated on abstract structures, but in more general parlance transformational rules explain the systematic relationships between various types of clauses and structures.

Typical transformations, based on transformational rules, produce questions from declarative sentences, passives from actives, and so on, and show the regular grammatical relationships between such pairs.

Perhaps a more important claim for transformational rules is that they can disambiguate structures that a purely surface grammar (such as Immediate Constituent Analysis) cannot. Thus two superficially identical structures, such as

They need helping
They like helping

can be shown to be derived, by different rules, from two different underlying abstract bases (which might very roughly mean 'they need + for someone to help them' and 'they like + when they help other people').

•• **transformational component**: see COMPONENT.

• **transformationalism**: transformational theory. **transformationalist.**

transformational grammar

A theory of grammar in which transformational rules form an essential part. Transformational rules as such were first introduced by Noam Chomsky in his *Syntactic Structures* (1957) (although he had already treated the subject in his doctoral dissertation of 1955). As this book also introduced the idea that the rules of a grammar should generate grammatical sentences, this type of grammar, contrasted with structuralism and other, more traditional grammatical models, is variously known as *transformational grammar*, *generative grammar*, and *transformational-generative grammar*, or *TG* for short.

TG is particularly concerned with DEEP and SURFACE STRUCTURES; later models that are less concerned with this, and more concerned with semantics, tend to drop the transformational label. But all three terms are somewhat loosely applied.

Compare GENERATIVE.

transition

Phonetics. A glide from one sound to another.

The concept of transition is somewhat technical, often being measured in terms of FORMANTS. In the detailed description of speech sounds, even a single phoneme can be analysed into several parts. Thus a plosive (or stop) consonant has three stages: the closing stage, the hold stage, and the release (or explosion) stage. In the first stage a *transition* (or *on-glide*) may link the preceding sound to the beginning of the plosive, and in the final stage another *transition* (this time an *off-glide*) may link the plosive to the following sound.

transitional

Of a word or words: indicating a change from one state, place, etc. to another.

This is not a widely used grammatical term, but is sometimes applied to conjuncts that semantically bridge a gap from the subject-matter of one statement to that of another; e.g.

meanwhile, in the meantime, incidentally

It is also applied to the meaning of a verbal form that indicates little or no duration, with a change of state about to result (e.g. *The bus was stopping*).

transitive

Of a verb: that takes a direct object. Contrasted with *intransitive*.

The division of verbs into *transitive* and *intransitive* is long established. (Both are to be distinguished from COPULAR verbs.)

Some verbs are virtually always transitive (e.g. *bury*, *deny*, *distract*). Others are virtually always intransitive (e.g. *arrive*, *come*, *digress*). But many can be both; e.g.

> I was cooking (breakfast)
> He lodged in Cambridge; He lodged a complaint

Even verbs that seem to be strongly transitive (e.g. *He made a cake/a mistake/a good job of it*) can have intransitive uses (e.g. *She made towards the river*), and similarly an intransitive verb such as *live* can be used transitively (e.g. *She lived a good life*). So for many verbs it is more accurate to talk of transitive and intransitive use.

Transitive verbs can be grammatically divided into three main types:

> MONOTRANSITIVE, i.e. simple transitive verbs, taking one object; e.g. *I've bought a new suit* (SVO)
> COMPLEX TRANSITIVE, e.g. with an object plus a complement or adverbial; e.g. *I found the story unreadable* (SVOC), *I put the book down* (SVOA)
> DITRANSITIVE, with an indirect and a direct object; e.g. *I've bought myself a new suit* (SVOO)

• **transitively**. **transitivity**.

transparent

Obvious in structure or meaning; that can be extrapolated from surface structure; (of a phonological rule) that can be extrapolated from every occurrence of the phenomenon, in which every context implies the rule. Contrasted with OPAQUE.

> 1977 I. DOWNING A compound may be highly transparent semantically when it is coined.

tree diagram

A diagram that shows the syntactic structure of a clause or other linguistic unit.

A tree diagram does not have to be tied to any particular theory of grammar. Thus the sentence *He looked up the words in his dictionary* could be diagrammed as

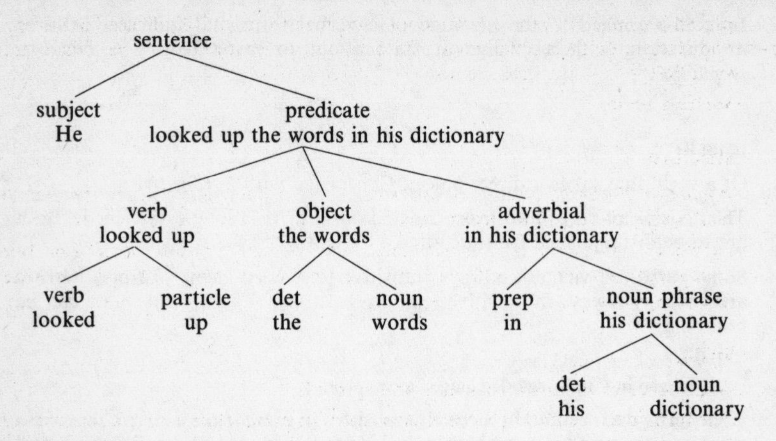

or the traditional division into subject and predicate could be omitted and the first split could be fourfold

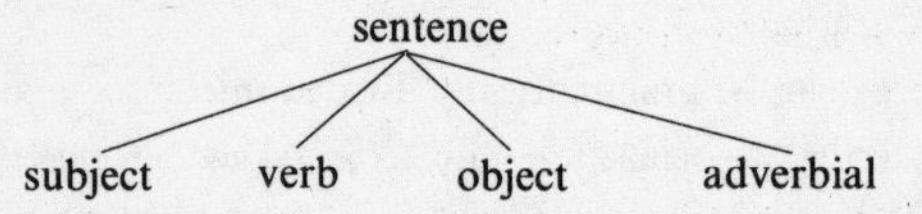

Tree diagrams are much used in Generative Grammar, where the labelling tends to be more abstract. *Looked*, for example, might appear as 'look + past', and the actual sentence would only appear at the bottom, having been generated by PHRASE-*structure rules*.

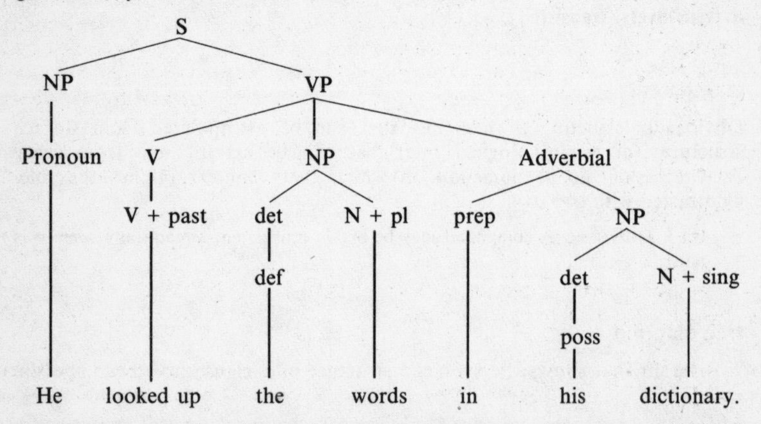

The root of the tree is at the very top, i.e. the S (for sentence); the other points from which lines branch off are NODES.

trigraph

Three letters representing one speech sound as in *manoeuvre*, where the *oeu* is pronounced /uː/.

trill

The same as ROLL.

triphthong

A vowel sound in which the vocal organs move from one position through a second to a third.

There are no triphthongs among the English phonemes, but such sounds occur when a closing diphthong is followed by /ə/. At least, they theoretically occur in a careful pronunciation of such words as

player /pleɪə/, shire /ʃaɪə/, royal /rɔɪəl/, slower /sləʊə/, hour /aʊə/

However, the glides between the elements of such triphthongs may be very slight, and the sounds actually articulated and heard are often more like diphthongs or even single long vowels.

• **triphthongal**.

trisyllabic

Having three syllables.

As with the related terms MONOSYLLABIC and DISYLLABIC, the term is particularly used of adjectives and adverbs. Trisyllabic or longer adjectives and adverbs have to take periphrastic comparison (e.g. *more delicious, most extraordinary, more hastily*).

tune

Phonetics. The pitch pattern heard over a whole tone unit.

With an utterance consisting of a single syllable (e.g. *Yes!*), *tune* and tone unit are the same, so the terms may be confused. A *tune*, however, depends on the overall pitch pattern and the height of any prehead or head (i.e. whether this is high or low).

two-part verb

A MULTI-WORD VERB consisting of a verb plus a prepositional particle (a PREPOSITIONAL VERB) or a verb plus an adverbial particle (a PHRASAL VERB). (Also called *two-part word*.)

U

ultimate constituent See IMMEDIATE CONSTITUENT.

umlaut

1 The same as MUTATION.

2 The diacritic ⟨ ¨ ⟩, which indicates mutation in German.

This same diacritic with another function is labelled DIAERESIS.

unacceptable See ACCEPTABLE.

unanalysable See ANALYSABLE.

unaspirated See ASPIRATED.

unattached participle See HANGING PARTICIPLE.

uncount

Designating a noun that has no plural form and cannot be used with numerical values; uncountable. Contrasted with COUNT. (Also called *uncountable*, *non-count*.)

Grammatically *uncount nouns* are distinguished by the fact that they can be used without any determiner or article and with certain determiners that are exclusive to them (e.g. *much*). Uncount nouns often refer in a rather general way to substances and abstract qualities, processes, and states (e.g. *china*, *petrol*, *poverty*, *rain*, *welfare*) rather than discrete units. But the *uncount* versus *count* distinction is grammatical, not semantic, and a number of English uncount nouns (e.g. *information*, *luggage*, *news*, *traffic*) have countable equivalents in other languages.

Uncount is generally synonymous with *mass*; but see MASS NOUN for a distinction sometimes made.

uncountable

(*n.* & *adj.*) (A noun that is) uncount.

See UNCOUNT.

underlying

Designating abstract 'deep' features posited to explain various relationships or meanings in the actual language.

In Generative Grammar, concepts such as *underlying form/structure/phrase-marker/string* are part of the apparatus of the grammar. Underlying structures are used, among other things, to explain why we sometimes have to interpret similar structures in different ways, while two very different structures may be understood to mean the same thing. See DEEP STRUCTURE.

Underlying structures are often said to contain elements not present in the 'surface' language. For example, in early Transformational Grammar the underlying structure of an imperative sentence like *Sit down* was supposed to be *You will sit down*. From this underlying structure the surface structure *Sit down* was derived by the application of transformational rules that deleted *you* and *will*.

In a less technical way (without involving transformational rules) various structures may be explained in terms of some underlying structure. Thus a postposed prepositional phrase, as in *the man in the iron mask* may be explained as having an underlying clause (i.e. *who was in the iron mask*). Or a genitive may be labelled subjective or objective on the basis of its meaning, e.g.

a hair of the dog (subjective: *the dog* has hair)
love of money (objective: someone loves *money*)

The concept of underlying structures can be seen as a more abstract extension of the UNDERSTOOD concept of traditional grammar.

• **underlyingly**.

understood

Of a word or words: deducible although not in the text.

The term *'you' understood* is sometimes used to describe the subject that is missing from most imperatives, but clearly implicit, as shown by the fact that it can be inserted or added in a tag, e.g.

(You) do as you're told
(You) be quiet
Don't (you) forget
Sit down, won't you

Understood is not entirely synonymous with RECOVERABLE. The latter is usually applied to words that could, with little or no change of form, be inserted in the text. *Understood* may relate more abstractly to UNDERLYING meaning. Thus a non-finite construction following a catenative verb has an *understood subject*: usually the same as the subject of the catenative, e.g.

They tried to telephone us

but occasionally someone else, as in the informal

We said not to worry

In neither of these cases could we insert the *understood* words (**They tried they to telephone us*, **We said they not to worry*).

unfulfilled condition

The same as *unreal* CONDITION.

ungradable See GRADABLE.

ungrammatical

Not conforming to the rules, deviant. Contrasted with GRAMMATICAL.

Grammaticality is judged in relation to what is considered a standard, but standards vary. *I never said nothing to nobody* is ungrammatical in standard English, but conforms to the rules of its own dialect.

Compare ILL-FORMED.

unheaded

The same as EXOCENTRIC.

unilateral

Phonetics. Of articulation: with the air released (rather unusually) around only one side of the tongue. Contrasted with BILATERAL.

uninflected See INFLECTED.

unique

Pertaining to something of which there is only one.

• **uniqueness.**

Unique reference is invoked to explain the use of the definite article (*the*) in various contexts where, although the referent has not been mentioned before, its definiteness may be assumed. Sometimes the referent really may be unique (e.g. *the Earth*); more often it is unique in the context of a particular place or time (e.g. *the Pope*, *the Queen*, *the moon*), or even in some much smaller situational context (e.g. *I'm going to the post office—please shut the windows*).

Uniqueness may also be due to grammatical or logical factors, e.g. *the day after tomorrow*; *May the best man win.*

unit

A discrete part of the linguistic stream at any level of analysis.

Linguistic unit is often used (as in this book) as a general term to cover word, phrase, phoneme, etc. without being more precise.

In Systemic Grammar, *unit* is one of the four theoretical CATEGORIES, along with STRUCTURE, CLASS, and SYSTEM.

See also INFORMATION UNIT, TONE UNIT.

unit noun

The same as a *partitive noun* (see PARTITIVE (*a*)).

A *unit noun* is a word that allows us to break up an uncountable noun into countable parts, e.g.

a *pat* of butter, two *pieces* of toast

universal

(*n. & adj.*) (A grammatical feature) that is common to all natural languages.

Features of English which seem basic to grammar, such as tense distinctions and prepositions, are non-existent in some languages. Theories of *universal grammar* and *language universals* which feature in all languages tend to be rather general; e.g. all languages use vowel and consonant sounds, and have ways of talking about time and place and distinguishing speaker and addressee.

unmarked

Of a linguistic feature: that is more basic, central, or usual than the MARKED form to which it is related.

1985 R. QUIRK et al. 'Measure' adjectives . . have two terms for the opposite ranges of the scale (*old/young*, *deep/shallow*, *tall/short*), but use the upper range as the 'unmarked' term in measure expressions.

For example, normal unmarked questions about age and height are

How *old* is the baby?
How *tall* is your little girl?

See MARKEDNESS.

unproductive See PRODUCTIVE.

unreal condition See CONDITION.

unrelated participle See HANGING PARTICIPLE.

unrounded See LIP POSITION.

unspecified *it* See IMPERSONAL.

unstressed See STRESS.

untensed See TENSED.

usage

Established and customary ways of using language.

Usage is a somewhat wider and somewhat vaguer term than *grammar* or *syntax*. In one sense, usage is what people generally say and write, how they actually use their language. Ideally therefore, *usage* should (*a*) include grammar and (*b*) be objective and descriptive, rather than prescriptive.

In practice, usage guides deal cursorily with consensual core grammar and pay most attention to areas of disputed usage, giving guidance that veers towards prescription (which is doubtless what most users of such books want). Grammatical usages discussed include such matters as *If I were* versus *If I was* or *used not to* versus *didn't use to*. Other areas are word formation and spelling (e.g. *blamable* versus *blameable*); pronunciation (*haRASSment* versus *HArass-ment*); and vocabulary (*flout* versus *flaunt*; *disinterested* versus *uninterested*; and the meaning of *decimate*).

Questions of usage are complicated by the fact that accepted usage may vary from one speech community to another, according to different national, regional, or social varieties, and such factors as who is writing or speaking to whom about what.

Many dictionaries employ *usage labels* to indicate whether particular senses, words, or phrases are formal, informal, British, American, dialectal, dated, slang, offensive, euphemistic, and so on.

A distinction is sometimes made between *usage* in the sense of what is grammatically and linguistically correct and acceptable (and of avoiding what is disputed) and *use*, i.e. what is appropriate to communication between people in an actual situation.

use See USAGE.

utterance

An uninterrupted sequence of spoken language.

Utterance is intended to be a more neutral term than the grammatically defined terms *clause* and *sentence*. *Utterance* is sometimes contrasted with TEXT, and sometimes included in it.

See also ECHO UTTERANCE, SPEECH ACT.

uvula

Phonetics. The fleshy extension of the soft palate hanging above the throat.

• **uvular**: (*Phonetics*) (of a sound) involving the uvula.

Uvular plosive consonants are heard in Arabic, and a uvular pronunciation of *r*, transcribed [ʀ] in the International Phonetic Alphabet, is standard in French. Uvular sounds are not a feature of standard RP, but uvular *r* may be heard in some northern accents.

V

V

1 Verb as an element in clause structure. See ELEMENT.

2 A symbol for a vowel in phonological structure.

valency

The number and kind of syntactic connections with other units that a linguistic unit can form.

> 1985 R. QUIRK et al. The term 'valency' (or 'valence') is sometimes used, instead of complementation, for the way in which a verb determines the kinds and number of elements that can accompany it in the clause. Valency, however, includes the subject of the clause, which is excluded (unless extraposed) from complementation.

value judgement

A judgement attributing merit or demerit to an action, event, etc.

One of many meaning categories assigned to DISJUNCTS. Thus in

> Foolishly, he didn't ask for a receipt

foolishly expresses the speaker's judgement of what happened (and does not describe the manner of asking).

Compare VIEWPOINT ADJUNCT.

variable

1 (*n. & adj.*) (A word) that has more than one form. Contrasted with INVARIABLE.

Variables include count nouns (with singular and plural forms); verbs (with third person singular present, present participle, past, and (in some verbs) past participle forms; e.g. *know*, *knows*, *knowing*, *knew*, *known*); and some gradable adjectives and adverbs (with comparative and superlative forms; e.g. *fine*, *finer*, *finest*; *soon*, *sooner*, *soonest*).

2 A word with variable reference, i.e. a word whose meaning is largely dependent on context.

For example, although personal pronouns contain meanings such as singular or plural, masculine or feminine, their referents are largely conditioned by the context. Thus the word *he* could in context refer to any male person (e.g. Shakespeare, the King, the man next door), but none of these 'meanings' are part of the dictionary meaning of the word *he*. And indeed *he* is not (outside politically correct circles) necessarily confined to male persons.

1984 R. HUDDLESTON *He* is used by many speakers as a 'variable' ranging over a set containing both males and females (normally human) as in . . *If any student wishes to take part in the seminar, he should consult his tutor*. The semantic distinction male vs female is here neutralised, and the fact that *he* is used, makes it the semantically unmarked member of the pair *he/she*. As we noted earlier, *they* has long been used as an alternative to *he* in this sense.

3 Any factor that contributes to differences between language VARIETIES.

e.g. *regional* or *social variables*.

4 **variable stress**: see STRESS (*c*).

variant See CONDITIONED *variant*.

variation

1 (The existence of) differences between VARIETIES of English.

Often with distinguishing word, such as *regional variation*, *stylistic variation*.

2 The existence of alternative linguistic forms within a single variety of English.

See FREE VARIATION.

variety

A distinct form of a language, used e.g. by a particular group or in a particular context.

The terms *variety* and VARIATION are particularly used in the analysis of different kinds of English. Thus we can talk of regional and social varieties (or variation); varieties according to the FIELD OF DISCOURSE; varieties consistent with spoken or written mediums; and 'stylistic' varieties, due to different degrees of formality, the attitude of the speaker, and so on.

Compare ATTITUDINAL.

velar

Phonetics. (*n. & adj.*) (A speech sound) made with the back of the tongue against the velum.

Standard English has three velar phonemes: the voiceless and voiced plosive pair /k/ and /g/ and the nasal /ŋ/, the sound at the end of *sing*.

The sound heard in the Scottish pronunciation of *loch* is a voiceless velar fricative [x].

See LABIO-VELAR.

velaric

Describing the air-stream mechanism used in the production of a CLICK.

In the velaric production of a sound, the tongue makes contact with the velum and is then drawn back, sucking air into the mouth.

velarization

The addition of a secondary, velar articulation to a speech sound.

See VELARIZE.

velarize

To add a secondary, velar articulation to a speech sound.

The so-called *dark l* allophone of the English /l/ is a velarized sound, articulated with the back of the tongue raised towards the velum. The symbol is [~] placed through the letter [ɫ]. Some English dialects have other velarized sounds. There are no phonemic distinctions of velarized versus non-velarized sounds in English.

velum

Phonetics. The soft palate.

The velum is the back part of the roof of the mouth, lying behind the bony hard palate, with the UVULA at its own back extremity. The velum is raised for ORAL sounds, and lowered for NASAL sounds.

verb

1 A member of a major WORD CLASS that is normally essential to clause structure and which inflects and can show contrasts of aspect, number, person, mood, tense, and voice.

In traditional grammar, the verb is sometimes defined notionally as a 'doing' word, but modern grammar prefers a more syntactical definition.

Verbs are usually subdivided first into:

(i) LEXICAL (or *full*) verbs
(ii) AUXILIARY verbs

Lexical verbs are further classified syntactically, depending on what accompanying elements are obligatory or permissible. The major types include TRANSITIVE, INTRANSITIVE, and LINKING (or COPULAR) verbs.

Auxiliary verbs are sometimes divided into PRIMARY and MODAL.

Compare MAIN VERB, MULTI-WORD VERB, PREPOSITIONAL VERB, REGULAR, TWO-PART VERB.

2 A major, and usually essential, element of clause structure.

In the representation of the functional elements of clause structure, *V* stands for the whole verb phrase (see VERB PHRASE (1)). Thus both *I bought oranges* and *I have been buying oranges* are SVO sentences. The verb element (V) is the only element that must always be filled by the same part of speech.

•• **verb of incomplete predication**: (an old-fashioned term for) a copular (linking) verb, especially the verb *be*, so called because such a verb is 'incomplete' without a complement.

See PREDICATOR.

verbal

1 Of, relating to, or derived from a verb.

•• **verbal adjective**: more usually called a PARTICIPIAL ADJECTIVE.

verbal conjunction: see CONJUNCTION.

verbal group: see VERB PHRASE.

verbal noun: the same as a GERUND.

2 Relating to words, particularly when spoken; ORAL (2). (As in *verbal ability*, *verbal abuse*).

Linguists usually try to avoid using the word *verbal* in this meaning, particularly when *oral* is more accurate (*oral communication*).

verbless

Of a clause: without a verb.

Verbless clauses are not usually recognized as such in much traditional grammar, where they are more likely to be dealt with as phrases. Some verbless structures, however, have some of the semantic and structural features of clauses. For example, some are introduced by a conjunction, e.g.

When in Rome, do as the Romans do
Come early *if possible*

Others have a subject introduced by *with* or *without*

With the exam behind her, she felt able to enjoy the holiday (compare *the exam being behind her . . .*)
Without you here, I don't know what I'd do (compare *if you were not here . . .*)

Others have neither conjunction nor subject, yet a paraphrase suggests a clausal rather than a phrasal interpretation, e.g.

Unhappy at the result, she decided to try again (compare *because she was unhappy . . .*)

•• **verbless sentence**: any type of MINOR SENTENCE lacking a verb.

Compare SMALL CLAUSE.

verb-particle construction

A phrasal or prepositional verb.

verb phrase (Abbreviated **VP**.)

1 A phrase consisting either of a group of verb forms which functions in the same way as a single-word verb or of a single-word verb on its own. (In Systemic-Functional Grammar called a *verbal group*.)

e.g.

has been thinking
must be leaving
was forgotten

having been urged
went

In a finite verb phrase, the first word is in fact the only word that is finite and indicates tense. The last word in both finite and non-finite verb phrases is the lexical verb. If a finite verb phrase consists of a single word, lexical word and tense are combined (e.g. *goes*, *went*).

Compare COMPLEX *verb phrase*.

2 (In Generative Grammar.) A sequence of words normally containing the verb and its complementation (i.e. virtually equivalent to the predicate), and forming, with the noun phrase that is its subject, a sentence. See TREE DIAGRAM.

> 1976 R. HUDDLESTON The . . forms will be classified as belonging to the category 'verbal group' (VGp), this term being distinguished on the one hand from 'verb', a category applying to single words, and on the other from 'verb phrase' (VP), which includes the object NP, etc., as well as the VGp. In *John will see Mary*, therefore, *will see Mary* is a VP, *will see* is a VGp, *will* and *see* are both verbs.

vibration See VOCAL CORDS.

viewpoint adjunct

An adverb qualifying the contents of a clause from a particular point of view. (Also called *viewpoint subjunct*.)

This is a subcategory of adverbials found only in some rather detailed classifications. E.g.

> *Morally*, the tax had much to commend it, but *politically* it was madness.

This kind of adverbial can be expanded by the addition of *speaking* (*Morally speaking* . . *but politically speaking* . .).

In classifications that call all adverbs *adjuncts*, adverbs of this kind are *viewpoint adjuncts*. In classifications which include the category *subjunct*, these are *viewpoint subjuncts*.

vocabulary

1 The entire set of words in the language.

2 A set of some of these.

(*a*) In Applied Linguistics, *vocabulary* is used in its everyday sense: there are books for foreign learners, for example, devoted to vocabulary, and most coursebooks have vocabulary sections.

(*b*) In Traditional, Structural, and Generative Grammar, vocabulary as such does not feature very much. The former two tended to concentrate on syntax and morphology, leaving word meanings to the dictionary. Early Generative Grammar also concentrated on syntax at the expense of meaning, but later versions sought to integrate word meaning into the theory of language, under such headings as the LEXICON or the SEMANTIC *component*.

There is an increasing trend towards the recognition of the interrelationship between vocabulary and syntax, and many modern dictionaries noticeably give a considerable amount of grammatical information.

Compare COLLOCATE.

(*c*) *Vocabulary size*. It is not easy to say how many words there are in the English language, partly because WORD itself is difficult to define, and partly because there are problems such as whether to include obsolete words, words in other internationally recognized varieties of English (e.g. Indian English, New Zealand English), slang or dialect words, and so on. A typical desk dictionary may define about 100,000 vocabulary items, while the *Oxford English Dictionary* lists rather more than 500,000, not all of which are fully defined.

As for how many words an individual English-speaking adult knows or uses, estimates vary greatly. Figures published in 1940 reporting tests on American college students claimed around 150,000, while some tests published in 1978 claimed no less than 250,000. Both figures have been challenged on various grounds, not least on what it is to 'know' a word. However the figure is likely to be many thousand.

> 1987 J. AITCHISON The number of words known by an educated adult, then, is unlikely to be less than 50,000 and may be as high as 250,000.

On the other hand, according to recent computer surveys, around 1000 common words account for over 70% of everyday speech and writing.

vocal cords

Two folds of muscle and connective tissue situated in the larynx, which are opened and closed during the production of speech. (Also *vocal folds*.)

Technically, the *vocal cords* are not cords or strings at all, but folds or bands of tissue (hence their preferred newer name). At the front of the larynx they are near the Adam's apple, and at the back they are attached to the arytenoid cartilages, which enable them to be opened and closed to allow or hinder the passage of air from the lungs. The space between the vocal cords is the GLOTTIS.

The main function of the vocal cords in the production of speech is to vibrate and produce VOICED sounds. This happens when they are held closely enough together for them to vibrate when subjected to air pressure from the lungs. See PHONATION.

When the cords are held rather wider apart they do not vibrate, and VOICELESS sounds are produced. Holding tightly together and then releasing the vocal cords produces a GLOTTAL STOP. The exact way in which the vocal cords work is debatable.

Vocal chords is a mistaken spelling.

vocalic

Phonetics & Phonology. Vowel-like; designating a sound produced with a comparatively free passage of air (i.e. with no major obstruction).

Various technical problems arise from dividing speech sounds into CONSONANTS and VOWELS (hence the introduction of the opposition CONTOIDS versus VOCOIDS). The contrast between *vocalic* and *non-vocalic* belongs to Generative Phonology and is another attempt at dealing with the problem.

vocal organs

The same as SPEECH ORGANS.

vocal tract

Phonetics.

1 The whole of the air passage above the LARYNX, including the ORAL tract (the mouth and pharyngeal area) and the NASAL tract (the air passage through the nose when the soft palate is lowered).

2 The entire area involved in the production of speech sounds, including the larynx, trachea, and lungs.

vocative

(*n. & adj.*) (An optional element in clause structure) denoting the person addressed.

In inflected languages the vocative is the case form taken by a noun phrase denoting an addressee. In English the vocative is marked not by inflection but by intonation. Vocatives can include: names (e.g. *Mary*, *Grandpa*), titles (e.g. *Sir*, *Mr President*, *Doctor*, *Waiter*, *Nurse*), and epithets and general nouns, both polite and otherwise (e.g. *darling*, *chums*, *bastard*, *friends*, *liar*, *mate*). Some of these can be expanded (e.g. *Mary dearest*, *my dear friends*, *you stupid idiot*).

Inanimate entities can also be addressed, but this tends to be in fairly formal contexts (and, in any context, to involve a degree of personification), e.g.

> I vow to thee, my country . . .
> Come, friendly bombs, and fall on Slough

vocoid (Pronounced /'vəʊkɔɪd/.)

Phonetics. A speech sound lacking any closure that causes friction. Contrasted with CONTOID.

Like *contoid*, the term *vocoid* was invented by the American linguist Kenneth Pike as a purely phonetic label, in the hope that VOWEL could be reserved as a phonological label.

Vocoid then includes not only most vowels as popularly understood (including many phonetic realizations of the letter *y*, as in *Mummy*, *city*), but also southern British [l], [r], and the semi-vowels [w] and [j].

voice

(*n.*)

1 A grammatical category which in English provides two different ways (ACTIVE and PASSIVE) of viewing the action of the verb.

Voice is applicable to verbs, verb phrases, and entire clauses or sentences. The names *active* and *passive* are linked to meaning in that the subject of an active verb is often the actor, or 'doer' of the verbal action, as in

The early bird caught the worm

while the subject of the passive counterpart is often shown as being 'passively' acted upon, as in

The worm was caught by the early bird

Some languages (e.g. Greek) also have a *middle voice*, which includes verbs of REFLEXIVE meaning.

2 *Phonetics.* Vocal cord vibration accompanying the production of speech sounds.

Various acoustically perceived types of *voice* are sometimes distinguished, e.g. breathy voice, murmur, creaky voice, tense voice, lax voice, and so on. Phoneticians explain these in articulatory terms. For example, in a creaky voice a large proportion, but not all, of the glottis vibrates.

(*v.*) Utter (a speech sound) with accompanying vibration of the vocal cords; cause to become voiced.

voiced

Phonetics. Of a speech sound: made with the vocal cords vibrating. Contrasted with VOICELESS.

In standard English, all the vowels are voiced, as are thirteen of the consonants and both the semi-vowels. The remaining nine consonants are voiceless.

voiceless

Phonetics. Of a speech sound: made without vibration of the vocal cords. Contrasted with VOICED.

There are nine voiceless phonemes in standard English, all of them consonants. (Some languages are said to contain voiceless vowels, but English is not among them.)

Compare DEVOICED.

voicing

1 *Phonetics.* The action or process of articulating a speech sound with vibration of the vocal cords.

Although *voicing* is part of the description of all vowel phonemes in English and of a majority of consonants, the amount of voicing in the production of a particular phoneme in a particular utterance may be affected by phonological context. See DEVOICED.

Some chiefly voiceless phonemes may be voiced in certain environments. Intervocalic [t] sounds more like [d] in many American accents, so that *betting* and *writing* may sound like *bedding* and *riding*. The same tendency occurs sporadically in colloquial British English in expressions like *get out!* /ge'daʊt/, *I've got it!* /aɪv 'gɒdɪt/, etc.

2 (In historical linguistics.) The change of a sound from voiceless to voiced.

Historically, probably around 1400, the voiceless fricatives /s/ and /f/ were voiced to /z/ and /v/ in certain positions, for example in final position in frequently unaccented words such as *as* (originally the same word as *also*), *his*, *was*, *of* (originally the same word as *off*). The voiceless fricative /θ/ was voiced to /ð/ in initial position in the frequently unstressed words *the*, *this*, *that*, *there*, etc.

volition

The action of willing something (one of the characteristic meanings of DEONTIC modality, together with *obligation* and *permission*).

Although *shall* and *will* sometimes have little meaning apart from the indication of futurity, they often express intention, promise, or other shades of volition, e.g.

We shall do all we can to help
I will not forget

• **volitional**.

Compare PREDICTION.

vowel

1 (*a*) A speech sound made with the vocal cords vibrating, but without any closure or stricture. Contrasted with CONSONANT.

(*b*) A speech sound that is central to a syllable and therefore SYLLABIC.

2 Any of the letters *a*, *e*, *i*, *o*, *u* representing such a sound.

As with the word *consonant*, the term *vowel* suffers from ambiguity. Vowel sounds, defined both phonetically as in (1*a*) and phonologically as in (1*b*) are usually represented by vowel letters as in (2). But there are discrepancies. Hence the invention of the terms VOCOID and CONTOID. But these latter remain specialist terms; and *vowel* and *consonant* remain ambiguous.

Vowels (or vowel-like sounds) in modern definition can therefore include syllabic consonants, such as the second syllable of *muddle* [mʌdl̩.].

Several vowel sounds are sometimes represented in writing by a combination of vowel and consonant letters, e.g.

ah, k*ey*, h*al*f, p*ar*t, l*aw*, n*ew*, d*ay*

The *vowel system* of English RP is usually analysed in terms of 12 PURE VOWELS (or MONOPHTHONGS), which may be long or short, and 8 DIPHTHONGS. Scottish English has only 10 monophthongs and 4 diphthongs.

Compare TRIPHTHONG.

•• **vowel alternation**: the same as ABLAUT.

vowel height: see HEIGHT.

vowel change

Morphology. A change in the internal vowel of related words.

Both ABLAUT and MUTATION (*umlaut*) can be covered by this more general term. But diachronically the two describe different phenomena. Ablaut has Indo-European origins, and explains the vowel changes in some current English irregular verbs (e.g. *sing*, *sang*, *sung*). Mutation is of Germanic origin and is a form of regressive assimilation. For example the vowel of *old*, /əʊ/, alternates with /e/ in *elder/eldest*; this reflects an original *a*-like vowel in the preliterary forms of all three words, which in the two inflected words was changed to /e/ under the influence of a vowel /i/ in the following syllable (now itself weakened to /ə/ and /ɪ/).

Synchronically, of course, the origin of these two kinds of vowel change is irrelevant, which is why both may be described by the same term.

vowel quality

The characteristics that distinguish one vowel from another.

vowel quantity

Length as a feature of vowel articulation.

vowel shift See GREAT VOWEL SHIFT.

VP

Abbreviation for Verb Phrase.

W

weak

I Of phonetic quality: obscured, lacking prominence. Contrasted with STRONG (I).

•• **weak form**: the pronunciation of a form word (grammatical word) word when unaccented and in a non-prominent position.

As grammatical words usually receive little stress or prominence, their weak forms (containing weak vowels) are their usual pronunciation.

Some words have more than one *weak form*. For example, the word *do* (strong form /du:/) may be

/dʊ/ before a vowel sound, as in *What do I care?*
/dʊ/ or /də/ before a consonant sound, as in *Why do we bother?*
/d/ in rapid speech, as in *D'you know what I think?*

Strong forms are used in sentence final position, even when a word is unaccented (*'You may not think so, but 'I do*).

Common words having weak forms are:

(determiners) *a, an, the, some, his, her, your*
(auxiliaries) *am, are, be, been, is, was, were, can, could, do, does, had, has, have, must, shall, should, will, would*
(nouns) *Saint, Sir*
(prepositions) *at, for, from, of, to*
(pronouns) *he, her, him, me, she, them, us, we, who, you*
(conjunctions & adverbs) *and, but, as, not, than, that, there*

1982 J. C. WELLS In many accents the pronoun *you* has a weak form /jə/ (conventionally spellable *ya* in the United States, but *yer* in non-rhotic-oriented England).

weak vowel: the vowel /ə/ (*schwa*) and certain other vowels occurring in unstressed, unprominent syllables (in contrast to *strong vowels*).

The vowel /ə/ occurs only in unstressed syllables and is the prime weak vowel of English. Many strong vowels lose their distinct quality and are replaced by /ə/ when they lose stress or prominence. E.g. /'lænd/ *land*, but /'ɪŋglənd/ *England*; /'fɔ:/ *for*, but /fə'get/ *forget*; /ʌn'taɪ/ *untie*, but /ən'les/ *unless*, and so on. In other unprominent contexts one or other of two other weak vowels may occur. One is somewhere in the region of /i:/ or /ɪ/; the other is somewhat like /u:/ or /ʊ/. On a phonemic analysis these two vowels are usually equated with the short vowels /ɪ/ (e.g. /'hæpɪ/ *happy*) and /ʊ/ (ɪnjʊ'endəʊ/ *innuendo*), since weak vowels tend to be short and no meaningful contrasts with /i:/ or /u:/, respectively, are involved. On a strictly phonetic analysis, these are distinctive weak vowels and other symbols may be assigned.

Weak vowels may be so unimportant that in some words the phonemic distinction between /ə/ and /ɪ/ is neutralized. Thus *believe* and *possible* may be pronounced (and so transcribed) as /bɪ'liːv/ or /bə'liːv/, /'pɒsɪb(ə)l/ or /'pɒsəb(ə)l/.

A weak syllable, in this sort of analysis, is not therefore any unstressed syllable, but a syllable containing one of the three weak vowels here described, or a syllable containing a syllabic consonant (e.g. ['bɒtl̩]).

2 (In some historical linguistics and in traditional grammar.) Of a verb or its conjugation: forming its past tense and past participle by adding a suffix (*-(e)d*/*-t*); in contrast to STRONG (2).

Regular verbs in modern English can be described as 'weak' since they require the 'help' of the suffix *-ed* to form the past simple and past participle (e.g. *liked*, *decided*). But some verbs now form the past with both an internal vowel change and an alveolar suffix (e.g. *sell/sold*; *sleep/slept*), while others form the past simple with a suffix and the past participle with vowel change and/or the 'strong' suffix *-(e)n* (e.g. *swell/swelled/swollen*; *show/showed/shown*). Hence this distinction is hardly useful now, and it is more usual to contrast regular and irregular verbs.

Compare LENIS.

weaken

Cause (a word, a vowel) to become weak; REDUCE (2).

well-formed

Capable of being generated by the rules of the grammar; contrasted with ILL-FORMED.

In Generative Grammar, a *well-formed* utterance is one generated according to the rules of syntax, semantics, and indeed phonology. It is a wider term than *grammatical*, which often has a more strictly syntactic sense. Hence many nonsensical or metaphorical utterances that fall foul of the selectional rules are excluded, even though they are grammatically acceptable and may well be interpretable.

• **well-formedness.**

1970 J. P. THORNE Between .. extremes of well-formedness occur sentences of varying degrees of grammaticalness . . . For example, a sentence like *The houses are asleep*, while clearly not ungrammatical in the sense in which *Asleep children the are* is ungrammatical, is nevertheless not completely well-formed.

Compare GRAMMATICAL.

***were*-subjunctive** See SUBJUNCTIVE.

wh

A symbol representing the interrogative or relative quality of the *WH*-WORDS. Probably introduced by Chomsky.

1957 N. Chomsky We shall have rules: *wh* + *he* → /huw/, *wh* + *him* → /huwm/, *wh* + *it* → /wat/ [/huw/ = *who*, /huwm/ = *whom*, /wat/ = *what*].

Wh is often prefixed to another word, indicating the interrogative or relative quality of the item which it denotes, e.g. *wh-clause*, *wh-question*, *wh-relative*.

wh-element

A phrase or clause containing a *WH*-WORD.

A *wh-element* may be a single word, just as a phrase may be. Thus in the sentence *What happened?* the word *what* here is the entire *wh*-element. But often the *wh*-element is longer, e.g.

Which of these pictures do you like best?
Tell me *what you decide*

Whorfian

Designating or characteristic of the theories of B. L. Whorf.

•• **Whorfian hypothesis**: the same as SAPIR-WHORF HYPOTHESIS.

wh-question

A question beginning with a *WH*-WORD.

The *wh-question* is one of the three main question types, the other two being the ALTERNATIVE QUESTION and the YES-NO QUESTION. They are sometimes called *information questions* because, unlike the other two types, they ask for information that is not contained in the question (e.g. *What happened? Who(m) did you tell? How do you know?*

Wh-questions tend to be spoken with a falling intonation.

wh-word

One of a small class of interrogative or relative words that begin with *wh-*, plus *how*.

The main *wh-words* are:

what, which, who, whom, whose; when, where, why, how

Wh-words are sometimes intensified by adding *ever*, often written as a separate word (and always so after *why*), e.g.

Who ever would have guessed?
Why ever didn't you say?

This usage is grammatically distinct from the similar-looking compounds used as subordinators (e.g. *Whoever may have guessed, nobody said anything*).

The *wh-* spelling does not correctly represent any present or past pronunciation. In modern RP the sound is of course /w/. An older (but never, since Old English times, universal) pronunciation, still current in Scottish and Irish English and in the United States and Canada, is in fact /hw/. In Old English the spelling *hw-* was used, but the letters were reversed in Middle English by analogy with *ch-*, *ph-*, *sh-*, and *th-*.

wishing

The verbal expression of a wish.

Wishing is sometimes singled out in grammatical description because the verb of a subordinate clause after an expression of wishing must be in the hypothetical past tense, e.g.

I wish I knew/had known what to say
I would that I knew/that I had known
If only I knew/had known

Such expressions when referring to present or past time denote what is contrary to fact; e.g. *I wish I knew/had known* implies *I don't/didn't know*. With reference to the future, wishing implies something that may be unlikely, but not necessarily what is impossible. For example

I wish you would come tomorrow

does not rule out your coming, though my remark is more diffident than *I hope you come/will come tomorrow*.

word

1 A meaningful unit of speech which is normally uninterruptable, and which when written or printed has spaces on either side (the *orthographic word*).

Native speakers intuitively recognize the *word* as a distinct meaningful grammatical unit of language. It is words whose meanings and very existence are catalogued in dictionaries, and which combine to form larger units such as phrases, clauses, and sentences.

Grammarians recognize smaller meaningful units in the grammatical hierarchy, such as morphemes, but the *word* has distinct characteristics. It is normally uninterruptable. It is also cohesive in the sense that its parts cannot be rearranged as words in a sentence can. Contrast *rearrangement* (in which no reordering is possible) with *Its parts can be rearranged* (which can be reordered, e.g. *Rearranged its parts can be*, *Can its parts be rearranged?*).

Another characteristic of the *word* which would probably seem obvious to the native speaker is enshrined in Bloomfield's definition of the word as a 'minimum free form', i.e. the smallest unit that can reasonably constitute a complete utterance, as in

Do you accept? *Yes/Maybe/Naturally*

But some words fail this test, e.g. *a/an* and *the*.

The characteristic of being 'complete in itself' is supported by the writing convention that separates one word from other, but there are problems with this. Opinions vary as to whether certain compounds are in fact one word or two (e.g. *half way*, *half-way*, *halfway*), and whether such forms as *don't* and *I'll* are single words or not.

2 A word, as (for example) listed in a dictionary, together with all its variants. Sometimes distinguished from sense (1) as LEXEME.

Although this is a more abstract sense of *word*, it is a common meaning (e.g. *see*, *sees*, *seeing*, *saw*, *seen* are all parts of the same 'word' *see*).

Compare CONTENT WORD, FORM WORD, ORTHOGRAPHIC WORD.

word blend See BLEND.

word class

A category grouping together words that broadly share the same syntactic characteristics.

The use of *word classes* is much the same as the more traditional classification of words into PARTS OF SPEECH, but favours less notional and more rigorous definitions. It also usually involves the use of some precise extra categories such as ARTICLES, DETERMINERS, and NUMERALS.

word ending See INFLECTION.

word form

Any variant of a lexeme. (Also called *form of a word*.)

The term is used as a way of avoiding the ambiguity of *word*. For example, *see*, *sees*, *seeing*, *saw* and *seen* are word forms of (or forms of) the lexeme *see*.

word formation

1 The whole field covered by MORPHOLOGY, including both INFLECTION and DERIVATION.

2 (More narrowly.) DERIVATION (the formation of new words other than by compounding).

In this model, inflection is handled as part of syntax.

word grammar

A grammatical theory which claims that the word—rather than, say, the clause, or levels of structure—is the most important element in language.

> 1990 R. HUDSON a. WG is *lexicalist* because the word is central—hence the name of the theory. Grammars make no reference to any unit larger than the word (except for the unit 'word-string', which as we shall see is used only in coordinate structures and is very different from the 'phrase' and 'clause' of other theories). b. WG is *wholist* because no distinction is recognized between the grammar 'proper' and the lexicon. The grammar includes facts at all levels of generality, all of which are handled in the same way.

word group

A HEAD (or *headword*) together with the other words that modify it. (Also called *word complex*.)

Word groups occupy a special place in Systemic-Functional Grammar, because GROUP is syntactically defined as the expansion of a headword, and is distinguished from PHRASE (a reduced clause). Word groups therefore have a distinct rank in this kind of grammar. The groups recognized are *nominal*, *verbal*, and *adverbial* (corresponding to *noun phrase*, *verb phrase*, and *adverb phrase* in other models), plus *conjunction group* and *preposition group* (which is distinguished from *prepositional phrase*).

word order

The order in which words come in clauses and sentences.

In inflected languages such as Latin, word order may be comparatively *free*, because a word's function is often indicated by its ending. English, having few inflections, has a much more *fixed* word order. The basic unmarked word order is SVO (Subject Verb Object). See ELEMENT.

Compare ADJECTIVE ORDER, INFORMATION STRUCTURE.

word stress See STRESS.

Y

yes-no question

A question capable of being answered by a straight 'yes' or 'no'.

The *yes-no question* is one of the three major question types, the other two being the ALTERNATIVE QUESTION and the *WH*-QUESTION. Syntactically, *yes-no questions* must begin with subject-verb INVERSION. The verb placed before the subject is the first auxiliary (or *be* or *have* if they are one-word main verbs); the lexical verb follows the subject.

Are you ready?
Are you sitting comfortably?
Have you any wool?
Have you quite finished?
Can anybody explain this?

This term was probably introduced (in the form *yes-or-no question*) by Jespersen (1924).

yod coalescence See COALESCENCE.

Z

Z element See ELEMENT.

zero

An abstraction, often symbolized by ø, representing the absence of any realization, where there could theoretically be, or in comparable grammatical contexts there is, a morphological or syntactical realization.

The concept of *zero* is used as a way of making rules more comprehensive and consistent than they would otherwise be.

It is not generally used in relation to ellipted structures, where the symbol [^] is preferred. Occasionally the missing word or words are referred to as a *zero anaphor* (or *null anaphor*).

•• **zero article**: a unit posited before an uncountable noun or a plural count noun when either is used with an indefinite meaning.

Thus ø alternates with *a/an* (used before a singular count noun) in a paradigmatic relationship, e.g. *ø food*, *ø vegetables*, *a cauliflower*.

zero genitive: the realization of the genitive inflection without an additional *s* in words that already end in *s*.

This is the usual genitive with regular plural nouns, as in *the athletes' achievements*, where the form is identical in speech with the ordinary plural form (*athletes*) and differs only in having an apostrophe added to the written form. *Zero genitive* also occurs with some singular words, particularly foreign names ending with /z/, e.g. *Aristophanes' plays*.

By contrast, irregular plurals not ending in *-s* show as marked a contrast of form between the plural common case (e.g. *men*) and the plural genitive (e.g. *the men's achievement*), as their singular forms (e.g. common case *woman*; genitive *woman's*).

zero plural: a plural form of a count noun that is not distinct from the singular.

Some count nouns have no distinct plural form (e.g. *sheep*, *cod*, *deer*). Other nouns for animals can have zero plurals (e.g. *fish/fishes*, *pheasant/pheasants*). Zero plural is often the norm with nouns of measurement and quantity (e.g. *He's six foot three*, *We need six dozen*) and is obligatory with some foreign plurals (*a series*, *two series*).

zero relative pronoun: the absence of a relative pronoun in a contact clause.

e.g.

The books ^ I bought yesterday
The girl ^ I was talking to

zero *that*-clause: a nominal clause or relative clause which could be introduced by *that*, but from which *that* is absent.

e.g.

He said he was sorry (= that he was sorry)
Here's the map I promised to lend you (= that I promised to lend you)

A *zero relative clause* (e.g. *I promised to lend you* above) is commonly called a CONTACT CLAUSE.

The use of *zero* in synchronic linguistics was originated by Bloomfield (1926).

List of Works Cited

ADAMS, V. (1988). *An introduction to Modern English word-formation* (edn. 2). London: Longman.

AITCHISON, J. (1987). *Words in the mind.* Oxford: Basil Blackwell.

ALGEO, J. (1990). 'It's a myth, innit? Politeness and the English tag question' in Ricks, C. & Michaels, L. *The state of the language*, 443–450. Berkeley and Los Angeles University of California Press.

ALLERTON, D. J., with French, M. A. (1975). 'Morphology: the forms of English'. In Bolton (1975), 79–134.

AUSTIN, J. L. (1955). Lectures published in: *How to do things with words* (1962). Oxford: OUP.

BASKIN, W. (1959). Translation of F. de Saussure's *Course in general linguistics.* New York: The Philosophical Library, Inc.

BELASCO, S., et al. (1966). 'Bibliography' in *American Speech* 41:225.

BENNETT, D. C. (1975). *Spatial and temporal uses of English prepositions: an essay in stratificational semantics.* London: Longman.

BENSON, M., BENSON, E., ILSON, R. (1986). *The BBI combinatory dictionary of English.* Amsterdam/Philadelphia: John Benjamins Publishing Company.

BLOCH, B. & TRAGER, G. L. (1942). *Outline of linguistic analysis.* Baltimore: Waverley Press.

BLOOMFIELD, L. (1926). 'A set of postulates for the science of language' in C. F. Hockett, ed. *Leonard Bloomfield anthology* (1970), 128–138. Bloomington: Indiana University Press.

BLOOMFIELD, L. (1933). *Language.* New York: Holt, Rinehart and Winston.

BOLINGER, D. & SEARS, D. A. (1981). *Aspects of language* (edn. 3). New York: Harcourt Brace Jovanovich.

BOLTON, W. F. (1975). *The English language.* Sphere history of literature in the English language Volume 10. London: Sphere Books.

BRINSLEY, J. (1612). *The posing of the parts.* London: Thomas Man.

BROWN, E. K. & MILLER, J. E. (1980). *Syntax: a linguistic introduction to sentence structure.* London: Hutchinson.

CASSIDY, F. G. (1961). *Jamaica talk: three hundred years of the English language in Jamaica.* Cambridge: CUP.

CATFORD, J. C. (1959). 'The teaching of English as a foreign language' in Quirk, R. & Smith, A. H., eds. *The teaching of English* (1959), London: Secker & Worburg 164–189.

CATFORD, J. C. (1988). *A practical introduction to phonetics.* Oxford: OUP.

CHALKER, S. (1984). *Current English Grammar.* London & Basingstoke: Macmillan.

CHOMSKY, N. (1955a). *The logical structure of linguistic theory* (microfilm, Massachusetts Institute of Technology).

CHOMSKY, N. (1955b). *Transformational analysis* (Ph.D. Dissertation, University of Pennsylvania).

CHOMSKY, N. (1957). *Syntactic structures*. The Hague: Mouton & Co.

CHOMSKY, N. (1965). *Aspects of the theory of syntax*. Cambridge, Mass.: The M.I.T. Press.

CHOMSKY, N. (1986). *Knowledge of language: its nature, origin and use*. New York: Praeger.

CHOMSKY, N. & HALLE, M. (1968). *The sound pattern of English*. New York: Harper and Row.

Collins Cobuild English grammar (1990). London: Collins.

CURME, G. O. (1931). *Syntax, a grammar of the English language, volume* 3. Boston: Heath.

DECAMP, D. (1968). 'The field of Creole language studies' in *Latin American Research Review* 3:35–46.

de SAUSSURE, F. (1916). *Cours de linguistique générale*. Paris: Lausanne.

de SAUSSURE. See Baskin (1959).

DOWNING, P. (1977). 'On the creation and use of English compound nouns' in *Language* 53:810–842.

EARLE, J. (1873). *The philology of the English tongue* (edn. 2, revised). Oxford: Clarendon Press.

FILLMORE, C. J. (1968). 'The case for case' in Bach, E. & Harms, R. T., eds. *Universals in linguistic theory*, 1–90. New York: Holt, Rinehart and Winston.

FIRTH, J. R. (1930). *Speech*. London: Benn's Sixpenny Library.

FIRTH, J. R. (1935a). 'The use and distribution of certain English sounds' (*English Studies*, 17:1) in Firth (1957), 34–46.

FIRTH, J. R. (1935b). 'The technique of semantics' (*Transactions of the Philological Society*) in Firth (1957), 7–33.

FIRTH, J. R. (1946). 'The English school of phonetics' (*Transactions of the Philological Society*) in Firth (1957), 92–120.

FIRTH, J. R. (1948a). 'Sounds and prosodies' (*Transactions of the Philological Society*) in Firth (1957), 148–155.

FIRTH, J. R. (1948b). 'Word-palatograms and articulation' (*Bulletin of the School of Oriental and African Studies*, 12. 3 and 4) in Firth (1957), 148–155.

FIRTH, J. R. (1951). 'Modes of meaning' (*Essays and Studies*) in Firth (1957), 190–215.

FIRTH, J. R. (1957). *Papers in linguistics 1934–1951*. London: OUP.

FOWLER, H. W. (1926). *Modern English usage*. Oxford: Clarendon Press.

FRANCIS, W. N. (1958). *The structure of American English*. New York: Ronald Press Co.

FRIES, C. C. (1940). *American English grammar*. New York: Appleton-Century-Crofts.

FRIES, C. C. (1952). *The structure of English*. New York: Harcourt, Brace & World.

GIMSON, A. C. (1962). *An introduction to the pronunciation of English*. London: Edward Arnold.

GREENBAUM, S. & QUIRK, R. (1990). *A student's grammar of the English language*. Harlow: Longman.

GREENWOOD, J. (1711). *An essay towards a practical English grammar*. London: Keeble, Lawrence, Bowyer, Bonwick & Halsey.

HALLIDAY, M. A. K., MCINTOSH, A., & STREVENS, P. (1964). *The linguistic sciences and language teaching*. London: Longman.

HALLIDAY, M. A. K. & HASAN, R. (1976). *Cohesion in English*. London: Longman.

HALLIDAY, M. A. K. (1985). *An introduction to functional grammar*. London: Edward Arnold.

HARRIS, Z. S. (1951). *Methods in structural linguistics*. Chicago: University of Chicago Press.

HAYS, D. G. (1967). *An introduction to computational linguistics*. London: Macdonald & Co. Ltd.

HOCKETT, C. F. (1958). *A course in modern linguistics*. New York: Macmillan.

HOEY, M. (1991). *Patterns of lexis in text*. Oxford: OUP.

HONEY, J. (1989). *Does accent matter?* London: Faber & Faber.

HOUSEHOLDER, F. W. (1952). Review of Z. S. Harris, *Methods in structural linguistics*, in *International Journal of Applied Linguistics* 18:260–268.

HUDDLESTON, R. D. (1976). *An introduction to English transformational syntax*. London: Longman.

HUDDLESTON, R. D. (1984). *Introduction to the grammar of English*. Cambridge: CUP.

HUDDLESTON, R. D. (1988). *English grammar: an outline*. Cambridge: CUP.

HUDSON, R. A. (1971). *English complex sentences: an introduction to systemic grammar*. Amsterdam: North-Holland.

HUDSON, R. A. (1980). *Sociolinguistics*. Cambridge: CUP.

HUDSON, R. (1990). *English word grammar*. Oxford: Blackwell.

JAKOBSON, R. (1973). *Main trends in the science of language*. London: Allen & Unwin.

JESPERSEN, O. (1909–49). *A Modern English grammar on historical principles* (I–VII). Copenhagen: Munksgaard.

JESPERSEN, O. (1924). *The philosophy of grammar*. London: George Allen & Unwin.

JESPERSEN, O. (1933a). *The system of grammar*. London: George Allen & Unwin.

JESPERSEN, O. (1933b). *Essentials of English grammar*. London: George Allen & Unwin.

JOHNSON, S. (1755). *A Dictionary of the English language*. London.

JONES, D. (1962). *An outline of English phonetics* (edn. 9). Cambridge: Heffer.

LABOV, W. (1968). *The study of nonstandard English*. Urbana: National Council of Teachers of English.

LADEFOGED, P. (1971). *Preliminaries to linguistic phonetics*. Chicago and London: The University of Chicago Press.

LADUSAW, W. A. (1988). 'Semantic theory' in Newmeyer, F. J. ed. *Linguistics: The Cambridge Survey* I. Cambridge: CUP.

LATHAM, R. G. (1841). *The English language* (edn. 3). London: Taylor, Walton, and Maberley.

LEECH, G. N. (1966). *English in advertising*. London: Longmans.

LEECH, G. N., DEUCHAR, M., & HOOGENRAAD, R. (1982). *English grammar for today*. London and Basingstoke: the Macmillan Press in conjunction with the English Association.

LODGE, D. (1972). *Twentieth century literary criticism*. London: Longman.

LONG, R. B. (1961). *The sentence and its parts: a grammar of contemporary English*. Chicago: The University of Chicago Press.

LYONS, J. (1968). *Introduction to theoretical linguistics*. Cambridge: CUP.

LYONS, J. (1970). *Chomsky*. London: Fontana/Collins.

LYONS, J. (1977). *Semantics*, 2 vols. Cambridge: CUP.

LYONS, J. (1991). *Chomsky* (edn. 3). London: Fontana.

MALINOWSKI, B. (1923). 'The problem of meaning in primitive languages' in Ogden, C. K. & Richards, I. A. *The meaning of meaning*, 451–510. London: Kegan Paul, Trench, Trubner & Co. Ltd.

MARSH, G. P. (1860). *Lectures on the English language*. London: John Murray.

MARSH, G. P. (1862). *The English language* (edn. 2, with additional lectures). London: John Murray.

MARTINET, A. See Palmer (1964).

MCCARTHY, M. (1991). *Discourse analysis and language teachers*. Cambridge: CUP.

MATTHEWS, P. H. (1974). *Morphology – an introduction to the theory of word structure*. Cambridge: CUP.

MATTHEWS, P. H. (1981). *Syntax*. Cambridge: CUP.

MATTHEWS, P. H. (1992). 'Bloomfield's Morphology and its successors' in *Transactions of the Philological Society* 90:2, 121–186.

MILLER, G. A., BRUNER, J. S., & POSTMAN, J. (1954). 'Familiarity of letter sequences and tachistoscopic identification' in *Journal of General Psychology* 50:129–139.

MITCHELL, T. F. (1975). 'Syntax (and associated matters)' in Bolton (1975), 135–213.

MORLEY, G. D. (1985). *An introduction to systemic grammar*. London and Basingstoke: Macmillan.

MORRIS, C. (1937). *Logical positivism*. Paris: Hermann et cie.

MORRIS, R. (1872). *Historical outlines of English accidence*. London: Macmillan & Co.

MORT, S. (1986). *Longman Guardian new words*. Harlow: Longman.

MURRAY, L. (1824). *English grammar* (edn. 38). London: Longman, Hurst, Rees, Orme, Brown & Green.

New English dictionary (1884–1928). (*Oxford English dictionary*, edn. 1). Oxford: Clarendon Press.

NIDA, E. A. (1948). 'The analysis of grammatical constituents' in *Language* 24:168–177.

NIDA, E. A. (1951). 'A system for the description of semantic elements' in *Word* 7:1–14.

NIDA, E. A. (1964). *Toward a science of translating*. Leiden: E. J. Brill.

NIDA, E. A. (1969). 'Science of translation' in *Language* 45:483–498.

NOREEN, A. (1904). *Vårt språk, Nysvensk grammatik i utförlig framställing*. Lund: C. W. K. Gleerup.

O'CONNOR, J. D. O. (1973). *Phonetics*. Harmondsworth, Middlesex: Penguin Books.

OHALA, J. (1975). Review of I. Lehiste, *Suprasegmentals* in *Language* 51:737

OLSSON, Y. (1961). *On the syntax of the English verb*. Gothenburg Studies in English 12, Göteborg: Acta Universitatis Gothoburgensis.

ONIONS, C. T. (1932). *An advanced English syntax* (edn. 6). London: Routledge and Kegan Paul.

PALMATIER, R. A. (1972). *A glossary of English transformational grammar*. New York: Appleton-Century-Crofts.

PALMER, E. (1964). Translation of A. Martinet's *Elements of general linguistics*. London: Faber & Faber Ltd.

PALMER, F. R. (1975). 'Language and languages' in Bolton (1975).

PALMER, F. R. (1976). *Semantics: a new outline*. Cambridge: CUP.

PALMER, F. R. (1984). *Grammar* (new edn.). Harmondsworth: Penguin Books.

PARTRIDGE, E. (1947). *Usage and abusage* (UK edn.). London: Hamish Hamilton.

PEI, M. A. (1966). *A glossary of linguistic terminology*. New York: Columbia University Press.

PIKE, K. L. (1943). *Phonetics: a critical analysis of phonetic theory and a technic for the practical description of sounds*. Ann Arbor: University of Michigan Press.

PLATT, J. T. (1977). 'The sub-varieties of Singapore English: their sociolectal and functional status' in Crewe, W. ed. *The English language in Singapore*. Singapore: Eastern Universities Press.

POTTER, S. (1957). *Modern linguistics*. London: Andre Deutsch.

PUTTENHAM, G. (1589). *The arte of English poesie* (ed. Arber, 1869). London: Edward Arber.

QUIRK, R. (1988). Introduction to: Greenbaum, S. & Whitcut, J. *Longman guide to English usage*. Harlow: Longman.

QUIRK, R. & GREENBAUM, S. (1973). *A university grammar of English*. London: Longman.

QUIRK, R., GREENBAUM, S., LEECH, G., & SVARTVIK, J. (1985). *A comprehensive grammar of the English language*. London and New York: Longman.

REINHARD, T. (1977). Review of T. A. van Dijk, *Some aspects of text grammars* in *Language* 53:247–253.

ROACH, P. (1991). *English phonetics and phonology* (edn. 2). Cambridge: CUP.

ROBINS, R. H. (1953). 'Formal divisions in Sundanese' in *Transactions of the Philological Society*, 109–141.

ROBINS, R. H. (1964). *General linguistics: an introductory survey*. London: Longmans.

ROSS, A. S. C. (1958). *Etymology: with especial reference to English*. London: Andre Deutsch.

SAMUELS, M. L. (1972). *Linguistic evolution with special reference to English*. Cambridge: CUP.

SAPIR, E. (1921). *Language, an introduction to the study of speech*. New York: Harcourt, Brace & Co.

SIEVERS, E. (1885). 'Philology' in *Encyclopaedia Britannica* 18:765–790.

STRANG, B. M. H. (1962). *Modern English structure*. London: Edward Arnold Ltd.

STRANG, B. M. H. (1970). *A history of English*. London: Methuen.

SWEET, H. (1874). *A history of English sounds*. London: English Dialect Society.

SWEET, H. (1888). *A history of English sounds* (edn. 2). Oxford: Clarendon Press.

SWEET, H. (1892a). *A short historical English grammar*. Oxford: Clarendon Press.

SWEET, H. (1892b). *A new English grammar*, Part 1. Oxford: Clarendon Press.

SWEET, H. (1898). *A new English grammar*, Part 2. Oxford: Clarendon Press.

SWEET, H. (1902). Letter to the Vice-Chancellor of Oxford University. In Firth (1946).

THORNE, J. P. (1970). 'Generative grammar and stylistic analysis' in Lyons, J. *New horizons in linguistics*, 185–197. London: Penguin.

ULLMAN, S. (1962). *Semantics*. Oxford: Blackwell.

VACHEK, J. (1964). 'Notes on the phonematic value of the Modern English [ŋ]-sound' in Abercrombie, D. et al., eds. *In honour of Daniel Jones: papers contributed on the occasion of his eightieth birthday*, 191–205. London: Longmans.

VANDERVEKEN, D. (1990). *Meaning and speech acts*. Volume 1. Cambridge: CUP.

WATSON, I. H. (1985). Letter. In *English Today* 4.

WEEKLEY, E. (1912). *The romance of words*. London: John Murray.

WEINREICH, U. (1953). *Languages in contact*. Publications of the Linguistic Circle of New York. New York: Linguistic Circle of New York.

WELLS, J. C. (1982). *Accents of English*. Cambridge: CUP.

WELLS, J. C. (1990). *Longman pronunciation dictionary*. Harlow: Longman.

WHATMOUGH, J. (1956). *Language, a modern synthesis*. London: Secker & Warburg.

WHORF, B. L. (*ante* 1941). 'The relation of habitual thought and behavior to language' in J. B. Carroll, ed. *Language, thought and reality: selected writings of Benjamin Lee Whorf* (1956), 134–159. Cambridge: M.I.T. Press.

List of Figures

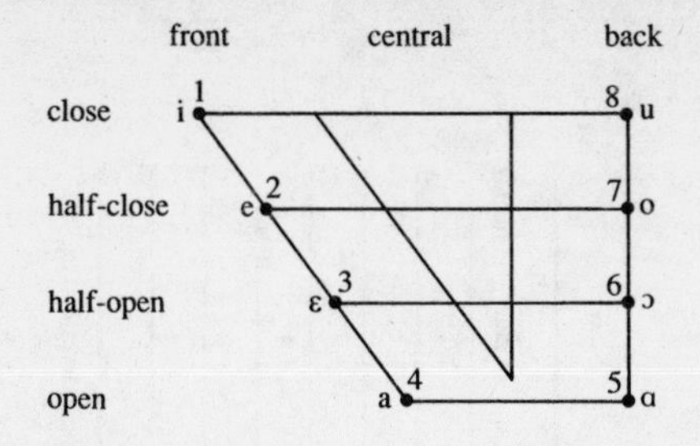

FIG. 1 The primary cardinal vowels

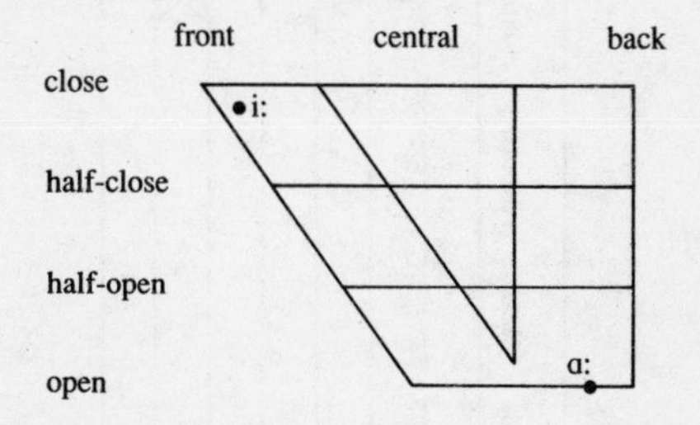

FIG. 2 Two English vowels compared with the cardinals

FIG. 3 The consonants of English

Manner of articulation	*Place of articulation* Bilabial	Labiodental	Dental	Alveolar	Palato-alveolar	Palatal	Velar	Glottal
Plosive	p b			t d			k g	
Fricative		f v	θ ð	s z	ʃ ʒ			h
Affricate					tʃ dʒ			
Nasal	m			n			ŋ	
Lateral				l				
Semi-vowel	w					j		
Frictionless continuant					r			

For the values of these symbols see FIG. 6

(/r/ is more accurately described as post-alveolar)

Fɪɢ. 4 Some IPA diacritics

ʰ	aspiration	as in /tʰɪn/ *tin*
◌̪	dental	as in /eɪt̪θ/ *eighth*
◌̥	devoiced	as in /pl̥eɪ/ *play*
◌̫	labialized	as in /k̫waɪt/ *quite*
◌̃	nasalized	as in /mæ̃n/ *man*
◌̩	syllabic	as in /bʌtn̩/ *button*
~	velarized	as in /fiːɫ/ *feel*
ː	length mark	as in /iː/ /ɑː/ /ɔː/ /uː/ /ɜː/ (the 'long' vowels)

Some of the phonetic features here are obligatory in the standard pronunciation of the examples given, while others are possible variants. Only the vowel length mark and the syllabic consonant mark are likely to be used in a broad transcription.

Note also:

Square brackets [ ] indicate a phonetic symbol, e.g. cardinal vowel 4 [a]
Slashes / / indicate an English phoneme, e.g. vowel 4 /æ/

Fɪɢ. 5 The elements of English clause structure

The five elements of clause structure are commonly found in these combinations:

SV	Time / will tell
SVO	Not many people / know / that
SVC	Small / is / beautiful
SVA	A funny thing / happened / on the way to the forum
SVOO	The world / doesn't owe / you / a living
SVOC	They / were painting / the town / red
SVOA	I / 'll forget / my own name / in a minute

Fig. 6 The phonemes of standard English

Vowels and diphthongs

1	iː	as in	**see** /siː/	11	ɜː	as in	**fur** /fɜː(r)/
2	ɪ	as in	**sit** /sɪt/	12	ə	as in	**ago** /ə'gəʊ/
3	e	as in	**ten** /ten/	13	eɪ	as in	**page** /peɪdʒ/
4	æ	as in	**hat** /hæt/	14	əʊ	as in	**home** /həʊm/
5	ɑː	as in	**arm** /ɑːm/	15	aɪ	as in	**five** /faɪv/
6	ɒ	as in	**got** /gɒt/	16	aʊ	as in	**now** /naʊ/
7	ɔː	as in	**saw** /sɔː/	17	ɔɪ	as in	**join** /dʒɔɪn/
8	ʊ	as in	**put** /pʊt/	18	ɪə	as in	**near** /nɪə(r)/
9	uː	as in	**too** /tuː/	19	eə	as in	**hair** /heə(r)/
10	ʌ	as in	**cup** /kʌp/	20	ʊə	as in	**pure** /pjʊə(r)/

Consonants

1	p	as in	**pen** /pen/	13	s	as in	**so** /səʊ/
2	b	as in	**bad** /bæd/	14	z	as in	**zoo** /zuː/
3	t	as in	**tea** /tiː/	15	ʃ	as in	**she** /ʃiː/
4	d	as in	**did** /dɪd/	16	ʒ	as in	**vision** /'vɪʒn̩/
5	k	as in	**cat** /kæt/	17	h	as in	**how** /haʊ/
6	g	as in	**got** /gɒt/	18	m	as in	**man** /mæn/
7	tʃ	as in	**chin** /tʃɪn/	19	n	as in	**no** /nəʊ/
8	dʒ	as in	**June** /dʒuːn/	20	ŋ	as in	**sing** /sɪŋ/
9	f	as in	**fall** /fɔːl/	21	l	as in	**leg** /leg/
10	v	as in	**voice** /vɔɪs/	22	r	as in	**red** /red/
11	θ	as in	**thin** /θɪn/	23	j	as in	**yes** /jes/
12	ð	as in	**then** /ðen/	24	w	as in	**wet** /wet/

Note: Many variants of these symbols have been used in the past, with consequent confusion. The symbols shown here have gained general acceptance in modern works. They are the symbols used in the *Concise Oxford Dictionary* (1990) and standard works from other publishers. A variant found in some modern books is the use of /ɛ/ for vowel 3, instead of /e/. This use affects vowels 13 and 19, which may appear as /ɛɪ/ instead of /eɪ/ and /ɛə/ instead of /eə/.

/ʔ/ represents the glottal stop (as in a Cockney pronunciation of butter /'bʌtə/ as /'bʌʔə/).

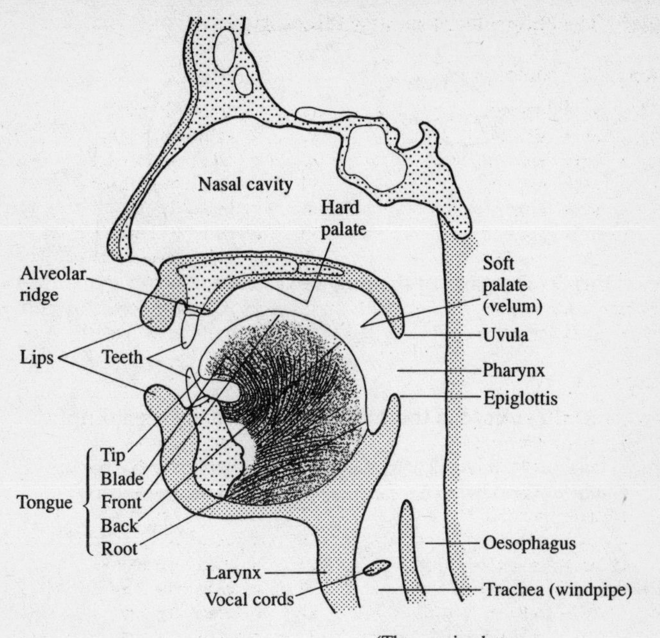

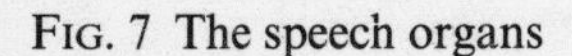
FIG. 7 The speech organs

(The opening between the cords is the glottis)

FIG. 8 Stress and intonation marks

ˈ	primary stress	as in graˈmmatical
ˌ	secondary stress	as in graˌmmatiˈcality

Tones

ˋ	high fall	
ˎ	low fall	
ˊ	high rise	
ˏ	low rise	
ˇ	fall-rise	
ˆ	rise-fall	
¯	high level	
_	low level	
ˈ	high stressed syllable in head	as in ˈTell me ˎmore
ˌ	low stressed syllable in head	as in ˌTell me ˎmore
‖	break between tone units	

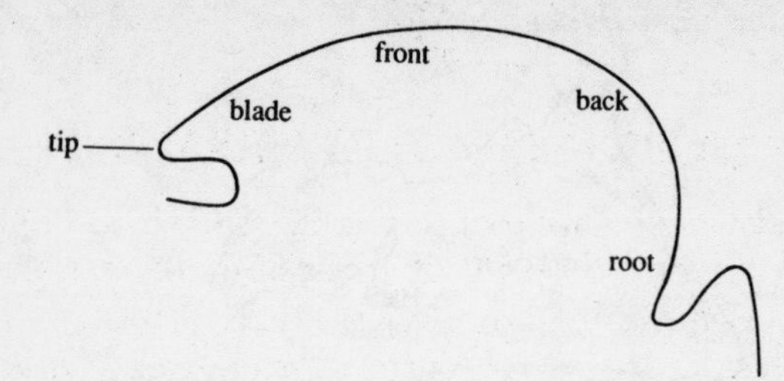

FIG. 9 The parts of the tongue

FIG. 10 A broad transcription /ə 'brɔːd træn͵skrɪpʃn̩/

'fɔːlshʊd ən dɪ͵luːʒn̩|| ər ə'laʊd ɪn'nəʊ 'keɪs 'wɒtsəʊ͵evə.|| bət 'æz ɪn ðɪ 'eksəsaɪz əv 'ɔːl ðə ͵vɜːtʃuːz|| ðər ɪz ən ɪ'kɒnəmɪ əv ͵truːθ

'edmən(d) ͵bɜːk

MORE OXFORD PAPERBACKS

This book is just one of nearly 1000 Oxford Paperbacks currently in print. If you would like details of other Oxford Paperbacks, including titles in the World's Classics, Oxford Reference, Oxford Books, OPUS, Past Masters, Oxford Authors, and Oxford Shakespeare series, please write to:

UK and Europe: Oxford Paperbacks Publicity Manager, Arts and Reference Publicity Department, Oxford University Press, Walton Street, Oxford OX2 6DP.

Customers in UK and Europe will find Oxford Paperbacks available in all good bookshops. But in case of difficulty please send orders to the Cash-with-Order Department, Oxford University Press Distribution Services, Saxon Way West, Corby, Northants NN18 9ES. Tel: 0536 741519; Fax: 0536 746337. Please send a cheque for the total cost of the books, plus £1.75 postage and packing for orders under £20; £2.75 for orders over £20. Customers outside the UK should add 10% of the cost of the books for postage and packing.

USA: Oxford Paperbacks Marketing Manager, Oxford University Press, Inc., 200 Madison Avenue, New York, N.Y. 10016.

Canada: Trade Department, Oxford University Press, 70 Wynford Drive, Don Mills, Ontario M3C 1J9.

Australia: Trade Marketing Manager, Oxford University Press, G.P.O. Box 2784Y, Melbourne 3001, Victoria.

South Africa: Oxford University Press, P.O. Box 1141, Cape Town 8000.

PHILOSOPHY IN OXFORD PAPERBACKS

THE GREAT PHILOSOPHERS

Bryan Magee

Beginning with the death of Socrates in 399, and following the story through the centuries to recent figures such as Bertrand Russell and Wittgenstein, Bryan Magee and fifteen contemporary writers and philosophers provide an accessible and exciting introduction to Western philosophy and its greatest thinkers.

Bryan Magee in conversation with:

A. J. Ayer
Michael Ayers
Miles Burnyeat
Frederick Copleston
Hubert Dreyfus
Anthony Kenny
Sidney Morgenbesser
Martha Nussbaum
John Passmore
Anthony Quinton
John Searle
Peter Singer
J. P. Stern
Geoffrey Warnock
Bernard Williams

'Magee is to be congratulated . . . anyone who sees the programmes or reads the book will be left in no danger of believing philosophical thinking is unpractical and uninteresting.' Ronald Hayman, *Times Educational Supplement*

'one of the liveliest, fast-paced introductions to philosophy, ancient and modern that one could wish for' *Universe*

PAST MASTERS

General Editor: Keith Thomas

SHAKESPEARE

Germaine Greer

'At the core of a coherent social structure as he viewed it lay marriage, which for Shakespeare is no mere comic convention but a crucial and complex ideal. He rejected the stereotype of the passive, sexless, unresponsive female and its inevitable concomitant, the misogynist conviction that all women were whores at heart. Instead he created a series of female characters who were both passionate and pure, who gave their hearts spontaneously into the keeping of the men they loved and remained true to the bargain in the face of tremendous odds.'

Germaine Greer's short book on Shakespeare brings a completely new eye to a subject about whom more has been written than on any other English figure. She is especially concerned with discovering why Shakespeare 'was and is a popular artist', who remains a central figure in English cultural life four centuries after his death.

'eminently trenchant and sensible . . . a genuine exploration in its own right' John Bayley, *Listener*

'the clearest and simplest explanation of Shakespeare's thought I have yet read' Auberon Waugh, *Daily Mail*

LAW FROM OXFORD PAPERBACKS

INTRODUCTION TO ENGLISH LAW
Tenth Edition

William Geldart

Edited by D. C. M. Yardley

'Geldart' has over the years established itself as a standard account of English law, expounding the body of modern law as set in its historical context. Regularly updated since its first publication, it remains indispensable to student and layman alike as a concise, reliable guide.

Since publication of the ninth edition in 1984 there have been important court decisions and a great deal of relevant new legislation. D. C. M. Yardley, Chairman of the Commission for Local Administration in England, has taken account of all these developments and the result has been a considerable rewriting of several parts of the book. These include the sections dealing with the contractual liability of minors, the abolition of the concept of illegitimacy, the liability of a trade union in tort for inducing a person to break his/her contract of employment, the new public order offences, and the intent necessary for a conviction of murder.

OXFORD POETS

Winner of the 1989 Whitbread Prize for Poetry

SHIBBOLETH

Michael Donaghy

This is Michael Donaghy's first full-length collection. His work has a wit and grace reminiscent of the metaphysical poets, and his subjects range widely, responding in unexpected ways to his curiosity and inventiveness. Among the varied pieces collected here are a number of love poems remarkable for their blend of tenderness and irony; a terse 'news item'; playful 'translations' of a mythical Welsh poet; and an 'interview' with Marcel Duchamp.

As the American critic Alfred Corn says:

'Michael Donaghy's poems have the fine-tuned precision of a ten-speed bike, the wit of a streetwise don, a polyphonic inventiveness . . . Poems so original, wry, and philosophical as these are hard to come by. Don't think of passing them up.'

Oxford Reference

The Oxford Reference series offers authoritative and up-to-date reference books in paperback across a wide range of topics.

Abbreviations
Art and Artists
Ballet
Biology
Botany
Business
Card Games
Chemistry
Christian Church
Classical Literature
Computing
Dates
Earth Sciences
Ecology
English Christian Names
English Etymology
English Language
English Literature
English Place-Names
Eponyms
Finance
Fly-Fishing
Fowler's Modern English Usage
Geography
Irish Mythology
King's English
Law
Literary Guide to Great Britain and Ireland
Literary Terms
Mathematics
Medical Dictionary
Modern Quotations
Modern Slang
Music
Nursing
Opera
Oxford English
Physics
Popes
Popular Music
Proverbs
Quotations
Sailing Terms
Saints
Science
Ships and the Sea
Sociology
Spelling
Superstitions
Theatre
Twentieth-Century Art
Twentieth-Century History
Twentieth-Century World Biography
Weather Facts
Word Games
World Mythology
Writer's Dictionary
Zoology